Microsystem Technology
and Microrobotics

Springer-Verlag Berlin Heidelberg GmbH

S. Fatikow U. Rembold

Microsystem Technology and Microrobotics

With 277 Figures

Springer

Sergej Fatikow
Ulrich Rembold
TU Karlsruhe
Kaiserstraße 12
76128 Karlsruhe

ISBN 978-3-642-08232-0

Library of Congress Cataloging-in-Publication Data

Fatikow, S. (Sergej), 1960-
Microsystem technology and microrobotics/S. Fatikow, U.Rembold.
p. cm.
Includes bibliographical references and index.
ISBN 978-3-642-08232-0 ISBN 978-3-662-03450-7 (eBook)
DOI 10.1007/978-3-662-03450-7
1. Electromechanical devices - Design and construction
2. Microfabrication. 3. Robotics. I. Rembold, Ulrich. II. Title.
TJ163.F38 1997
621-dc21 96-53992
CIP

Originally published by Springer-Verlag Berlin Heidelberg New York in 1997
Softcover reprint of the hardcover 1st edition 1997

Cover Design: Atelier Struve & Partner, Heidelberg
Typesetting: Camera ready by authors
SPIN 10518788 33/3142 – 5 4 3 2 1 0 – Printed on acid-free paper

To our wives who inspired
and supported this work

Preface

Within recent years, the microsystem technology (MST) has become a focus of interest to industry and has evolved to an important field with a great application potential. Many industrial countries, e.g. the USA, Germany, France and Japan, have initiated major programs to support and coordinate developments in MST. Despite of the growing interest in this new technology, there is hardly any publication that treats MST in a coherent and comprehensive way. There are some books describing partial aspects of MST, however, in general they are unsuited for a beginner trying to understand the basic concepts of this field; often they even deter the reader from being more inquisitive. This book is an attempt to give the reader a systematic introduction to MST and microrobotics; it discusses all important aspects of this rapidly expanding technology. Numerous technical and economical innovations are increasingly being influenced by MST. In the near future, MST will play a decisive role for industry to be competitive in many areas such as medicine, biotechnology, environmental technology, automotive products, appliances etc. Various branches of industry are already including MST components in their new products; they will outperform the old ones and will capture a large section of the market.

It is the purpose of this book to inform the practicing engineer and the engineering student about MST, its diversity of products and their fields of application. This will help them to get quickly familiar with the miniaturization of sensors and actuators and with the integration of these components into products. The book contains an overview of numerous ideas of new devices and the problems of manufacturing them. The reader not familiar with the subject will find the text and the many illustrations easy to understand. The book will not go into details of theory, however, a basic knowledge of engineering, physics and chemistry may help the reader to better understand the working principle of the presented devices.

The idea of the book originates from the lecture "Microsystems Technology and Microrobotics" which has been given for three years to students of engineering and computer sciences of the University of Karlsruhe. These students are the experts of tomorrow who will be designing and manufacturing microsystems and microrobots and who will be integrating them in advanced products.

It virtually is impossible to explain every idea and research work on MST that had been discussed in the literature. A representative selection was made of them and the authors believe that all important devices, operating principles and problems of application have been covered. Maybe at a later date there will be a chance to go into more details. We are sure that not everything presented will be free of errors, and for this reason we welcome any suggestion the reader wants to give to the authors.

Many of the prototypes of the devices presented in the book are in the forefront of technology, eventually they will be reaching maturity and they will be opening up a mega-market for MST products. It is expected that the market volume will be following that of microelectronic products. The market penetration and success will be belonging to the innovators who are currently experimenting with MST and the many facets of applying its products. It is the wish of the authors that this work will help to generate an awareness for this new technology and that it will serve many readers as a guide to conceive improved products and processes.

This work was not created over night, it took several years of effort to collect the material and to put it into a proper form. There were numerous authors who presented their research results in papers and conferences and who eventually contributed with their publications to the success of the book. Many of them provided us with graphs and pictures which became an integral part of the work and which made it much easier to understand the text. We also had much help in preparing the manuscript and in drawing the artwork. Mirko Benz, Jörg Seyfried, Michelle Specht and Jing-Jing Zhang deserve our sincere thanks for their time and efforts. The final proof reading was done by Esther Freyer and we are indebted to her numerous suggestions and corrections. Last but not least we would like to thank our wives and children for their patience and understanding.

Karlsruhe, October 1996

Sergej Fatikow
Ulrich Rembold

Contents

1 Ideas and Problems of Microsystem Technology and Microrobotics

1.1 Introduction

In recent years, the microsystem technology has become an important source for sensors, actuators and entire control modules. So far there is no generally accepted definition for MST. Most MST researchers characterize a microsystem as the integration of miniaturized sensors, actuators and signal processing units, enabling the overall system to sense, decide and react. MST can be defined as the functional integration of mechanical, electronic, optical and any other functional elements, using special MST techniques.

The goal of this technology is to fabricate intelligent monolithic or integrated chips which can sense (with sensors), plan, make decisions (signal processing units) and actuate (with actuators). Compared with conventional systems, these highly reliable microsystems provide the user with previously unheard functions. Figure 1.1 roughly shows how MST is structured. There are numerous possible applications. E.g. the whole field of robotics could be revolutionized by the development of highly miniaturized actuators and sensors. Today, intelligent multi-purpose microrobots become a reality [Ishi95], [Aoya95], [Fati95]. Microrobots will open up new fields of applications, achieve a great reliability (by a high degree of integration) and promise a long-term cost reduction for many handling and assembly operations.

There is also no general agreement concerning the dimensions of a microsystem or microcomponent. Some researchers refer to a microsystem as having dimensions of a few centimeters, others feel that only a system having dimensions in the micrometer range should be called a *micro*system. One compromise is to define a microsystem as being a system in which as many functions as possible are realized within a very small amount of space and which contains at least one micromechanically manufactured component. Figure 1.2, which depicts the size of an electrostatic micromotor compared to a human hair (diameter of 50 to 100 μm), gives an idea about typical dimensions of a microsystem component.

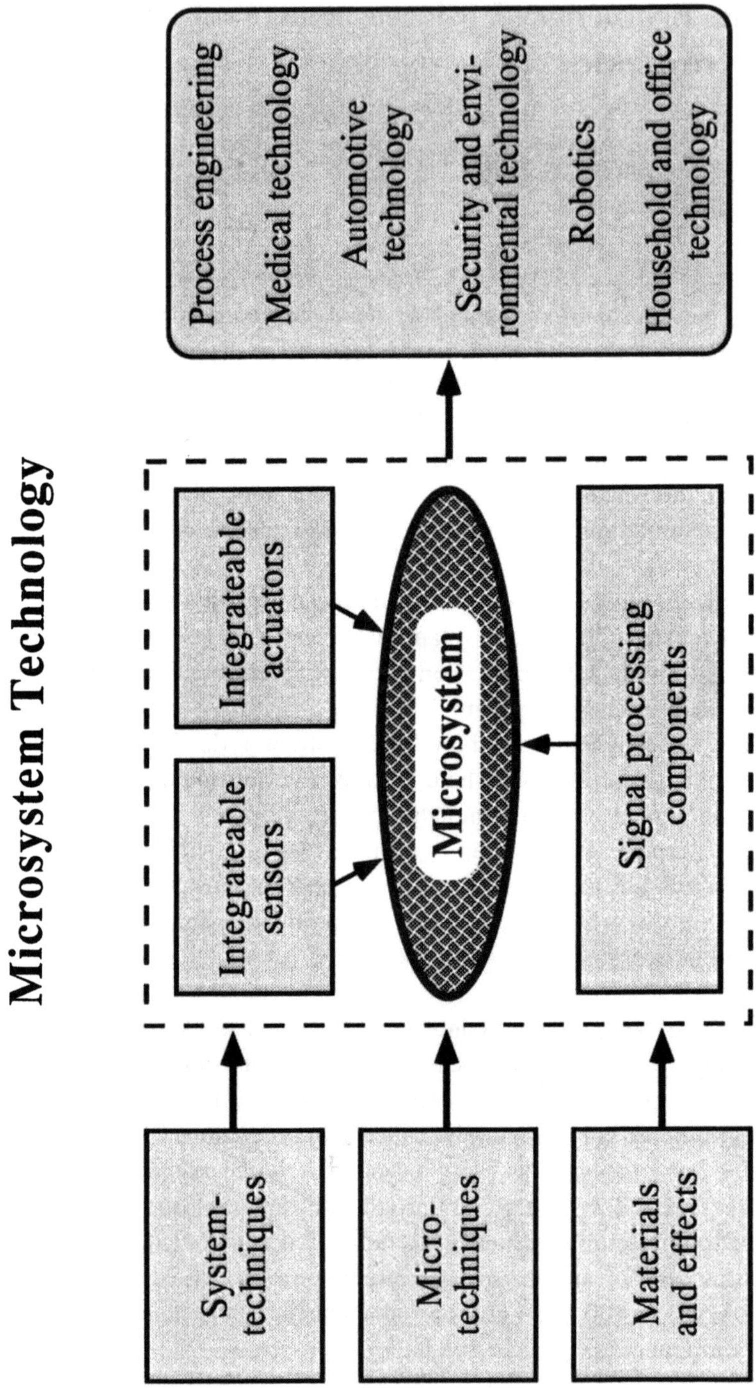

Fig. 1.1: Overview of the MST

Fig. 1.2: Dimensions of a polysilicon micromotor compared to the diameter of a human hair. Courtesy of the Berkeley Sensor and Actuator Center, University of California, Berkeley

One main goal of MST is to make a technical copy of living creatures of this world. E.g. researchers at Toyota are concentrating on duplicating the functionality of a mosquito – a natural microsystem which searches for blood cells with sensors, punctures skin and sucks and pumps blood. Theoretically, such a device can be used as a micromachine for blood diagnosis. In the future the nanotechnology will attempt to make systems even tinier and to build up electronic and mechanical systems atom by atom [Hata95]. One day, nanodevices will control tiny wheels, manipulators, actuators and sensors. E.g. human phantasy even goes so far that invisible nanorobots working in the upholstery of a pneumatically controlled armchair will form the seat to the owner's shape, or in another application, that nanorobots in a mouthwash spray are sent to scrub off tooth plaque [Drex91]. Presently such ideas are far-fetched, but these visions could quite possibly become reality.

Numerous technical and economical innovations are increasingly being influenced by MST. In the near future, MST will play a decisive role for industry to be competitive in many areas such as medicine, biotechnology, environmental technology, automotive products, appliances etc. Various branches of industry are already including MST components in their new products; they will outperform the old ones and will capture a large section of the market. An overview of the potential microsystem market is symbolically presented in Figure 1.3 [Hund94].

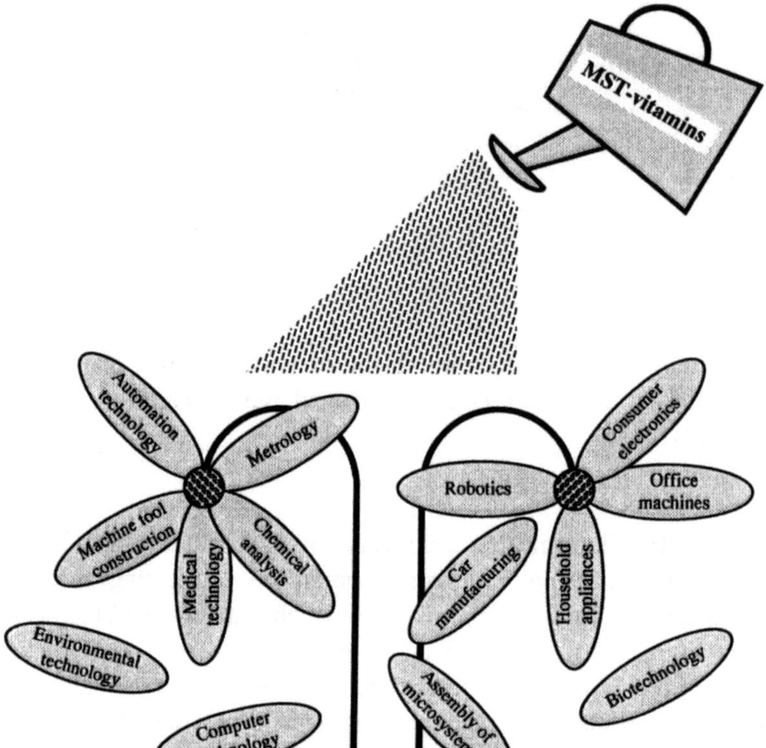

Fig. 1.3: Symbolic presentation of microsystems market potential. According to [Hund94]

Many microcomponents, such as different kinds of microsensors and -actuators, can already be realized for specific applications, but they are very expensive. Microsystems will be able to replace conventional systems when they are available at low cost. This, however, will not happen before they will be mass-produced and available in high quantities, which on the other hand demands large markets. This necessitates the availability of special micromechanical techniques, especially silicon-based ones. With these techniques, which allow the production of many identical components at the same time on one chip (batch process), it will be possible to lower the processing cost for each individual element. The manufacturing methods used in today's MST practice are discussed in detail in Chapter 4.

One major problem of present innovative MST developments is the transformation of research results into practical industrial applications – in other words, the transition from the technical fascination to the economic success. The main reason for this problem are the high initial investment costs for the production of microtechnical products, since they might be a financial risk for newcomer firms, especially for small and medium-sized companies. The costs for the equipment of special clean rooms which are necessary for microproduction are very high. A single speck of dust landing in the wrong place on the microcomponent can cause considerable functional defects. The risk to individual companies would be lowered by a well-planned government support, since it would encourage a greater portion of industry to get involved in creating microdevices. Due to the high investment costs, presently only large companies can work independently on MST product development. Most of the smaller potential MST manufacturers have to work together closely to pool their resources, which is often slow and inefficient. Unfortunately, the decision makers of these companies often do not realize how quickly MST will establish itself and in which way the technology will influence products.

Although in almost every application-specific project, problems come up which must be solved by basic research, there is often the unwillingness of cooperation between industry and institutes. Typical problems are material compatibility, scalability of actuation principles or the unavailability of algorithms to intelligently control microsystems (artificial neuronal nets, fuzzy logic). Therefore, cooperation between industry and science is important – from the planning phase, through the development and manufacturing phases, to the product realization. In this procedure, the final product should be the main focus. Standardization should also become a more important point for MST; if standardized components and flexible interconnection technology are available, a further cost advantage can be obtained. This would also encourage smaller and medium-sized companies to react quickly to individual customer needs.

1.2 Microsystem Structure

A complete microsystem should detect, process and evaluate external signals, should make decisions based on the obtained information, and finally should convert the decisions into corresponding actuator commands. The microsystem can then perform task-conditional manipulations. The principle components of a microsystem are shown in Figure 1.4. The special feature of microcomponents compared with conventional ones is that sensors and actuators are compatible with the microelectronics in terms of size and in component cost. In this sense, MST takes over where traditional precision machining technology has its limit. For example, parts made with microtools consisting of hard materials (e.g. diamond) in connection with CNC controlled machines become increasingly expensive with the degree of miniaturization.

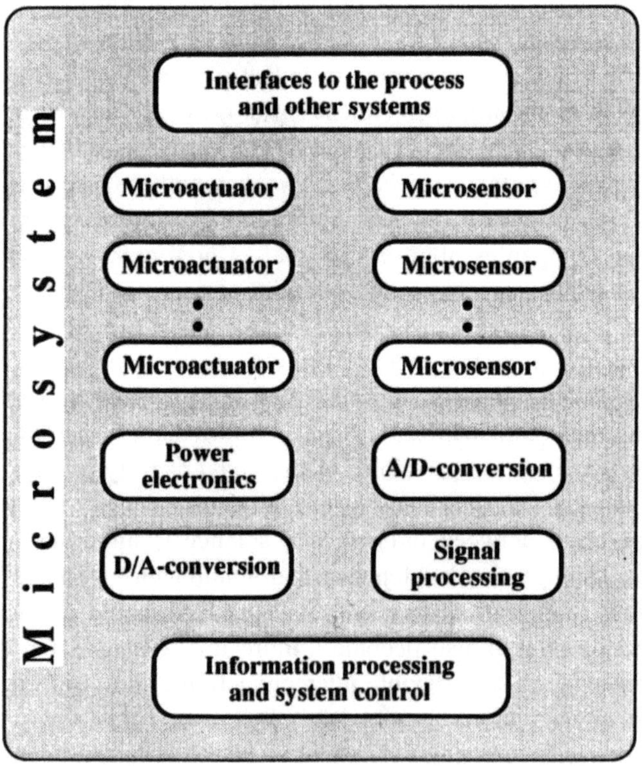

Fig. 1.4: Components of a microsystem unit

Individual sensors, which make up a sensor module, are no longer individually balanced components, but can be mass-produced on a small substrate with relatively low production costs. Several microsensors can be integrated together to form a sensor array. This considerably increases the system reliability and a failure of single sensors is no longer critical. The measurement range can also be laid out optimally. Depending on the application, these sensors can use mechanical, thermal, magnetic, chemical or biological principles.

The actuators are the active microsystem components which allow the system to react to a stimulus. They are small motors, pumps, valves, grippers, switches, relays, and special microsystem actuators, which are usually produced micromechanically. As we will see soon, various tasks of microrobotics can be performed by microactuators, such as causing a displacement or manipulating a tiny object. As opposed to actuator miniaturization, the sensor miniaturization is already quite advanced. Most micro-products available today are sensors. According to the source [Schwe93], in 1993 about 74% of the MST products were microsensors, 17% microactuators and 9% microoptics.

The development of MST signal processing components is also very demanding since the tasks are complex and the systems are limited in size and power. The control algorithms should be tailored to the need of MST, which means they should utilize the microprocessor's full computing power.

Many of the unsolved MST problems involve the interfaces. Microsystems have to keep in contact with their surroundings in order to be able to exchange energy, information and substances with other sytems, Fig. 1.5. The feasibility and marketability of future microsystems mainly depend on whether practical micro-macro interfaces can be developed. So far, electric interfaces for transmitting information and energy are the furthest developed. Various possibilities are being investigated, such as optical, thermal and acoustic interfaces. Up to date, only substances can be conveyed using microfluidic methods. Examples are the release of medicine in drug delivery systems or the suction of organic material in medical applications. A summary of possible interface ideas is given in [Kohl94a].

A/D and D/A converters are often part of an electric interface. They allow the conversion of analog sensor signals for digital processing and the control of analog actuators using digital control commands which are generated by

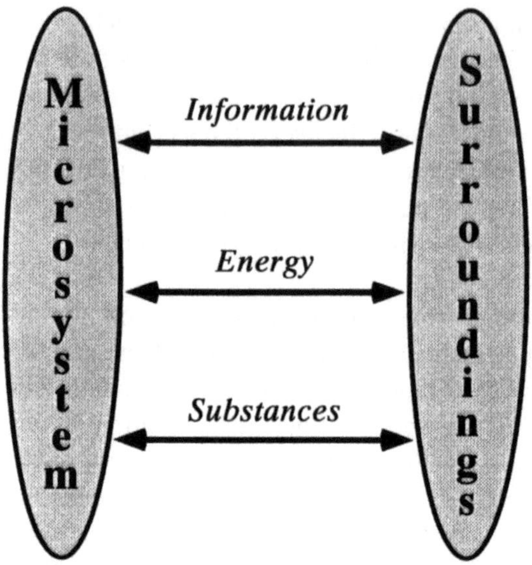

Fig. 1.5: Interfaces between a microsystem and its surrounding

the microcontroller. In a decentrally structured microsystem, in which intelligent sensors and actuators are equipped with their own microcontrollers, the A/D and D/A converters can be integrated directly onto the microsensor or microactuator chip. Power electronic components are essential for almost any microsystem, they often lead to thermal or electromagnetic problems. Such problems should be taken into account during the design phase of a system (Chap. 7).

1.3 MST Techniques

Microsystems can be designed with special micro- and system techniques using the appropriate materials and physical effects. Various techniques and processes have been developed which can often be used in connection with others, Fig. 1.6. This section gives the reader the basic information needed to understand the following discussion on MST techniques. The most important MST techniques will be treated in detail in Chapters 3 and 4.

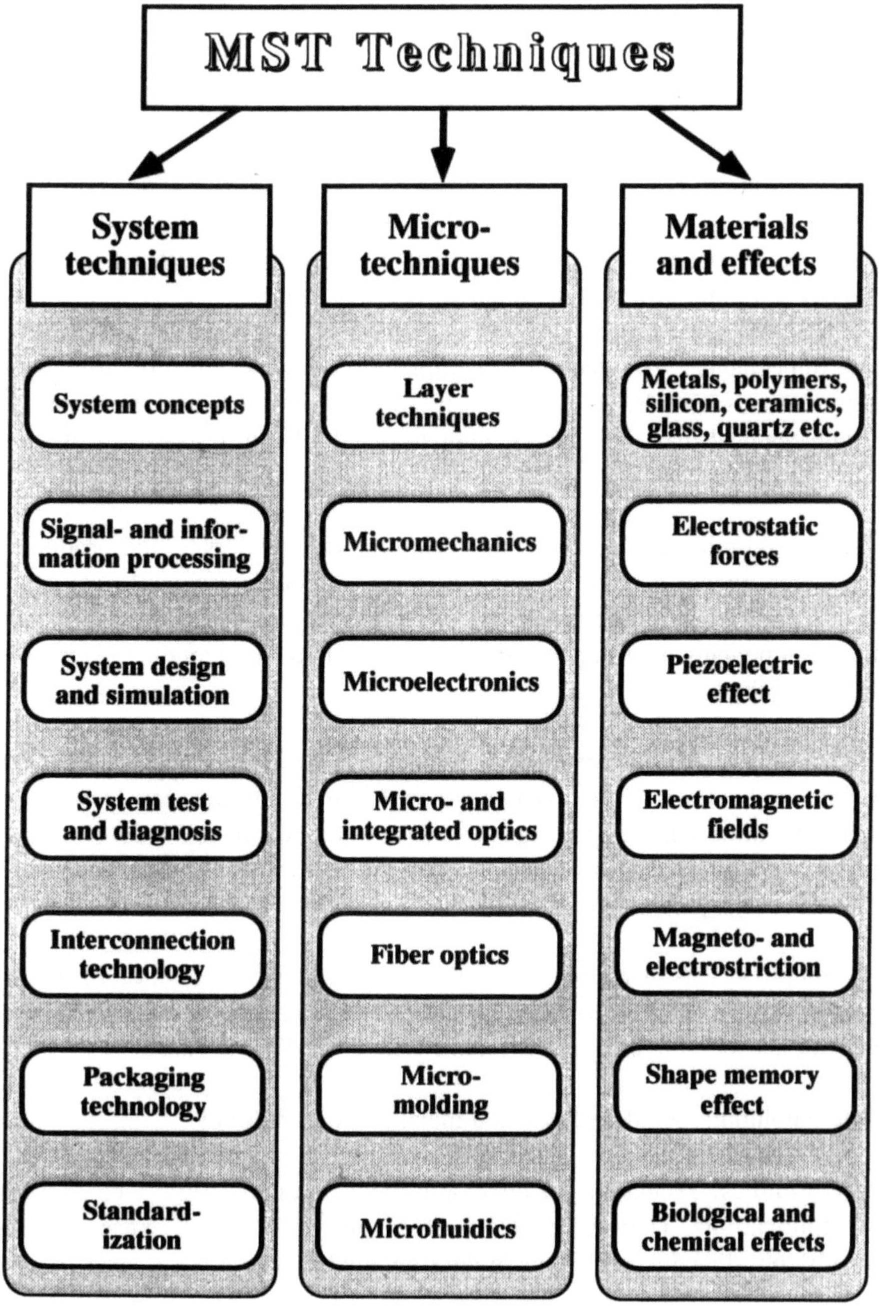

Fig. 1.6: Fundamental MST techniques

Microtechniques

In this section we will define the various techniques used to manufacture microsystem components.

- Layer techniques: Methods for producing layers of different materials on the surface of a substrate. Depending on the deposition method, the layer thickness can range from a few hundred μm to a few nm.

- Micromechanics: This technique comprises in general the three-dimensional structuring of solids, with at least one dimension in the micrometer range. Micromechanical materials include single-crystal silicon, polysilicon, metals, plastics and glass.

- Integrated optics: Definition of the technique for developing and producing miniaturized, planar optical components, such as modulators, switches, couplers, etc. Analogous to microelectronics, the aim is to integrate all the named optical components onto one substrate, such as glass, semiconductor material or lithium niobate.

- Fiber optics: Used to transmit optical signals in light-conducting media. The principle applications of fiber optics are first, the communication technology utilizing an almost noise-free signal transfer and second, the sensor technology for detecting changes of light to determine various physical, chemical and biochemical parameters.

- Microoptics: This technique deals with the design and production of miniaturized optical image processing elements such as mirrors, lenses, filters, etc., which are needed in hybrid microsystems with optical functions.

- Micromolding: Includes plastic and metal powder molding

- Microfluidics: Technique for developing and producing fluid elements for many applications. They have high performance, are free of wear and are relatively robust against pollutants in the flow media.

System Techniques

- System concept: Defines the microsystem architecture and interface concepts for the different MST techniques.

- Signal and information processing: Describes the receiving and processing of primary electric sensor signals, the execution of algorithms, the transformation of output information into control signals. It is also concerned with the management of data storage and retrieval.

- Design and simulation tools: Defines the tools for computer-based microsystem analysis, simulation and design.

- Test and diagnosis of microsystems: Methods and tools to test the functionability of microsystems

- Interconnection technology: Deals with the technological operations needed to physically integrate components within a small amount of space.

- Casing technique: Design of the casing for a microsystem, which usually is an essential part of the system and may influence the overall system function and size

- Standardization: As in many other branches of industry, very important for developing microsystems. It often can lead to the economical success of a research result.

Materials and Effects

- Biological materials and effects: Mainly used in biosensors to selectively measure concentrations of substances in fluids and to determine biological parameters, such as toxicity and the effect of allergens.

- Chemical materials and effects: Used almost exclusively in chemical sensors. These sensors can detect a specific component in a foreign substance as well as its concentration in this substance.

- Piezoelectric effect: The changing of the geometry of a piezocrystal when applying an electric voltage to it. This effect is used in actuators.

- Electrostatic force: Appears between two parallel metal plates when an electric voltage is applied between them.

- Electromagnetic field: Generated when current flows through a conductor or coil. This effect is often used for magnetic actuators.

- Magneto- and electrostrictive effect: Magnetostriction (Electrostriction) is the deformation of a ferromagnetic (ferroelectric) material under the influence of a magnetic (electric) field.

- Shape memory effect: Describes the property of a shape memory alloy. When permanently deformed under a certain temperature it remembers its original shape and returns to it when heated above this temperature.

- Silicon, silicon oxide, silicon nitride, ceramics, quartz, metals (nickel, gold, aluminum, copper, etc.), polymers, glasses and other materials are designations for materials used for MST.

Most of the above-named subjects are in various stages of development, resulting in different stages of microsystem integration. Production lot sizes and costs are relevant data.

1.4 Worldwide MST Activities

All the major industrial countries have recognized the significance of microsystem technology. In Germany, Japan and the USA, many MST programs have been implemented, still the individual aims of these countries are very different. In 1991, worldwide about 300 companies and institutes were actively working on MST projects [Tschu92]. Now, five years later, this number has been exceeded several times since many academic and research institutes and companies in Europe have joined the various MST activities. In 1995, more than 8,000 companies have participated in the development and production of MST products. Another 11,000 want to start within the next few years [Bras95]. Most MST projects (considering the number of participating companies and institutes) are going on in the USA, followed by Japan and Germany. The distribution of the market share between the USA, Europe and Japan is expected to be about 40:35:25 in the year 2002 [NEXU95]. The estimated worldwide market sales volume of MST products and products based on this technology will be between 90,000 to 100,000 million Dollars.

1.4.1 MST-Activities in Europe

In Europe, especially in Germany and France, there is thorough application know-how in MST. In many MST areas Europe has a leading international position and therefore has a good chance of securing its competitiveness.

1.4.1.1 Promotions by the European Community

Characteristic for Europe are the many independent research and development plans on national and European levels, which should be coordinated and organized by a central organization. For this reason, the EU commission has gained a significant position in Europe since it is able to coordinate the international developments and use of MST Europewide. An example is the *Network of Excellence in Multifunctional Microsystems* (NEXUS) which has the important task of expanding and intensifying information exchange between institutes, universities and companies even outside the EU boundaries. In 1996, NEXUS had more than 250 European members, and it is still growing. Further, the EU supports programs as the *University Enterprise Training Partnerships* (UETP), the *Joint European Submicron Silicon Initiative* (Jessi) and the *Joint Analogue Microsystems Initiative of Europe* (Jamie). The project *Microsystem Usage Strategy and Technologies* (MUST), with the aim of a broad transfer of technologies is also supported by the EU.

Also the 4th EU research program (1994–1998) supports MST work. In particular, the system-specific technologies are being sponsored. One focus of work is the multidisciplinary design and manufacture of microsystems through the integration of already existing basic technologies. An intense professional cooperation between the grants recipients is anticipated. Furthermore, an increased orientation towards practical applications is demanded. Obviously the program management is interested in convincing application examples of operating microsystems. Most European MST activities are concentrated in the academic and research facilities whereas industrial companies of all sizes are very hesitant to become active in this new technology. The main applications of the program are for the automotive, medical and environmental products. Altogether, Europe gives about $55–200 million a year, with most of it going toward projects of the automobile industry [Bryze94].

1.4.1.2 Germany

As mentioned earlier, a complete microsystem can only occasionally be produced by one company. This fact gives German industry a competitive advantage since it has many small and medium-sized companies that can pool technologies. Germany competes very well with the USA and Japan; all current German activities stress cooperation. Between 1990 and 1993, the program "Microsystem Technology" was founded by the German Ministry for Research and Technology (since 1995: German Federal Ministry for Education, Science, Research and Technology, German abbreviation: BMBF) with a total contribution of $180 million [BMFT90]. Industry played a major role in choosing the R&D topics. One major topic was the ability of small and medium-sized companies to manage and handle the micro and system techniques, which are the basis for future microsystem integration. The ongoing program "Microsystem Technology" logically builds on the results of the first program and is planning the MST developments in Germany up to the year 1999 [BMFT94]. Emphasis is laid on cooperative projects with predominantly industrial participation. The projects are usually initiated by the companies themselves; considering the intensive competition in MST this is undoubtedly the only way. Besides the fundamental investigation of system techniques, the program also tries to encourage the development of exemplary prototypes in various areas of application, such as health, environment and transportation. An aim of the cooperative projects is also to standardize microsystem components and interfaces, which is very important for economic development as the established R&D areas such as microelectronics and computers show.

Within the framework of the institutional sponsorship of the German Research Foundation (DFG), research institutions are directly supported to do fundamental research. The Karlsruhe Research Center (FZK) and the Fraunhofer Society for Applied Research in Munich (FhG) received the most resources to date and have a high scientific MST potential in Germany outside the universities. Besides to federal government sponsorships of MST, in particular BMBF, several German states have set up regional programs for MST.

A major prerequisite for training new MST scientists is the inclusion of various MST subjects in the university curricula. This should help in establishing a broad base of qualified scientists and engineers. The ideal MST engineer should have specialized knowledge in all areas of MST, good knowledge in related research fields, and should be able to work interdisciplinarily. Universities are aware of their role in MST. Several universities in Germany are offering many courses in MST now and have institutes doing basic research on this topic. Large German companies which are working on MST, estimate that the demand for engineers with a graduate degree in MST will make up 5% to 20% of the entire need for engineers and basic scientists. In Germany, there are presently about 6,000 engineers graduating every year from electrical engineering and physics; for MST this means a training of 300 to 1,200 engineers per year. These numbers are only intended to give a basic idea about future need for people of this new technology.

1.4.1.3 Other European Countries

Switzerland also has a strong position in Europe with its many advanced MST research facilities. At the *Swiss Centre for Electronics and Microtechnology (CSEM)* in Neuchâtel and at the ETH institutions in Lausanne and Zurich, important activities for sensor research, the development of micromechanical components and the investigation of microrobots had been established. At the Institute of Microtechnology of the University of Neuchâtel, successful fundamental and application-oriented research has been carried out on silicon sensors and actuators. Many MST activities are being coordinated through the *Swiss Foundation for Research in Microtechnology* and the *National Foundation for Scientific Research.* However, Swiss industry does not receive any federal grants, making it difficult to transform research results into products.

Recently, *France* has become very active especially in R&D work of microsensors. In 1993, a call for MST-oriented research proposals was published as a cooperative initiative by three ministries as well as the *Agence Nationale de Valorisation de la Recherche (ANVAR)*. At the same time, the *Centre National de la Recherche Scientifique (CNRS)*, the *Microtechnology Cooperative Center - Toulouse* and the *Research Group on Microsystems* were founded to help coordinate the R&D work of various French research institutes and companies. This endeavor should help to realize MST products up to the prototype phase. The most prestigious institutes include the *Laboratoire d'Automatique et d'Analyse des Systemes (LAAS)*, the *Laboratoire d'Electronique, de Technologie et d'Instrumentation (LETI)*, the *Institut des Microtechniques de Franche-Comte (IMFC)* and the *Centre Technique de l'Industrie Horlogère (CETEHOR)*. An entire spectrum of MST problems are being investigated by these institutes, the main ones being microsensors, microelectronics and system technology. At IMFC, microrobots including controllers, architectures and energy sources are being developed as well as high-resolution positioning systems. In order to encourage the cooperation between research institutes, universities, the atomic energy commission (CEA), CNRS, and industry, the *Groupement Microcapteurs Chimiques (GMC2)* was founded a few years ago by the *French Ministry for Research and Technology*.

Industry-oriented research is done in the *Netherlands*. The most significant MST activities are being pursued at the University of Twente, where a new institute for microelectronics, materials research and sensors and actuators (MESA) was founded, which deals with a broad spectrum of MST problems. The *Netherlands Study Centre for Technology Trends* has published a study report on microsystem technology, which can serve as an introduction to MST. It includes new ideas and describes existing technologies for the realizing of structures and systems. Everybody concerned with this technology and research institutes can use this report to become familiar with MST, which hopefully may lead to European-wide cooperation on this subject.

Special topics are being investigated in the *United Kingdom*, such as optical microsystems and biomedical sensors, with the nanotechnology as the main focus point. Interesting results were achieved at the *University of Cambridge*, *University of Oxford* and the *Imperial College* in London. The *Microengineering Common Interest Group* was formed under the *Department of Trade and Industry (DTI)* for the purpose of spreading technical and economic information concerning MST. Over 100 companies, research institutes and universities participate in this group. The group also tries to organize larger programs.

In *Belgium*, MST activities are mostly concentrated at the University of Leuven. Among others, research is going on in the application of shape memory actuators for microsystems. In *Sweden*, MST studies were halted a few years ago due to the financiers' opinion that the work being done was unrealistic. Today, only scattered projects are being pursued at the Universities of Uppsala, Linköping, Lund, Stockholm, Göteborg and at the *Industrial Microelectronic Center (IMC)*. Although Italian industry is becoming increasingly interested in MST, in *Italy*, there have only been a few MST activities to date. The *University of Pisa* plays an important role in Italy with numerous MST research projects. Research institutes in *Austria, Spain, Portugal, Denmark, Greece, Ireland* and the *Baltic countries* only work on a few exemplary projects. *Russia* and a few former Soviet Republics have a high MST potential, resulting from their experience in high-tech areas such as their space and military programs.

1.4.2 MST Activities in Japan

In Japan, the strive for intensive cooperation between large companies and universities has been a tradition, with public organizations playing a coordinating role. For this reason, there are clear government strategies and commitments to targeted projects, which is how MST research in Japan differs from that in Europe and the USA. In 1991, the *Ministry of International Trade and Industry (MITI)* began the program "Micromachine Technology" and helped found the Micromachine Center (MMC) in Tokyo in 1992. Many industrial firms (among which are the largest firms of Japan) are also involved in this center as well as the *New Energy and Industrial Technology Development Organisation* and the *Agency of Industrial Science and Technology (AIST)*. The latter has a patronage over three national research facilities: the *Mechanical Engineering Laboratory (MEL)*, the *Electrotechnical Laboratory (ETL)* and the *National Research Laboratory of Metrology (NRLM)*. All of them are more and more involved in MST projects.

Within the framework of this program, fundamental research, component and system development will be done over a period of 10 years. The main goal is to manufacture very small machines and instruments which can operate with high accuracy in a tiny space. These machines can be used in medicine, biotechnology and industry, Fig. 1.7. In medicine, the conceived microrobots should be able to independently carry out examinations, analyses and treatments at otherwise inaccessible locations during minimal-invasive surgical operations. Many operations which now endanger the patient's health (due to anesthesia and blood loss) could be done by micromachines one day. In the long-term, robots having a diameter under 1 mm are to be developed which can also move inside a blood vessel. Small industrial microrobots should go into complex machines and locally repair defects. Thus, the machines would no longer have to be disassembled for repair. Such tools would especially be practical for maintaining airplane motors and pipelines. Another research goal is the development of a bullet-shaped "intelligent pill". As soon as a patient swallows it, it is expected to collect information concerning the patient's health problem, analyze it, and finally release liquid medicine to cure it. This analysis and delivery system is anticipated to have a diameter under 1 cm.

From 1991 to 2000, MITI will be supporting the research plan with about $25–35 million a year. These amounts are regarded necessary to develop new medicine technologies in Japan. According to a prognosis, in the year 2020 about 24% of the Japanese population will be 65 years and older. Due to the overaging population, which poses difficulties for the health care program of the government, great financial investments are required. In particular, microsensors have a special importance in Japan's MST research. Private companies and public institutions are working together to maintain and improve Japan's international competitveness. In the past 10 years, more patents for biosensors were filed in Japan than in Europe and the USA together.

Besides the government sponsored research, industry is pursuing a government independent precompetitive program with its own money, it is organized by the *Advanced Machining Technology & Development Association (AMTDA)*. Japanese industry, including companies such as Nissan, Hitachi, Nippondenso, Toyota and many others, contributes about $45–70 million per year to this endeavor. Traditionally, the academic institutes of Japan are not very active in high-tech R&D activities. Nevertheless, wide-ranging MST research is being done at the Universities of Nagoya, Tokyo and Tohoku. Altogether, hundreds of millions of dollars are spent by industry and government on the research and development of MST in Japan every year.

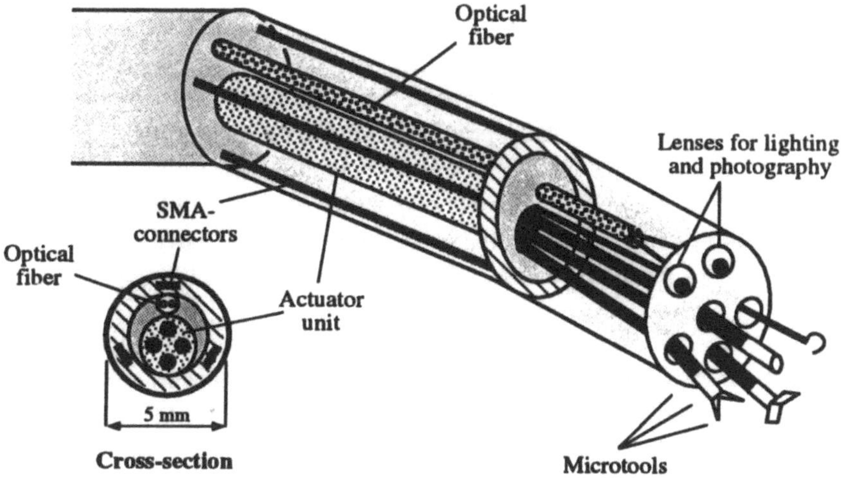

a

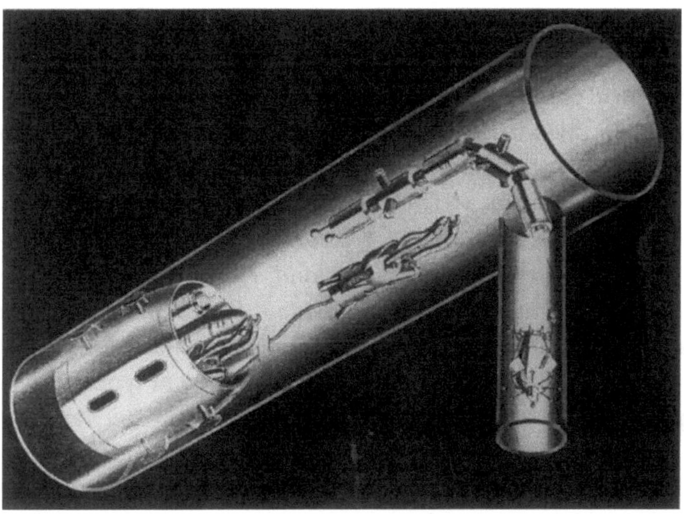

b

Fig. 1.7: Goal of the MITI program – active micro machines
a) medical catheter; b) industrial microrobots. According to [MITI91]

1.4.3 MST Activities in the USA

The USA are also very active in pursuing MST projects. Most of them are application-oriented, as opposed to projects in Europe and Japan. The government support is lower than in Europe or Japan. The *Engineering Research Center Program (ERC)* supports and educates new engineers with the help of application-oriented projects. At present, the ERC sponsors 18 centers in various MST activities; industry helps to finance this program.

The *National Science Foundation (NSF)* has sponsored and coordinated the development of MST for more than 10 years. Researchers from various disciplines are subsidized with more than $2 million per year by the *Solid State and Microstructures Program* [Haze93]. Among others, this program supports the *National Nanofabrication Facility* at the Cornell University, which is a unique interdisiplinary facility. This facility is available to all institutes working in MST, such as universities, research institutes and industrial firms. It offers possibilities to do practical experiments as well. Further, the NSF supports two other programs, the *Presidential Faculty Fellow (PFF)* and the *NSF Young Investigator program (NYI)*, for young engineers and scientists who want to establish themselves in the MST business. A sponsorship can last for 5 years with a funding of up to $100,000 per year. Another program is the *Research Initiation Award (RIA)* with a duration of 3 years and with a funding of $100,000 a year. Besides these programs there are a few others which mostly concentrate on smaller applications. Important MST research facilities have been installed at the *University of California, Berkeley, Massachussetts Institute of Technology (MIT), University of Wisconsin, University of Michigan, Case Western Reserve University at Cleveland, North Carolina State University, Carnegie Mellon University* and *Louisiana Tech University*. The nanotechnology is a new discipline which is being pursued at the *Institute for Molecular Manufacturing* in Palo Alto. It involves many different types of systems and their components, which have dimensions in the nanometer range.

At the federal level, MST development is being supported by the *Chamber of Commerce* and the *National Institute of Standards and Technology (NIST)*. NIST directs the *Advanced Technology Programm (ATP)*, which helps to sponsor special high-tech projects in small and medium-sized companies. The entire program has a budget of $1.4 billion for the period 1993–1997. There are also many MST military applications which are being converted to consumer products. The *Advanced Research Projects Agency (ARPA)* coordinates many such projects under the direction of the *Secretary of Defense*. The development of fluidic components, to be used for measuring moments of inertia, controlling electomagnetic and optical beams, is supported with about $8 million a year [Bryze94]. This program also subsidizes the development of design and simulation tools and production facilities as well as a micromachining service for universities and industry. Also NASA, the *National Health Foundation* and the *Federal Coordinating Council for Science, Engineering and Technology (FCCSET)* support MST projects.

In terms of total investment, MST is mostly funded by industry in the USA and in Japan. The largest contributions are given by small and medium-sized companies which have recently gone from microelectronics to MST and which are developing various components for microsensors and -actuators. Their advantage is deep knowledge and experience in silicon technology and microelectronics. Large companies, such as *IBM, AT&T, Ford, Texas Instruments Honeywell, Motorola, Nova-Sensor, Hewlett-Packard* and *Analog Devices*, founded their own MST research groups and have already developed their first products [Mast93], [Stix92], [Fan93], [Furu93]. There is much competition among the American companies. Almost all MST plans are kept secret. There is few cooperation between companies and the research organizations, compared to Japan and Europe. It is therefore difficult to estimate the amount of money spent on American MST activities.

1.4.4 International Market Prognoses

Up til now, there have not been any reliable international studies which compare the worldwide MST markets. It is clear, however, that MST has a great market potential. This was also confirmed by the investigation report of the Batelle-Institute [Tschu95]. According to this study, there already exists a market for microsensors and -actuators in the USA, Europe and Japan. The market prognoses estimate a total worldwide sales volume of $12 billion by the year 2000; the Western European countries will hold a share of 2.5 to 5 billion Dollars share of it. It is obvious that new, non-existing MST products are not taken into consideration yet. In Figure 1.8, the results of a MST prognosis of the Advanced Technologies Research Group of the System Planning Corporation by the year 2000 are presented. The results are based on polls taken from representatives from government and industry. This study, which evaluates many technologies, concentrated on five areas which will be expected to make the most progress over the next 5 years and which will have the greatest influence on the development of microsystems. The products include pressure and inertia sensors, fluid-technical components, optical switching elements and data memory components.

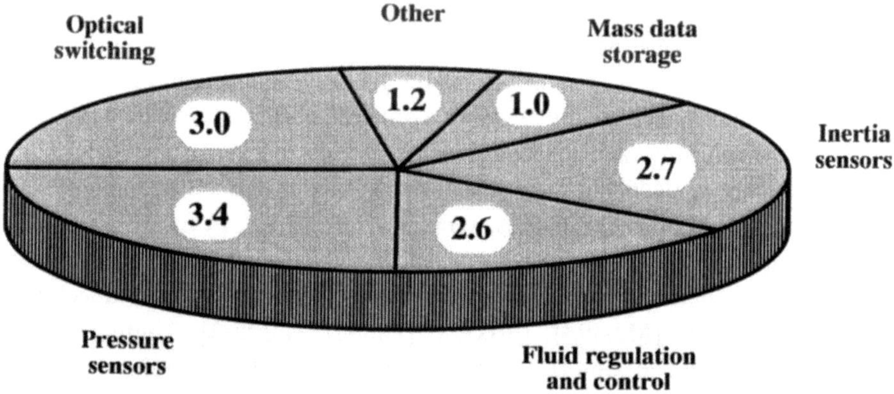

Fig. 1.8: The MST market by the year 2000. According to [Burns94]

There are estimates for various areas of application. In the automobile sector, MST will make up 2% to 4% of the total product value of a car. In Europe, the value of MST components in the automobile production will be between 2.5 and 4 billion Dollars in the year 2000. In 1993, the market for airbag sensors was about 4 million units and it is growing due to legal requirements [Bryze94]. MST is also becoming more and more relevant in the control of the environment. In Europe, the amount to be spent on MST components for water treatment will be around 50 billion Dollars in the year 2000. The growth rate in this area is expected to be 20% yearly [Bene95]. Not quite as spectacular, but still impressive, is the expected growth rate of the micromotor market; about 5.8% in the year 2000 [Prod95]. The expected sales volume in Europe will be more than $4 billion. The estimated worldwide market potential for biochemical sensors will be $1 billion in the year 2000 [Gard94].

There is also an important market to be expected for medical chips which will take on many measuring and evaluation functions. The experts believe that there will be a distribution of the market potentials for environmental and medical microsensors of 1:2.4.

Hundreds of thousands of silicon pressure sensors are produced every year for the monitoring and control of industrial production. Sensors and actuators based on MST will be used for controlling hydraulic systems, compressors, cooling aggregates, air conditioners, etc. The consumer products constitute the most important market for microsensors, it is expanding rapidly. MST components will be used as pressure meters for scuba diving gear, digital tire pressure meters, barometer and depth meters, bicycle computers to measure inclination, for hygrometers etc. Also, sensors will be installed in vacuum cleaners, washing machines, dryers, toasters or even in sport shoes for damping control.

An economic breakthrough of MST can only be achieved when Europe, the USA and Japan follow a cooperation based on partnership. The market for process analysis and control devices, for production control and for medical instruments will greatly benefit from MST, in particular in Europe [Lind93]. The USA and Japan will dominate in other areas, such as in air and space applications, and in the consumer and household technologies.

2 Microsystem Technology Applications

2.1 Introduction

In practically all technical areas, the performance of components and systems is greatly improved by their miniaturizing. The present state of the development of microsystem technology is not very advanced yet, but it allows us to get an idea of what will happen in the future. In a few years, microsystems will be able to perceive many events happening in their environment, evaluate them and convert the result into actions by means of intelligent control algorithms. Inexpensive mass production, a lower material content and low energy consumption will make the microproducts competitive economically; they will become more reliable with increasing degree of integration, and new functions will be created by using MST.

The most economically interesting application areas of MST are the use of sensors for measuring physical or chemical parameters, such as temperature, acceleration, pressure, or the concentration of certain substances. Other areas include actuators which interact with the real world. At this time, there have already been successes with integrated microsensors, but today most other applications are only ideas placed on paper. When one speaks of MST, it is important to distinguish between the products of MST which are increasingly being used and the developments of the nanotechnology. Figure 2.1 shows the spectrum of microsystems and their corresponding manufacture and measuring methods. This classification covers a dimensional range from 10 mm down to a few nm and includes all relevant mechanisms. MST clearly covers a broad range of applications. Most applications, which have been realized to date or are considered realistic by experts, can be found in the upper two boxes of this table. The corresponding application areas are illustrated in Figure 2.2.

New MST developments for medical applications should revolutionize present diagnosis and therapy methods and make it possible to recognize illnesses reliably and quickly and to treat them gently. Automotive technology has a demand for intelligent microsystems. On the one hand, automobiles are beco-

	Observation Methods	Parts	Manufacturing Technology
Milli-machines (10 mm – 1 mm)	Visible	Miniaturized parts	Precision manufacturing
Micro-machines (1 mm – 1 μm)	Optical microscope	Microparts	Silicon process LIGA-process
Nano-machines (1 μm – 1 nm)	Scanning electron microscope Scanning probe microscope	Molecular parts	Protein engineering

Fig. 2.1: The place of MST compared with other technologies. According to [Fuji92]

ming smaller and smaller with less energy consumption. On the other hand, there are such high demands for safety, control and stability which conventional manufacturing methods cannot meet under present economic conditions.

An effective environmental protection requires the availability of many different microsystems, especially inexpensive ones, in order to guarantee reliable monitoring and control. In the future, sensor networks for monitoring large areas, such as waste dumps, may be in operation if suitable microsensors will be developed. Intelligent sensors will find applications in production control and in aviation and space technology.

Microsystems for the latter two technologies are of particular interest to the military. Typical applications include the display of oil, fuel or hydraulic pressures, indication of speed and altitude, or monitoring the functions of the ejection seat. Civil developments in these areas have an increased demand for very efficient fuel and safety systems, which can only be obtained with new sensor and control concepts. MST also has a high potential in the areas of household appliances, consumer electronics and office technology, just to name a few. In the following, the most important microsystem applications will be introduced in more detail with a discussion of state of the art examples.

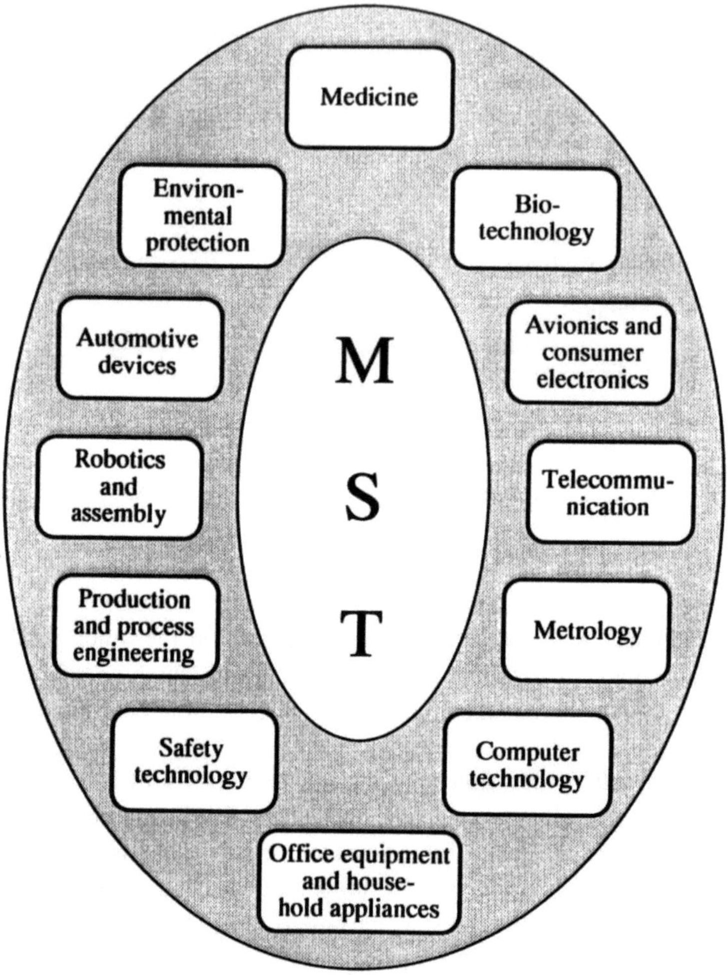

Fig. 2.2: MST applications

2.2 Medical Technology

Presently, the medical field is an excellent candidate for MST. Many parts of the traditional medicine will be undergoing radical changes since new and often unusual MST-based methods and tools are being built. They will allow the introduction of new and more effective diagnosis techniques (like endoscopy), implantable dosing systems, tele-microsurgery methods, neuronal prosthesis, etc.

A new concept based on MST developments is the minimal-invasive therapy, which is increasingly being used. Very small incisions and even natural orifices are used to reach the focus of an illness. Various concepts are being developed for carrying out this therapy, from improving existing instruments, to active endoscopy and to the design of microrobots. The latter can monitor, measure and operate. Major advances are expected in surgical practices, where up til now patients had to endure relatively large incisions. Endoscopic principles, which appeared in the late 80's and were initially only used for diagnostic, are most promising for minimal-invasive surgery. No other medical instrument has changed and advanced medicine as much as the endoscope. E.g. a thin fiber cable with an integrated cold light-source and video camera (endoscope) can be inserted into a natural bodily orifice or a small incision to make the surgery environment visible to the surgeon. If necessary, other tiny incisions are made to insert other miniaturized instruments, such as clamps, needles, slings, scalpels and rinsing/suction tubes. MST will make it possible to integrate various instruments needed for an operation within an active endoscope. The design of such an endoscope was presented in Figure 1.7a.

Minimal-invasive surgery is economical and will be very important to the future development of medicine. Less pain and fewer scars, and with this faster recovery and briefer stay in hospital (many endoscopic operations can today be carried out as out-patient treatments) will reduce costs for health insurance companies and employers. Endoscopic inspection with subsequent laparoscopic removal of the gall bladder, appendix and ovaries is a standard procedure now. Today, 80% of all gall bladders are removed with the help of minimal invasive surgery [Conra94]. Other potential applications are being more closely considered, too, Fig. 2.3. Brain surgery will also be possible by operating through a very small hole in the skull-cap. Thereby a special endoscope and a monitor will be employed, and the surgeon will be able to cut and remove tissues with a microscalpel.

Although the surgeon can observe the operation on the monitor, there are still many problems which accompany minimal-invasive surgery. Accidental injuries or hemorrhaging often force the surgeon to quickly resort to conventional, large-incision surgery. Endoscopic operations are also more difficult due to orientation difficulties with a remote tool.

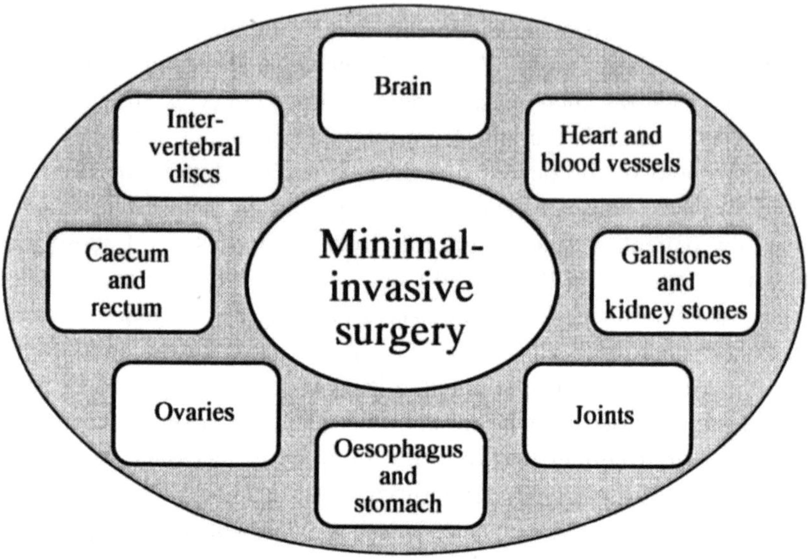

Fig. 2.3: Minimal-invasive surgery spectrum

A neurosurgeon described the difficulties in operating with endoscopic methods: "It is like trying to sew a button onto a blanket cover in the bedroom through the keyhole of the front door with a pair of tweezers. In addition, the rooms are filled with a lot of furniture around which you have to lead the tweezers. But never knock anything over!" Due to a potential risk, some necessary operations cannot be performed. Therefore, surgeons need very delicate miniaturized tools with tactile microsensors for carrying out these minimal-invasive operations. The development of such instruments is an expanding research area. Some flexible tools developed can be seen in Fig. 2.4.

Angioplasty is another future technique where MST will play a part. Here, sick organs can be reached by going through veins and arteries without having to make large incisions in the body. Today, this technique is mainly used to inspect pathological spots in blood vessels. Active microcatheters and microrobots open up new methods, whereby instruments can be remotely cont-

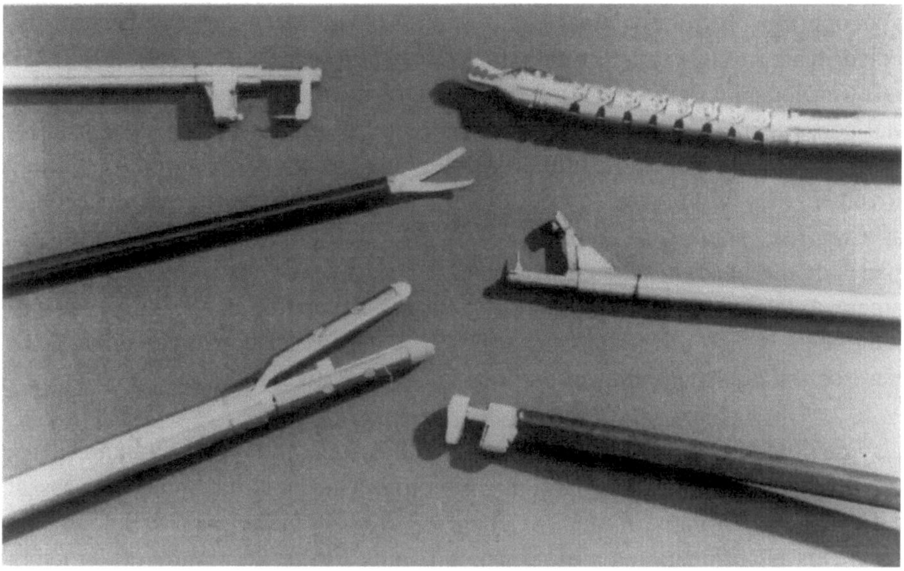

Fig. 2.4: Various miniaturized surgical effectors. Courtesy of the Karlsruhe Research Center

rolled when travelling to the sick organ and operate. They may revolutionize traditional scalpel surgery. Such devices are no longer a science fiction idea, but really form a basis for technique development now; they already have shown practical results. A similar concept is the basis of a national research project in Japan, planned for a period of 10 years until 2000 (Section 1.4.2). MST could also offer a potential in the treatment of coronary heart disease, the most common cause of death in Germany [Erbel94]. Many balloon catheter techniques have already been developed and practiced which allow the enlargement of restricted arteries. The balloon catheter is inserted into the leg artery, pushed to the constriction of the blood vessel with a wire and is inflated there. Future catheters will be equipped with integrated systems, such as cutting tools and might be driven by micromotors and control electronics in the catheter itself. These intelligent mechanisms may also search for fat deposits and clots in blood vessels and remove them with their integrated tools.

Another interesting application of modern R&D results in medicine is telemicrosurgery, such as a surgical operation using a stereo microscope with special microstructured tools. With a large magnification, the surgeon can even operate on minute pieces of tissue. The greatest success can be expected in neurosurgery and ophthalmology where access to a space is extremely limited and damage to microscopically small blood vessels may have fatal consequences. Plastic surgery has already used this technology. Manual microsurgery, however, is limited by the inaccuracy of the doctor's hand movement. There are high expectations from "virtual reality", which could, for example, support remotely controlled microoperations. Stereographic images of the location of the operation are dynamically transmitted to a CCD camera display worn on the doctor's head. The doctor can change the position of the stereo microscope lenses by moving his head and can control the microrobot via telecommunication lines by moving his finger.

It is imaginable that artificial prosthetics will replace almost all human organs, but there are often miniaturization problems. The smallest most successful artificial organ of modern medical practice certainly is the pace maker for the heart. This is an integrated system which can be further miniaturized by using MST techniques. The application potential of MST is very high here; in the USA alone, 120,000 of these systems are needed per year [Tana95]. There are many applications for sensor prostheses, such as hearing aids. The biggest problems with prostheses cause the microactuators and energy supply. Human body organs use chemical energy to operate, but artificially made systems require other energy sources.

One important goal of medical technology is the development of implantable drug delivery systems, such as artificial pancreas, which should contain glucose sensors and controllable dosing pumps. It is important that these organs can be removed without operating on the patient. A possible solution is the use of capsules which can be destroyed by external signals without hurting the patient. For treating intestinal infections, a so-called torpedo pill is being developed which the patient simply has to swallow [Quick92]. The pill's coating protects the sensitive drug from the aggressive stomach acid and the drug is discharged exactly at the inflamed part of the intestine. The pill's position is located by ultrasound or an X-ray monitor. As soon as the capsule has reached the infected area, the active ingredient is released, for example by means of a radio signal. The pill is about 2 cm long and contains a discharge mechanism and a container for the drug; there is also an antenna and possibly a transmitter. The pill will leave the body through the natural way, via the digestive track. After sterilization, it could be used again.

Implantable sensor systems may also be necessary for permanent monitoring of various physical parameters inside the human body, such as blood pressure, ion concentration, temperature, etc. Obviously, all the components of such intelligent microsystems must be in the submillimeter range for *in vivo* applications and should also be inexpensive, reliable and biocompatible. There are only a few technologies which can meet these high demands, and they are still in the experimental or development phases. Today, only disposable pressure sensors are mass-produced in the USA [Axel95]. Especially fiber optic sensors are important since they make it possible to monitor various body parameters *in vivo*. An implantable glass fiber dosimeter is being developed for radiation therapy, making it possible to monitor the radiation dosis in the body of a patient in real-time [Büker93]. The dosimeter is made from two fibers, one made of radiation-sensitive lead glass and the other used as a medium for the signal transfer, Fig. 2.5. The radiation causes attenuation in the light fiber which can be detected at the end of the fiber.

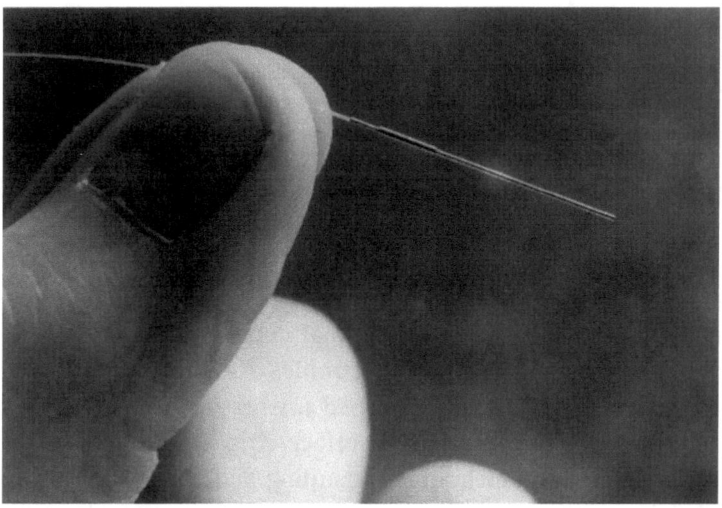

Fig. 2.5: Implantable fiber-optic dosimeter. Courtesy of the Jülich Research Center

An integrated micro-measurement system is being developed for the continuous measurement of various blood parameters [Mokw93]. The sensor chip contains a pressure and temperature sensor, as well as signal processing electronics and can be fitted into a catheter (chip dimensions of 0.7 mm by 7 mm). The oxygen content of blood, which takes about 2 hours to determine today, is another important parameter. Since it is a very sensitive process, the doctors must exactly adhere to the analysis steps and times. A new microsensor, which will be integrated into the above microsystem additionally, will determine the blood parameters in real-time.

A concept of an implantable sensor chip of a few mm^2 in size for blood pressure measurement was introduced in [Zache95]. The chip should have no physical connection to the outer world, neither for control nor for energy supply. It contains a capacitive pressure sensor and an interface for telemetric information and energy transfer. A schematic diagram is shown in Figure 2.6.

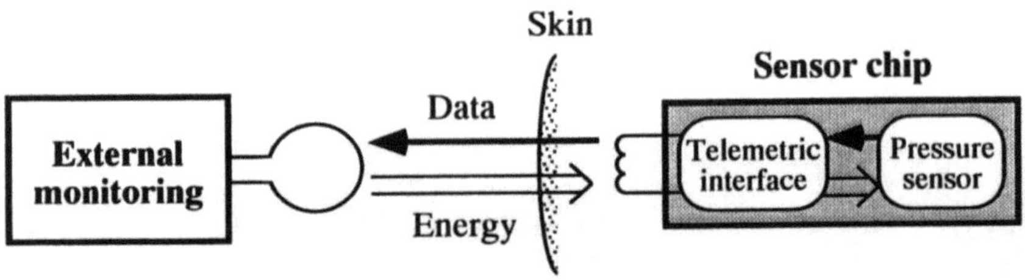

Fig. 2.6: Concept of a telemetric microsensor system. According to [Zache95]

A future application of MST is neurotechnology, where researchers investigate the implantation of microstructured neuronal connections between regenerating peripheral nerves and external electronics for the restoration of certain neural functions, Fig. 2.7. A sieve-shaped microstructure serves as a neuronal interface. The holes in the silicon interface chip are surrounded by recording/stimulating electrodes which allow recording and stimulation of individual axons. Thereby researchers can gain important information about the complex organization of the peripheral nervous system. The depicted electrode arrangement was succesfully implanted between the cut ends of peripheral taste fibers of rats and nerve fibers functionally regenerated through the microholes.

Such neuronal interfaces are needed for a new generation of motorized/-sensory prostheses. They will make it possible to control a prosthesis by the

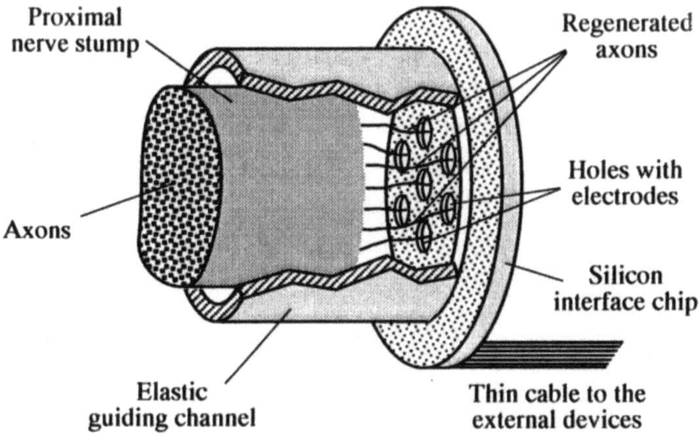

Fig. 2.7: Principle of neural prosthetics. According to [CISC91]

so-called functional electrical stimulation. For spinal cord injuries involving severed nerves it should be possible to bridge over the interrupted nerve paths, thereby allowing affected body parts to be directly stimulated. First, the axon signals of the proximal nerve stump are decoded and processed by external electronics, then the associated stimulation is propagated to the distal nerve stump of the restricted body part via a silicon interface. The development of such microsystems, which function as bidirectional connectors between the peripheral human nervous system and external devices, is an active research area today; here the skills of medicine, computer science and microsystem technology are combined [Cocco93], [Dario94] and [Meyer95]. The importance of these endeavors is clear: in Germany alone, one thousand people are paralysed by accidents, another thousand lose their hearing and 2,500 their sight each year [Wein94].

Initial research results show that the main problems in creating a suitable connection between living cells and artificial microstructures are negative effects such as the rejection of a device. Electronic and biological systems can be coupled with micromechanical chips, a prototype of which is shown in Figure 2.8 [Akin94]. This neurochip is made of silicon (the technologies will be discussed in Chapter 4) and contains several holes of different diameters, down to 1 μm in size. When it is implanted, neural cells grow through the holes, establishing a stable mechanical and electrical connection between the neurons and the control electronics. The interface chip must be integrated into a special guiding channel made of a biocompatible and flexible material such as a polymer or silicon rubber. This channel allows the guided growing of axons as well as the fixture of the chip.

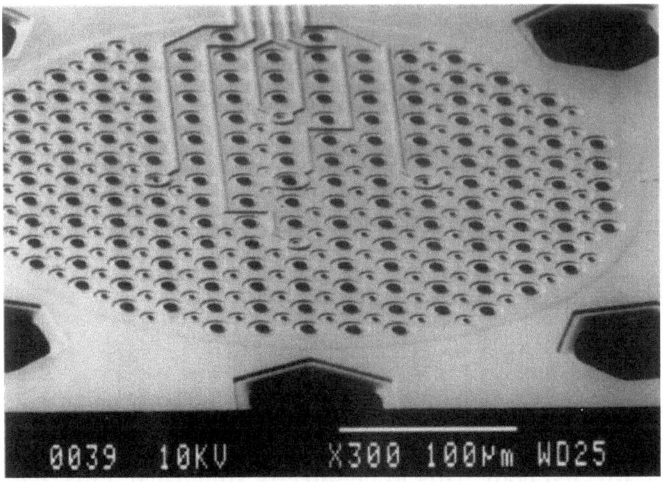

Fig. 2.8: Silicon interface "axon-control electronics". Courtesy of the University of Michigan (Solid-State Electronics Laboratory)

To generate control signals for prosthetic devices, the next step after establishing the mechanical connection between the electrodes and nerves is transforming or decoding the neural signals into electronic data and vice versa. An electronic interface must be developed for this to process data flows in both directions. The final step is the development of an intelligent control system which can correctly interpret biological signals and act on them. This type of neurotechnology is still at its very beginning.

2.3 Environmental and Biotechnology

The analysis of chemical substances is a main focus of environmental technology. The goal is to prevent the creation and release of contaminants and to quickly and accurately detect any contamination of the environment so that an emission can be kept to a minimum. The present environmental problems can only be solved by continuous and reliable on-line analysis, since this is the only way to avoid possible damage in time. It is also hoped that this technology will reduce the risks of dangerous processes and their impact on the environment. This is only possible if the danger is quickly detected.

An analysis system which can fulfill the above-mentioned requirements must integrate many functions, such as sample taking, parameter measurement, signal processing and performing data output and subsequent control. The weakest link in this chain, which determines the reliability of the entire system, is the measurement part, which includes various chemical microsensors and microprobes; they are discussed in detail in Chapter 6. Also of great interest to environmental technology is the measurement of effects spread over a wide surface and the simultaneous testing for various environmental toxins or contaminants. This is only possible, however, if inexpensive analysis instruments can be developed. Here it is important to use very small samples, since the substances being analyzed are often not abundantly available, are expensive or are toxic. Often micropumps must be integrated in a chemical sensor system for sample collection. The minute samples are then delivered by the micropump in an exactly defined amount to the place of measurement.

The development of a spectral microanalysis instrument containing a microspectrometer and micropumps made by the LIGA method, chemical microsensors and data processing units was described in [Menz93a] and [Schom93]. This microanalysis instrument is only a few mm in size and its reflection grid has dimensions in the submicrometer range, Fig. 2.9. When in operation, the micropump takes a sample of the substance to be analyzed, such as contaminated river water and it transfers the sample via a valve to an inspection chamber where it is spectroscoped. An integrated photodiode array translates the spectrogram into electrical data which is then analyzed by a microcontroller. The value of the measured concentration of the sought substance is transferred to an external display over an interface. After the measurement, the sample is discharged, the sensor rinsed by means of a second micropump and recalibrated with a reference liquid. This analyzer requires a very small amount of expensive chemicals compared with conventional ones. The micropumps used here are described in more detail in Section 4.2.6 after the introduction of the LIGA-technique. The microspectrometer, which is to be integrated in the above-described microanalysis instrument, is described in [Müll93] and [Hage95]. The principle function of the optical measurement system is depicted in Figure 2.10.

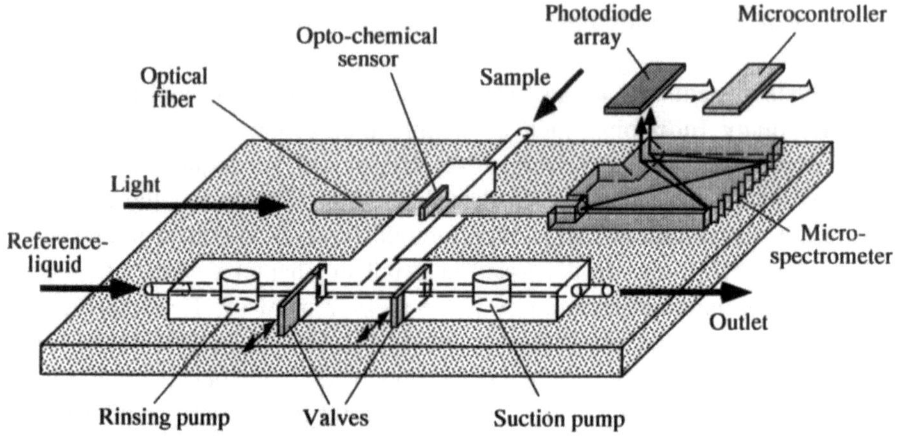

Fig. 2.9: Contaminant analyzer using an optical principle. According to [Schom93]

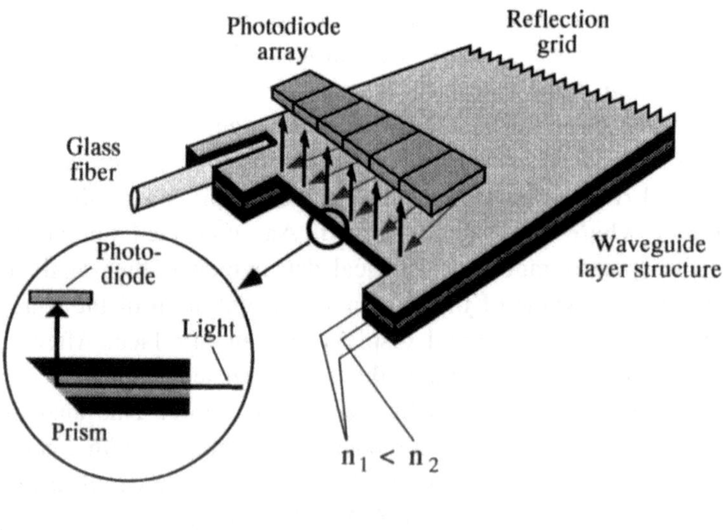

Fig. 2.10: Microspectrometer (LIGA method). According to [Menz93a]

The microspectrometer consists of several polymer layers and was made with the LIGA method. A glass fiber which is positioned within a shaft, transmits the incoming light to the sensor base where it is directed toward a serrated, self-focussing reflection grid. There, the light is split into its spectral constituants and retransmitted through the sensor base. Inside this base the light is deflected by a 45° prism and brought to a photodiode array for detection. The grid is made of PMMA and has about 1,200 serrations, which are about 3 μm wide and 0.5 μm high. The spectral range of this device is between 400 nm to 1100 nm and the spectral resolution is about 10 nm.

In the biotechnology, special microtools are needed to do micromanipulations like handling, holding, transporting, sorting, slicing and injecting under a microscope. Such cell manipulations are important for many biotechnical applications, such as genetic research and breeding. Environmental investigations also need microscopic operations, since cells are a good indicator for the presence of dangerous substances. It is often necessary to partially or completely automate such sensitive operations since the human hand is too coarse a tool. Both numerous stationary micromanipulation systems and universal mobile microrobots, equipped with delicate tools, are being developed; such systems will be introduced in Chapter 8. Besides the use of mechanical gripping systems, other non-contact methods are being investigated which will allow precise and gentle handling of individual cells.

2.4 Automotive Technology

Every year, more than 35 million cars are manufactured worldwide; they are equipped with increasingly complicated control and safety devices. High customer demands on comfort and safety has led to an increasing number of electronic systems being installed in motor vehicles. The BMW 750 e.g. contains 73 electric motors, 50 relays, 1,567 plug-connectors and 25 control devices. Anti-blocking system, automatic stability control, power steering, motor performance management, air conditioning, central locking systems, diagnosis systems, etc. are all controlled electronically. Soon the driver can imagine that he is driving in a huge chip rather than in a car [Nach92]. Since the average car size may not increase, more and more functions have to be integrated within the same space as before. Today, about 10% of the automobile's weight and 15% of the cost are made up of the electrical and electronic components [Ehle92]. This means that electronic devices have to be further miniaturized and that eventually MST will be used.

Especially important for the future of the automobile industry is a drastic miniaturization of sensor elements. In 1970, practically no sensors were installed in automobiles whereas today the number ranges from 20 to 80, depending on the make and model. This trend will go up as the demand for travel comfort and safety increases. Another factor to consider is the continuous political discussion concerning the ban of cars which exceed a certain level of emission gases and fuel consumption. New concepts for reduction of pollution and fuel consumption are needed. The monitoring of car functions has become important now. There are sensors which monitor the motor, control emission levels and supervise brakes functions and soon they will have many other tasks to take care of. Figure 2.11 shows a system model of all automobile sensors.

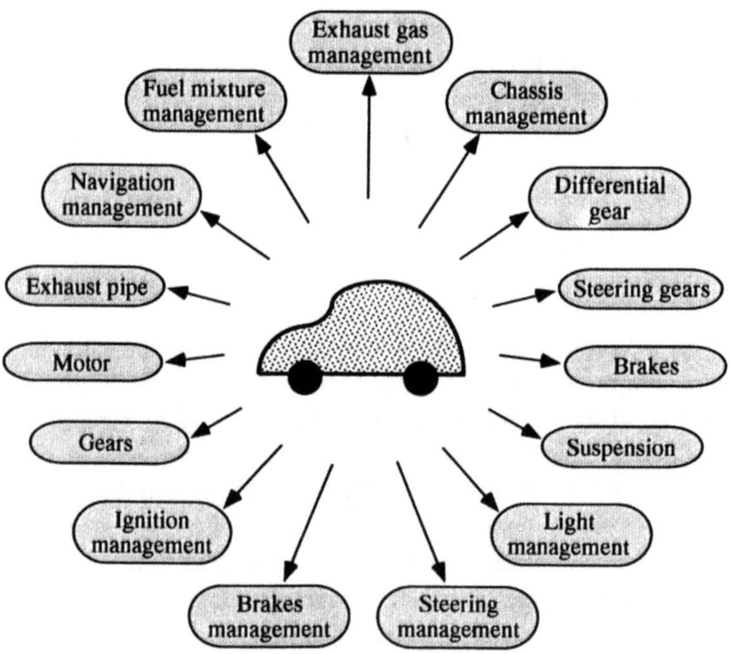

Abb. 2.11: Possible use of microsensors in a car. According to [Klei92]

The final goal is to have the automobile control completely networked via a bus system. Such a data network will make the vehicle data representation more compact and will supply the driver with wireless interfaces for future traffic information systems.

Presently, microsensors (acceleration sensors) for airbag control have the greatest market chances today; there will be a great demand for them in the near future, Fig. 2.12.

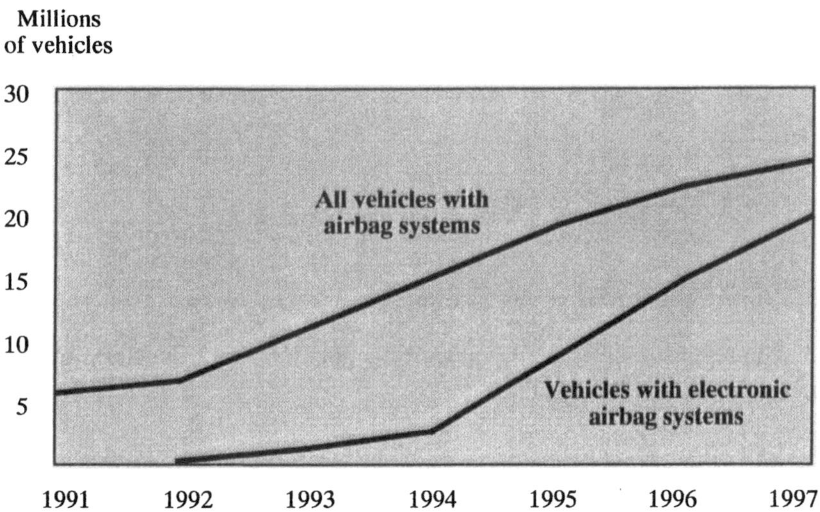

Fig. 2.12: Market study for airbag systems. According to [Ajlu95]

Advanced airbag control systems have already existed for a long time in the automobile industry, but the micro-acceleration sensors which are manufactured using silicon-based batch processes or the LIGA-technique are expected to lower their price to under $10 each and to eliminate the bulky electromechanical switches of today. The mass-production of silicon crash sensors started in the USA and is now also done in Europe [Axel95]. A typical microsystem for measuring two-dimensional acceleration was described in [Menz93a], Fig. 2.13. The device has three capacitive acceleration sensors each for the x and y directions (Chapter 6). The sensors are made by the LIGA method using nickel and are coupled to analog microchips for collecting data. A 3D accelerometer with an additional sensor to measure acceleration in the z direction is also being developed [Menz95], [Kröm95]

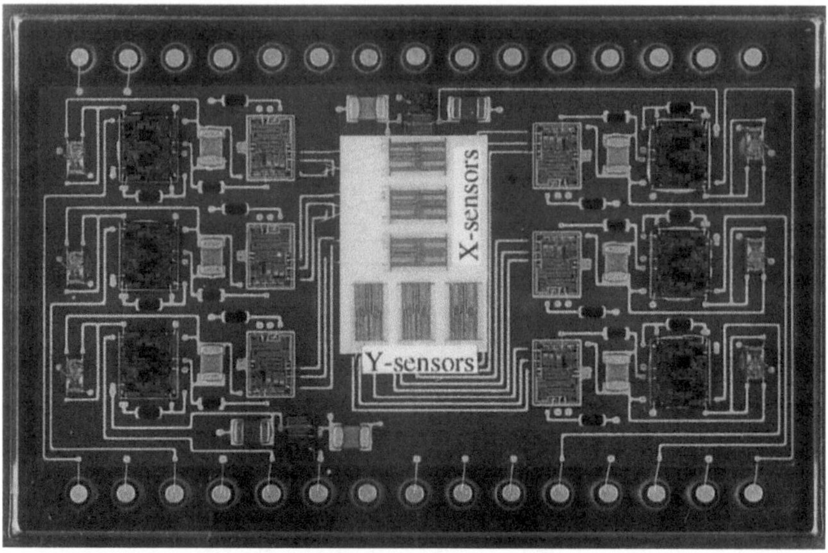

Fig. 2.13: Microsensor system for measuring acceleration. Courtesy of the Karlsruhe Research Center, IMT

Miniaturized angular speed sensors will also have a successful future. They make up an essential part of the intelligent navigation and control systems [Ajlu95]. The entire American market for acceleration and angular speed sensors is expected to grow from 17 million sensors in 1995 to 114 million in the year 2000. Companies like Motorola and Analog Devices, plan to have micromechanical gyroscopes as a standard feature of every car by the year 1998. Many major automobile manufacturers have recently started to install various silicon micro-pressure sensors for optimizing fuel consumption of their cars. These companies also plan to have microsensors for measuring the pressure in the fuel, oil, air and hydraulic systems of engines and transmissions. Since these sensors will open up important markets, many automotive suppliers have already started development activities in this field.

The initial successes of MST gave rise to enthralling future expectations. Developers expect that one day tiny microdiagnosis systems equipped with sensors will be able to move autonomously or by remote control through fuel lines to find leaks; others will optimize the combustion process of low-pollution motors by measuring operating parameters directly in the combustion chamber. Distance sensors installed in the front and back bumpers will not only help with parking (as they already do in the larger models of luxury cars), but should also help to prevent body dents. Instead of using individual conventional sensors, a network of microsensors will be installed, making the failure of a few sensors not critical. A market prognosis estimates that automobile microsensors will be a $500 million market by the year 2000 [Bryze94].

2.5 Manufacturing and Metrology

Important developments in the quality control of object surfaces are conventional scanning electron microscopy and laser scanning microscopy, where the object surface is scanned with a laser beam with a diameter of 1 μm, and a newcomer, the scanning probe microscopy (SPM). This technology allows access to minute material structures in the subnanometer range. Conventional light microscopy is often insufficient since atomic defects in crystals or defects in integrated circuits cannot be detected. These structures have typical dimensions of just a few nm. Scanning probe microscopy has an extremely high resolution, which was unattainable until now and which can go down to the atomic level. This technology opens up new doors for the localization of defects. SPM instruments can be used as *Scanning Force Microscopes* (SFM), (also called *Atomic Force Microscopes* (AFM)) or as *Scanning Tunneling Microscopes* (STM). The measurements are non-destructive and allow three-dimensional, non-contact examination of any surface pattern with dimensions of μm to nm; the tolerances are in the nm range. This will lead to greater understanding of the physical and chemical properties of the materials and surfaces, and in the end it will improve the quality of microstructures. These microscopes have a very high resolution of about 0.3 nm and can even follow the manipulations of an individual atom [Torii93].

In order to work in the nano-world, local micromechanical probes are needed which have a tip of just a few μm in length. These probes scan the surface of the microstructure at a nm distance away from it. The STM measurement principle uses the tunnel effect known from quantum mechanics. The tunnel effect occurs between two current conducting media under a defined voltage; the surfaces are separated from each other by an insulating layer or vacuum, Fig. 2.14. The tunnel current flowing between the probe tip (attached to a 3D piezotransformer) and the scanned surface is measured. It is held constant by keeping the measurement gap constant, which is obtained by energizing the piezo transducers as needed for the positioning of the tip. The displacement of the individual piezotransducers gives an indication of the surface structure.

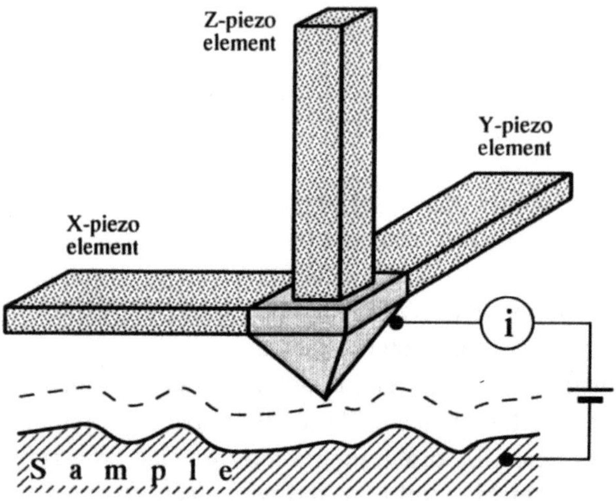

Fig. 2.14: The piezotransformer and measurement principle of a scanning tunnel microscope

The SFM instruments make use of the attractive forces between the probe tip and the atoms on the surface of the microstructure. Instead of a piezotransformer, the probe in the SFM instrument is fastened to a tiny cantilever beam having a small spring constant. The cantilever is made to vibrate and is guided over the structured surface. The resulting attractive forces cause the beam to deflect. This deflection is measured directly by either using piezoresistive or piezoelectric force sensors integrated in the probe or by indirectly using capacitive or optical measurement principles. The motion of the cantilever e.g. can be registered by a laser sensor, Fig. 2.15. The position change of a laser beam which is reflected by the cantilever is measured. The angle of

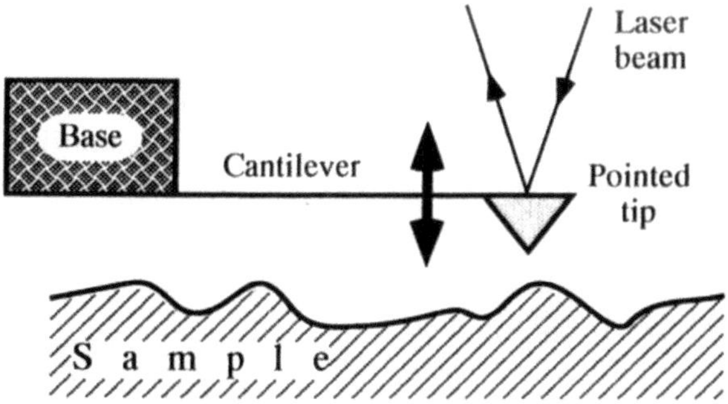

Fig. 2.15: The measurement principle of a scanning force microscope using a laser. According to [Torii93]

reflection can be exactly correlated to the probe movement. A high resolution image of the surface structure can be produced electronically from this information. In Figure 2.16 a silicon probe of a scanning force microscope using the capacitive measurement principle is shown. Here, the motion of the cantilever beam is registered by the change of the capacity between the beam and the other electrode placed above it.

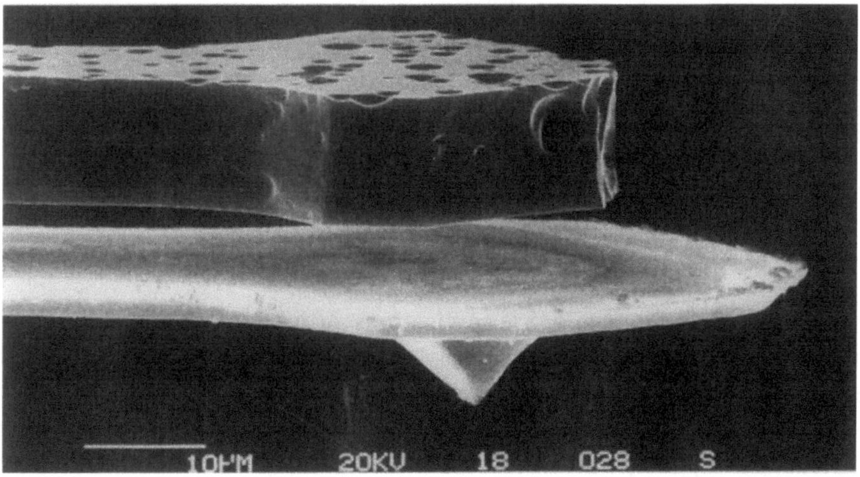

Fig. 2.16: Photo of the silicon probe of a scanning force microscope using the capacitive measurement principle. Courtesy of the University of Neuchâtel (Institute of Microtechnology)

These methods can be used to find defects directly after a manufacturing step, which will prevent defects from being propagated and with this increase product quality and lower costs. The performance of such instruments in quality control can be increased by using a parallel, time-saving sampling method. Rows of probe tips, each having its own spring, can be integrated in one instrument, whereby each probe can still addressed individually, Fig. 2.17. Figure 2.18 shows a scanning force microscope image of a surface of a PMMA microstructure made by the LIGA method.

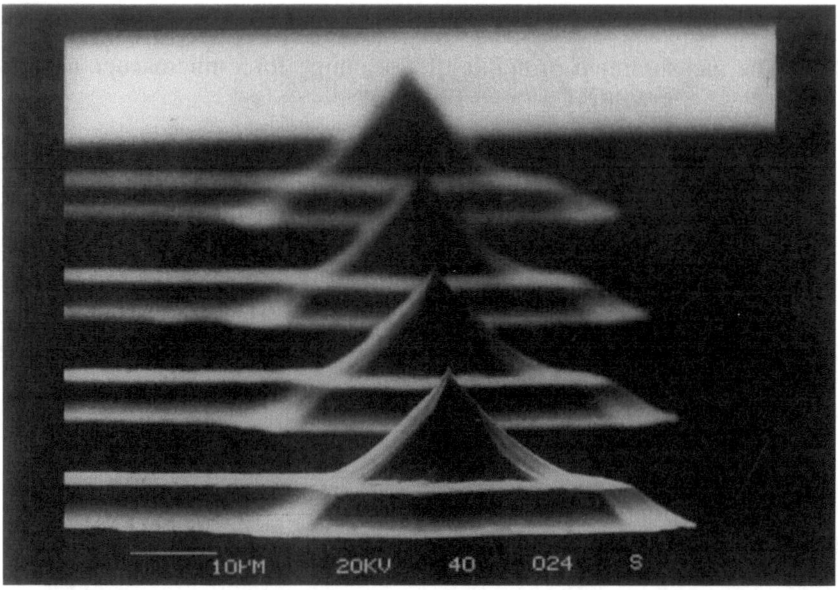

Fig. 2.17: Rows of SFM probes. Courtesy of the University of Neuchâtel (Institute of Microtechnology)

There are other conceivable MST applications in production technology. In many production processes high precision is demanded by sensor-guided assembly tools and machine tools with low vibrations. In the case of a grinding machine, for example, there are often chatter vibrations between the workpiece and the tool, which drastically reduces its performance. This may cause the surface of the workpiece to produce ripples with differences in height of a few micrometers. To solve this problem, microactuators can be used which help to actively dampen the vibrations.

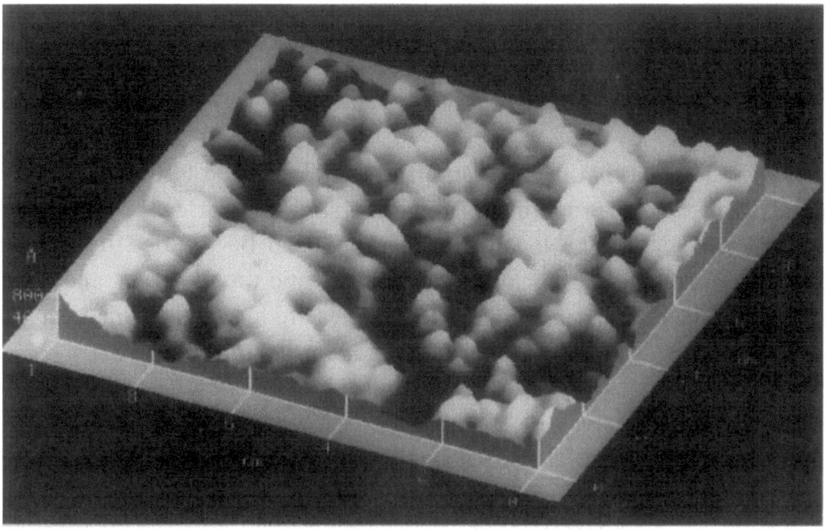

Fig. 2.18: Example of a measurement using the SPM method. Courtesy of the Institut für Mikrotechnik GmbH, Mainz

There are many more potential MST applications that can be used in products from almost all sectors of human life. For example, consumer electronics and computer technology offer many applications for microsystems. Inkjet heads for printers or drive heads for disks are only a few good examples for microcomponents which are already being mass-produced [Axel95]. The whole field of microrobotics, which is currently being investigated, may lead to possible applications in medicine, precision manufacturing, metrology, biology and numerous others. Many of them will be covered in detail in Chapter 8. In the following chapters we will also discuss many MST applications, in which microactuators and microsensors play an important role, most of them are application-specific.

3 Techniques of Microsystem Technology

In Chapter 1 we learned that in MST an effort is being made to combine microtechniques with special properties of materials and their effects to fabricate functional microdevices. The techniques of MST can be divided into three classes: the microtechniques, the system techniques and the materials and their effects, Fig. 3.1. The borderlines of these techniques are not well defined. However, the classification is useful when discussing the principles of MST.

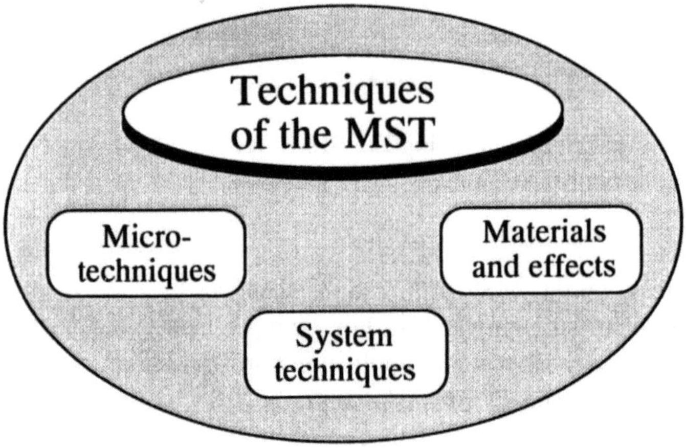

Fig. 3.1: Classification of the MST techniques

3.1 Microtechniques

The availability of the fundamental microtechniques shown in Figure 1.6 is the basic prerequisite for the miniaturization and integration of micro components into a microsystem. The majority of the microtechniques originate from microelectronic and conventional sensor and actuator technologies. They are being adapted to the special requirements of MST. A more detailed discussion of the most important microtechniques can be found in [Heub91] and [Menz93].

3.1.1 Layer Techniques

These techniques comprise methods of producing and structuring layers of materials in the micro and nanometer range. With these methods, layers of conductive or non-conductive materials of various thicknesses are deposited on a substrate. They allow the production of devices for a large variety of applications. Metal layers are mainly used as electric conductors or resistors, whereas insulation layers prevent the flow of electricity between conductors. Passivation or sacrificial layers are frequently employed and there are numerous materials which serve as sensitive layers of sensors. The special role of the layer techniques in MST will be explained in detail in Chapter 4.

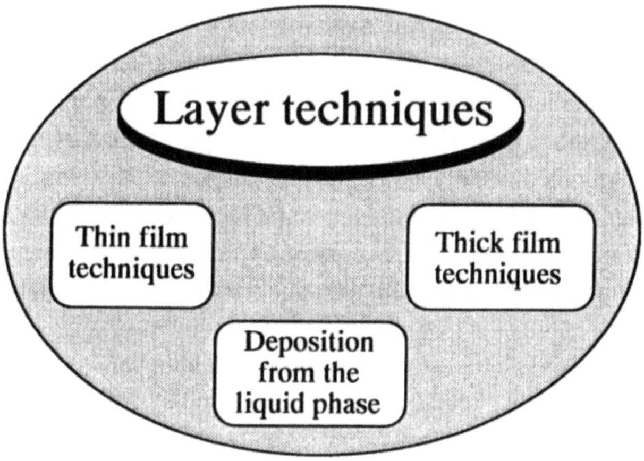

Fig. 3.2: Structure of the layer techniques

The layers can be attached by numerous techniques; here the cleanliness of the material and the adhesion to the substrate are of great importance. The layer techniques can be divided into three methods: the thin film techniques, the thick film techniques and the deposition techniques from the liquid phase, Fig. 3.2. Originally, the thin film and thick film techniques were used to produce semiconductor elements and to miniaturize electronic circuits to reduce cost and material. Deposition methods from the liquid phase are not widely used and are still under development. The LIGA method applies the latter method in one of its processing steps (Chapter 4).

3.1.1.1 Thin Film Techniques

A thin film layer has a thickness in the range of a few nanometers to a few micrometers. They have two important functions to perform; first, the generation of functional layers and second, the structuring of the chip. The processes used to generate layers employ thermal, chemical and physical deposition techniques. Important processes to structure thin film layers are the lithographic and etching methods as well as doping and microforming. They are covered in Chapter 4 under the basic micromachining technologies.

- **Thermal deposition**

Silicon is a semiconductor material which is quickly covered by a layer of oxide when exposed to air. Chemically resistant silicon oxide layers are often used for masking in connection with the chemical etching processes to obtain a defined surface structure. Thereby, certain areas of the substrate may not be exposed to the process during etching and are masked. An important aspect with complex microsystem devices is that these layers also serve as electrical insulators. Oxidation can significantly be accelerated by exposing the silicon in a reactor containing oxygen at temperatures of up to 1200°C. Such thermal oxidation is widely employed by industry.

- **Physical layer deposition**

Electrical conductors of aluminum, chromium, nickel and platinum are deposited on a substrate by sputtering or vapour deposition. With the vapour deposition method, the material to be deposited is heated up in a vacuum chamber to high temperature, Fig. 3.3. The atoms of the evaporated material condense on the substrate to be processed. Vapour deposition is a conventional well understood method. However, it is rarely used today, because of poor layer adhesion.

The sputtering process is also done in a vacuum chamber whereby the cathode is made of the metal to be deposited, Fig. 3.4. Inert gas ions (e.g. argon) which are generated by the plasma in the chamber are used to bombard the cathode, they tear loose metal atoms which are then condensed on the closeby substrate. This method is suitable for various materials. The adhesion of the layers is much

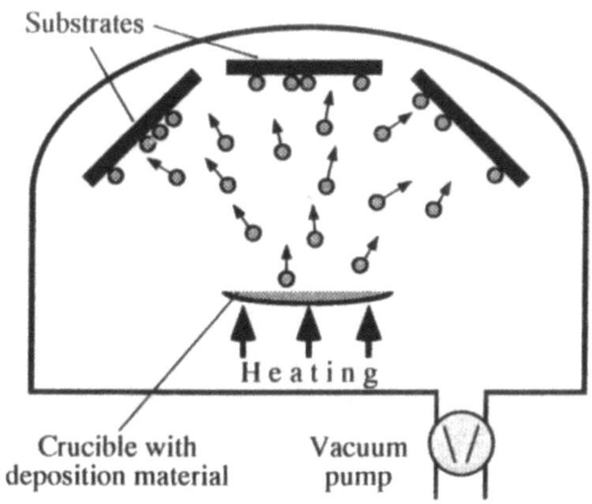

Fig. 3.3: Principle of vapour deposition. According to [Müll95] and [Menz93]

greater than that obtained by vapour deposition. The basic sputtering process can be altered in many ways. With magnetron sputtering, for example, it is possible to control the trajectories of the electrons to increase the sputter rate. By adding oxygen or other reactive gases, a chemical component can be added to the mechanical process of sputtering, which is then called reactive sputtering. The composition and properties of layers can be exactly controlled by the chemical reaction that occurs in chemical sputtering.

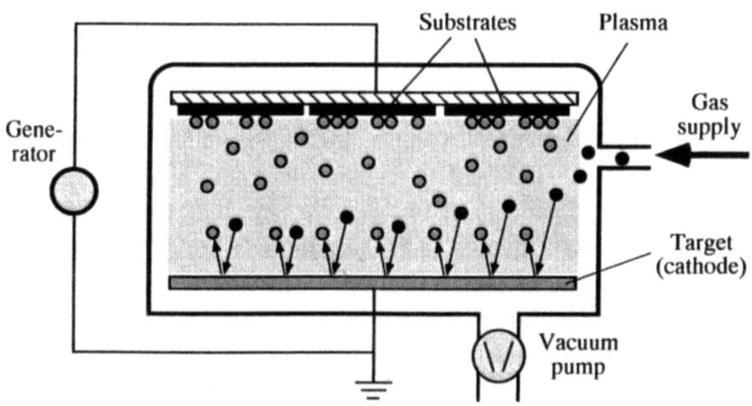

Fig. 3.4: Principle of sputtering. According to [Müll95] and [Menz93]

- **Chemical layer deposition**

Presently, chemical vapour deposition from a gaseous phase is the predominantly used thin film technology. With this method, a substrate is placed in a reactor and exposed to thermally instable gas which contains the material to be deposited, Fig. 3.5. Under a high reactor temperature (up to 1250°C), a chemical reaction on the substrate surface decomposes the gas into a gaseous and a solid component. The latter is deposited on the surface of the substrate as a very thin and uniform film, whereas the gaseous component is sucked away. By this method, SiO_2 and Si_3N_4 layers as well as amorphous poly- and mono-crystalline materials may be deposited.

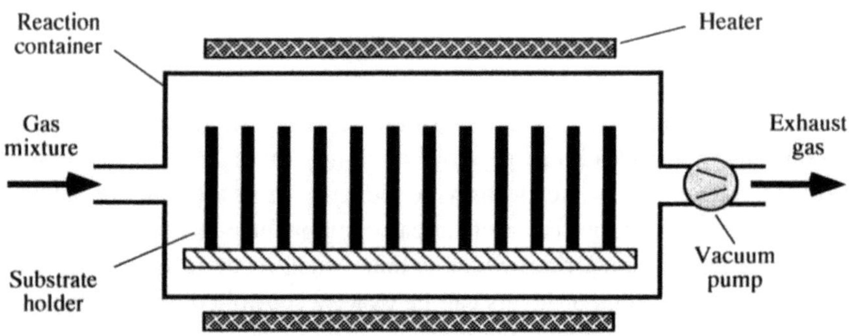

Fig. 3.5: Principle of the chemical vapour deposition method. According to [Müll95] and [Menz93]

Depending on the reactor pressure employed, the method is called atmospheric pressure or low pressure CVD. The disadvantages of this method are relatively high process temperatures and the use of toxic gases. Another version of the CVD technique is the Plasma Enhanced CVD method (PECVD). It uses a plasma discharge process at temperatures between 150–350°C; high energy electrons break up the chemical bonds and ionize the molecules of the layer material. These ions are then deposited onto the surface of the substrate. According to the frequence of the plasma-initiating electrical fields, there exist different versions of the PECVD process. The widely used epitaxy process for the deposition of a single crystalline layer of silicon, known from microelectronic technology, is a special version of the CVD technique. With this method, the processing temperature of the substrate has to be kept at a very exact value over 1000°C, which may be very difficult.

The chemical layer deposition methods have important advantages in comparison with other methods. The deposition of a layer can be exactly dosed in respect to its thickness and physical properties and the purity of the used material is extremely high. The CVD and PECVD methods produce very durable layers of polycrystalline silicon and passivasion layers of silicon nitride and silicon oxide. They are widely used by industry.

An improvement of the thin film method may be obtained by the use of various laser methods, taking the advantage of the excellent focusing ability and the power density of laser. Other advantages of laser are high resolution and minimal damage to the material being processed. The microstructures made by this method do neither require lithographic preparation nor etching. The physical laser method is of pure thermal nature, it uses the direct interaction between a laser beam and the surface of the substrate. This method has been used for a long time for piercing holes, welding, cutting and soldering. With the chemical laser method, the substrate to be machined is immersed into a liquid or gaseous substance. When the laser beam is applied, a pyrolytic or photolytic reaction takes place between the surface atoms of the substrate and the molecules of the surrounding substance, thereby drastically increasing the speed of reaction. The laser structuring methods have a limited processing speed, making them infeasible for mass production of complex microstructures. Presently, one of their main applications is for the repair of masks. A detailed description of numerous laser based thin film techniques can be found in [Heub91].

3.1.1.2 Deposition from the Liquid Phase

Methods of deposition from the liquid phase use galvanic, spin-coating, catalytic and other principles. They will be briefly described:

• Galvanic techniques have been used for many years by industry, e.g. for the application of the primer coating on an automobile. The microgalvanic method permits the deposition of various metals on a microstructured substrate surface. The substrate is immersed into an electrolyte, and a voltage is applied between the part and a counter-electrode, making the material deposit to form the desired surface structure. The obtained metal layers have a very high aspect ratio. The term "aspect ratio" is of importance for MST and will accompany the reader during his further studies. Its meaning can be visualized from Figure 3.6. Of importance is the so-called maximum aspect ratio. It considers the minimum width W and the maximum height H that can be obtained with a process.

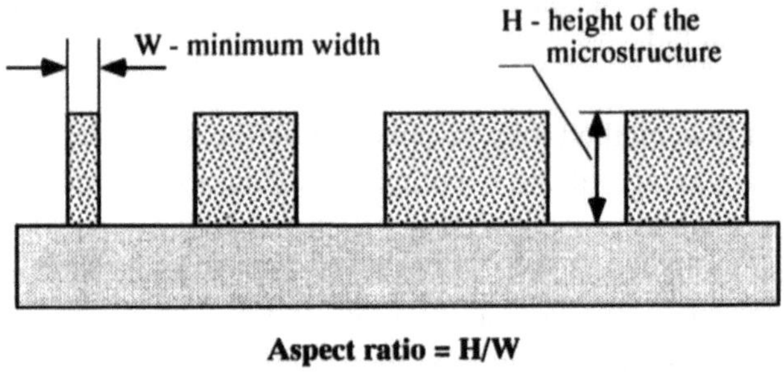

Fig. 3.6: The meaning of the term *aspect ratio*

Galvanic techniques are used by the LIGA process on an industrial basis. Of disadvantage is the slow deposition speed and the impurities often contained in the metal layer, making the process unsuitable for some applications.

- The catalytic method only allows the deposition of metals. With this method, at first, a catalytic active metal (e.g. Pd, Pt) with a layer thickness of several nm is deposited on the substrate. Thereafter, the layer is photolithographically patterned to obtain a desired surface structure and it is immersed in an electrolytic solution. A reaction with the catalyte takes place, and the metal which is in the solution is deposited on the areas covered with the catalyte.

- The spin-coating method is mainly used for the manufacture of semiconductors to obtain a photosensitive coating. With this method, a small amount of liquid coating material is deposited on a substrate which is spun at a constant high speed. The liquid spreads equally over the surface and is hardened in a drying process forming a thin coating.

There are numerous other methods experimented with. However, they have adhesion problems with the layers they produce and are not of commercial value.

3.1.1.3 Thick Film Techniques

These techniques are a standard in microelectronics to produce printed circuit conductors and insulation layers. In the case of screen printing, a special paste is applied on the substrate via a mask. The layer is dried and burned in at high temperature. The mask is made from a very fine screen and the desired microstructure is obtained by covering up certain areas of the screen to make the mask impervious to the paste at these places. The resolution which can be obtained is limited to 50 μm, making the method useless for many MST devices.

3.1.2 Micromechanics

The three-dimensional structuring of solids is understood by micromechanics which is an attempt to design devices in the micrometer range. Typical design elements are membranes, cantilevers, reeds, gears, grooves etc.; they are used to build miniaturized sensors and actuators [Fati94]. Fabrication methods for micromechanics have the advantage that the devices can be easily fabricated in small batch sizes and, when mass-produced, the manufacturing cost goes down considerably [Scha94], [Bier93]. Micromechanics starts where precision mechanics stops. Fig. 3.7 shows a rough classification of the basic micromechanics technologies.

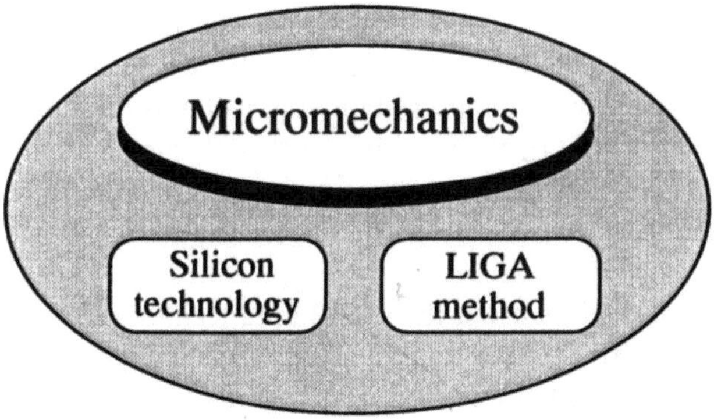

Fig. 3.7: Structure of the micromechanics

Since micromechanics was partially derived from microelectronics and the properties of silicon are well known, several successful results had been obtained by using silicon machining. Today's R&D activities in MST focus (i.e. more than 80%) on silicon-based technologies. Various micromechanical pressure and acceleration sensors made from silicon are being marketed.

The other process shown in Figure 3.7 is the LIGA (Lithographie, Galvanoformung, Abformung) method, which can be applied for manufacturing 3D parts with a high aspect ratio. Various metals are being used. Possibilities of designing complex components which can be made by mass production principles are quite numerous. A real breakthrough with this method was obtained when X-ray lithography became available.

Both of the above mentioned methods are key technologies for MST. The question remains which of the two technologies will be the prime candidate to produce high precision microparts. Because of their great significance, both methods will be described separately in Chapter 4; there will also be a discussion on various components produced by these techniques.

3.1.3 Integrated Optics

The integrated optics is used to produce planar miniaturized optical components for switches, interfaces, modulators, splitters etc. Such an optical component may conduct light signals or convert light to electrical signals and vice versa. In general, several optical components are integrated on one substrate. Communication and microsensors are typical areas of application. The methods of producing integrated optics are characterized by the underlying substrate material onto which the optical components are constructed or into which they are integrated. Figure 3.8 shows a classification scheme of the substrates used for microoptics.

- Numerous integrated optical devices using the glass based technology have already successfully been produced. The spectrum of components includes light wave guides, splitters, couplers, grid filters and spectrometers. In order to conduct light through a fiber optical cable, the conductors have across their diameter a different light refraction index, which is produced by ion exchange or a special implantation method. The fabrication methods of structuring optical components have been borrowed from the semiconductor technology. They include photolithography, etching etc.

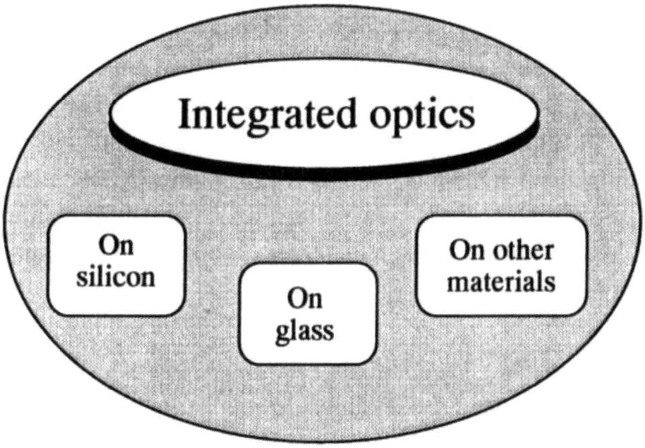

Fig. 3.8: Structure of integrated optics

- Silicon is used for the most part as a base material for integrated optics. For this purpose, thin layers are applied to a silicon substrate and are structured by using photolithography and etching techniques. Very precise microgrooves can be realized to be used for adjusting optical wave guides. Layers made of SiO_2 and SiON can also be used as wave guides (Chapter 6). Numerous products with integrated optical components on silicon are already available.

- Lithium niobate ($LiNbO_3$) is also a very interesting material for integrated optics due to its good electro-optical properties; passive as well as active integrated optical components (such as lasers and photodetectors) can be realized on it. The use of III–V semiconductor materials is also very promising for monolithically integrated optoelectronic devices. Passive and active optical components, such as photodiodes and laser diodes can be realized on GaAs. Some polymers having good optical properties, both linear and nonlinear, can also be used for producing inexpensive integrated optical circuits and nonlinear optical components. The fundamentals of the polymer technique, however, are still being researched.

3.1.4 Fiber Optics

Fiber optics is useful for coupling, guiding and decoupling of optical signals in light-conducting media (usually glass). Thin glass fibers with their extremely low attenuation are gaining importance in MST and are an indispensable part of many optical components. Fiber optics can be used in illumination sources such as laser diodes and LEDs or for decoupling optical signals, e.g. as silicon photodiodes or InGaAs diodes. In communication technology, multi- and mono-mode fibers are already being used. However, inexpensive methods for mass producing the connectors are presently not available. There are many other optical components used in MST such as filters, polarizers, and light modulators, which also depend on optical fibers.

In MST, these components are generally used for fiber optical sensors, making it possible to miniaturize optical measurement systems. Many physical and chemical effects such as speed, pressure or pollutant concentration can be transformed into optical parameters like wave amplitudes, phase shifts, spectral distribution, frequency or time signals and can then be evaluated. Fiber optical sensors are characterized by their high sensitivity and their employability in dangerous or relatively inaccessible surroundings. The latter is very important in control engineering. Fiber optics can transmit signals and data free of disturbance, even over long distances. There is also a great potential in the medical, aviation and space technologies and as energy supply media for microsystems. Special fibers for microsensors are still in the development phase. A disadvantage of fiber optics is that devices made from this technology are very expensive. Therefore, fiber optical elements must be especially justified compared with other solutions before they become interesting to a user.

3.1.5 Fluidic Techniques

Fluidic devices are well suited for producing forces or movements in a simple and safe manner with minimum pollution. When used in a microsystem, fluidic drives can take over a wide range of tasks. They are particularly suited for MST, because they have a high power density despite of their tiny dimensions. Various types of fluidic components are being used in medicine for microdosing systems, suction catheters etc. and in environmental protection for chemical analysis. The components include microvalves, micropumps and fluidic switches. In order to integrate such elements in a microsystem, proper interfaces must be developed to connect them to sensor and signal processing units.

3.2 System Techniques

Complete microsystems are obtained when the individual system components such as microsensors, signal processing units and microactuators are integrated to realize a desired overall system behavior; this is done by applying system techniques. System technologies use system concepts and architectures as well as signal processing methods for the integration of system components. As an example, a system may have to handle information, energy and mass flow. In general, computer-supported tools for designing and simulating a microsystem and its components are necessary for developing complex products.

Well thought-out interconnection methods are required for the integration of microcomponents in a microsystem. Further, a microsystem should be packaged properly to protect it from environmental influences. Application-specific test and diagnosis methods are needed to assure that the system and its components function properly. So far, most products achieved with MST have been partial solutions, e.g. certain microsensors or microactuators. Several microsystems which already exist as prototypes were fabricated as a hybrid combination of previously developed components; they were conceived according to a bottom-up design. However, a microsystem can only be optimally adapted to an application using a top-down design. A fine-tuning to the application using the above-mentioned system techniques is decisive here. An economic breakthrough for MST can therefore only be reached after these techniques have been better understood and developed.

To define the information-technology tools for developing microsystems, all system techniques for solving information-related tasks are grouped together under the term “information techniques”. It is not a completely accurate term, but it helps the reader to distinguish information technology from system techniques used for interfacing and packaging. Information techniques will be discussed in the next section and more of it can be found in Chapter 7.

3.2.1 Information Techniques

System concepts and development tools are essential to make optimal decisions in the design phase concerning the type of information to be processed and transmitted and the energy to be supplied. They are also essential in defining the interfaces to the surrounding, the number of subsystems needed and the structure of the system architecture. It must be determined how the intelligence built into the microsystem will be distributed over the entire microsystem.

Information processing involves the handling of many different types of system and sensor parameters, thereby producing signals to control the system actuators and the self-diagnosing components. The processing of the information for the communication with external systems is also an important issue. Microsystems increase in complexity with increasing number of components. Complexity is an inherent problem with multisensor and multiactuator systems and for them the layout of the information flow circuitry can no longer be managed with conventional model-based techniques. In practice, a microsystem should be able to function reasonably well in an unstructured environment where the sensor information is incomplete and noisy. The system complexity can be handled best by using advanced signal processing and control methods e.g. neural networks and fuzzy logic (Chapter 7).

The design of a microsystem is not without problems. The interaction of physical, biological and chemical parameters must be well understood. There must be knowledge about mechanisms and the processing of technical parameters. Modern design tools mainly try to optimize the individual system design steps, whereas manufacturers try to optimize a solution to a technical problem. But well-defined design concepts neither exist for the conception of an overall system nor for subsystems like microactuators and microsensors. For a functional device an effort must be made to control the entire design process from the conception and simulation of the microsystem up to its production. MST design methods will profit a lot from the further development and adaptation of existing computer-supported design methods.

In order to be able to guarantee the proper operation of a complex system through the entire process of design, production and usage, suitable test and diagnosis methods must be provided and used. Model-based test methods and hybrid test procedures from existing components and simulated models may be applied. Often, the problem is that the test strategies and the used devices are application-specific and are difficult to standardize. Microsystems which find their entrance in safety devices, e.g. as medical *in-situ* systems or as inspection robots in power plants, often must be able to perform on-line tests during operations and any occuring error must quickly be diagnosed, reported and eliminated. During the design of a microsystem, internal test sensors and the corresponding signal processing units must be integrated into a system. As soon as the first microsystems will appear on the market, the demand for test and diagnosis methods will sharply increase.

3.2.2 Interconnection Technology

The interconnection technology is important for the development of microcomponents such as sensors and actuators. This is a key technology in MST: it allows microcomponents to be integrated into a very small space. Good interconnection methods are a technological prerequisite for coupling microelectronic and non-electronic elements together to form one complete microsystem. To reduce development costs, the system limitations and the applicability of materials and components, which strongly depend on the interconnection technologies, should be taken into account early in the planning phase. The biological compatibility of a microsystem for medical applications, the long term stability in aggressive environments and the material compatibility in a complex system are just a few examples. Individual microcomponents are suitable for a MST production process if they are compatible with the process steps of the microsystem and if the techniques needed to connect them with the rest of the system are available at reasonable cost. A survey has shown that over 40% of all companies engaged in MST in Germany find that the most important technology for the system development is the interconnection technology [Schwe93].

An essential task of the interconnection technology is the manufacture of mechanical, electrical and optical connections for clamping, coupling and supplying of energy. Examples include the alignment of glass fibers in an optical component with an accuracy in the nm range, the connection of two microstructures having different heat expansion coefficients and the creation of a dependable and inexpensive electrical contact. At the moment, microelectronic bonding methods such as gluing, soldering, welding, wire bonding, etc. are being modified for MST and are being continually improved. To increase the integration density of a microsystem, new contact techniques are being used, such as TAB (Tape Automatic Bonding), the Flip-Chip technique and the so-called Embedding Techniques. A detailed description of these techniques, which is beyond the scope of this book, can be found in [Kohl94a], [Menz93] and [VDI93].

An interconnection technique tailored especially to the demands of MST is anodic (electrostatic) bonding. It is simple and inexpensive and offers new possibilities which can not be realized with conventional methods. Electrically conductive or semiconductive materials (e.g. silicon) and insulators the conductivity of which increases with temperature (e.g. special glasses containing alkali) can be hermetically bonded under the influence of an electric field. A schematic sketch of a silicon-glass bond can be seen in Figure 3.9.

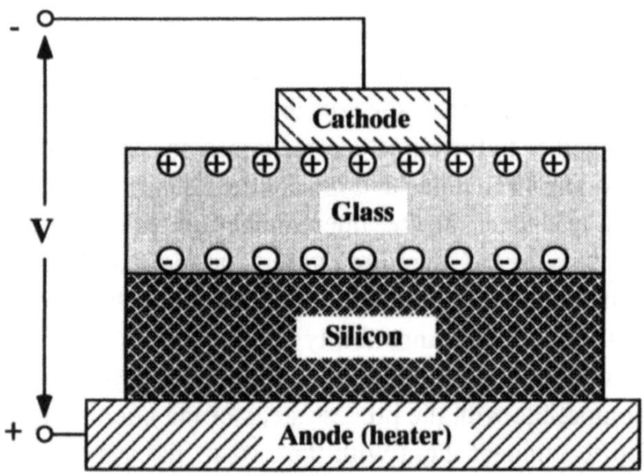

Fig. 3.9: Anodic bonding of silicon to glass

With this method, the two materials to be bonded are heated up by a hot plate to a temperature between 200°C and 500°C. Simultaneously a voltage of about 1000 V is applied. The heated glass becomes conductive and positive Na^+ ions move to the cathode. The non-mobile negative O^- ions stay in place. This results in a negative space charge in the glass-silicon boundary line, producing a strong electric field causing high electric attraction forces. Under a voltage of 600 V the electrostatic pressure can reach 34200 MPa [Schu95]. By carefully preparing the substrate surface to a smoothness of under 0.1 μm, the physical contact between the two materials is very close. Depending on the two mating materials, an electrochemical reaction will occur here. In the case of a silicon-glass bond, oxygen ions from the glass migrate into the silicon. The resulting silicon dioxide layer forms a strong and hermetic connection between the silicon and glass. This technique is often used for constructing microcomponents and for packaging microsystems (see next section). Many different methods have also been developed to connect two silicon structures. In all methods, a close contact is established between the silicon parts, resulting later in a strong chemical bond. Anodic bonding can be used here by first depositing a thin pyrex glass layer on one of the silicon substrates [Heub91], [Schu95].

The interconnection technology will have an increasing importance for this new technology since MST applications are using more and more smaller and complex components and need new types of materials, like alloys and polymers. There is also research work concerned with the development of new biological integration methods for MST.

3.2.3 Packaging

Packaging concerns the encasing of the microsystem so that it can be used for a practical application. MST can make use of many known methods of chip packaging. The techniques are rather advanced due to the long development history of the microelectronic industry. However, not every solution can be automatically adapted to an application. The system package and the microsystem interior cannot be considered separately. A casing is indispensable for a microsystem; its task is to protect the mechanically sensitive parts of the microsystem from damage and environmental influences. It should also be equipped with task-specific interfaces with its surrounding to obtain a secure contact with the system environment. Often, brittle and disturbance-prone microsensors must be exposed to a harsh environment to obtain a desired function. The package therefore greatly influences the development of a microsystem. In Figure 3.10 various packaging factors to be considered are shown.

There is a high demand for inexpensive multifunctional packaging materials which can be used for various microsystem designs. An important material is

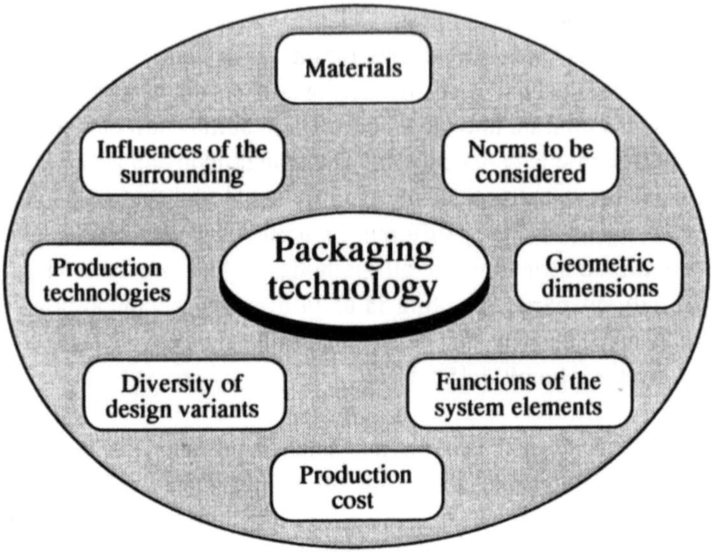

Fig. 3.10: Determining factors for packaging. According to [M2S293]

ceramics. Due to its high dimensional stability, its gas tightness and its ability to withstand many aggressive environments, ceramics is well suited for packaging MST devices. Also plastics, silicon and glass casings are often suitable components; concerning silicon and glass the anodic bonding technique discussed above can be used.

3.3 Materials and Effects

MST tries to integrate many components with various functional properties, e.g. chemical, mechanical, fluid-dynamic, optical, biological, electrical and electronic, etc. Thus a multitude of different materials is needed to build them. It is also necessary to develop materials for monolithic and hybrid microstructures and to sufficiently investigate their properties under various design conditions, e.g. form, dimension, usage, etc. so that they can be widely used. In addition, it is necessary to develop new interconnection techniques and to do further material research.

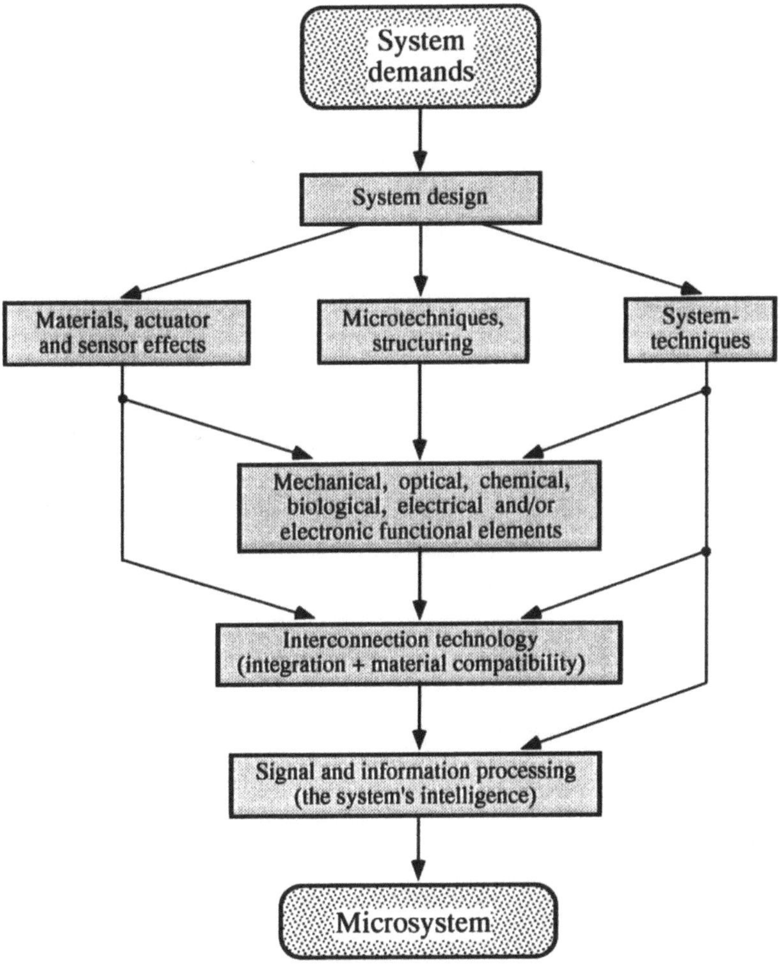

Fig. 3.11: The development cycle of a microsystem. According to [Zum93]

Figure 3.11 shows the development cycle of a microsystem. It is getting clear that the development is strongly influenced by the properties of the used materials. The most important requirement is the "MST compatibility", which means that the materials must be inexpensive, that they can be microstructured, and that they can be batch processed. Suitable microsystem materials include: semiconductor materials, which are in an advanced stage of development; metals such as nickel, titanium, aluminum, gold, platinum, etc. and their alloys, which are preferred due to their good conductivity and the ability to deposit them easily by galvanic methods; ceramics, which have advantages due to their dielectric properties, stiffness and chemical stability; polymers, which have light weight and an excellent formability as well as resistivity to corrosion; special glasses and quartzes; diamond due to its excellent mechanical properties etc.

Although the LIGA method is being used more frequently due to the number of materials that can be processed, silicon is still mostly used for MST (Chapter 4). It is especially important to use well-controllable and inexpensive techniques. The III–V compound semiconductors containing elements of the III and V groups of the periodic table (like GaAs or AlGaAs) are of interest to optoelectronics and to microsensors due to their excellent physical properties [Hjort95]. Their restrictions are their high material and production costs, however.

This group of MST techniques also comprises various energy transformation effects which are relevant to microsystem applications. The effects may be of mechanic-, thermal- and chemical-electrical as well as thermal- and magnetic-mechanic nature and are used in constructing microsensors and actuators, Fig. 1.6. A more exact description of the most often used energy transformation effects will be found in Chapters 5 and 6. Also several sensor and actuator prototypes will be introduced to show the use of proper materials and their effects.

3.4 Outlook

This concludes the overview of microsystem technology, and we will now turn our attention to the special areas of MST which are relevant to microrobotics. In the next chapter, the two key microstructuring methods will be discussed, namely silicon-based micromechanics and the LIGA method. We will give many examples of components that have already been realized. In later chapters, typical microactuator and microsensor principles will be explained in detail, also accompanied by many examples. Information processing problems in microsystems will be treated and some advanced solutions will be suggested for intelligent systems. Then we will discuss future developments of microrobotics, and finally, we will describe microrobots which already exist. Based on the various MST techniques and applications introduced in the first three chapters, it is now possible to give an overall view of MST, Fig. 3.12.

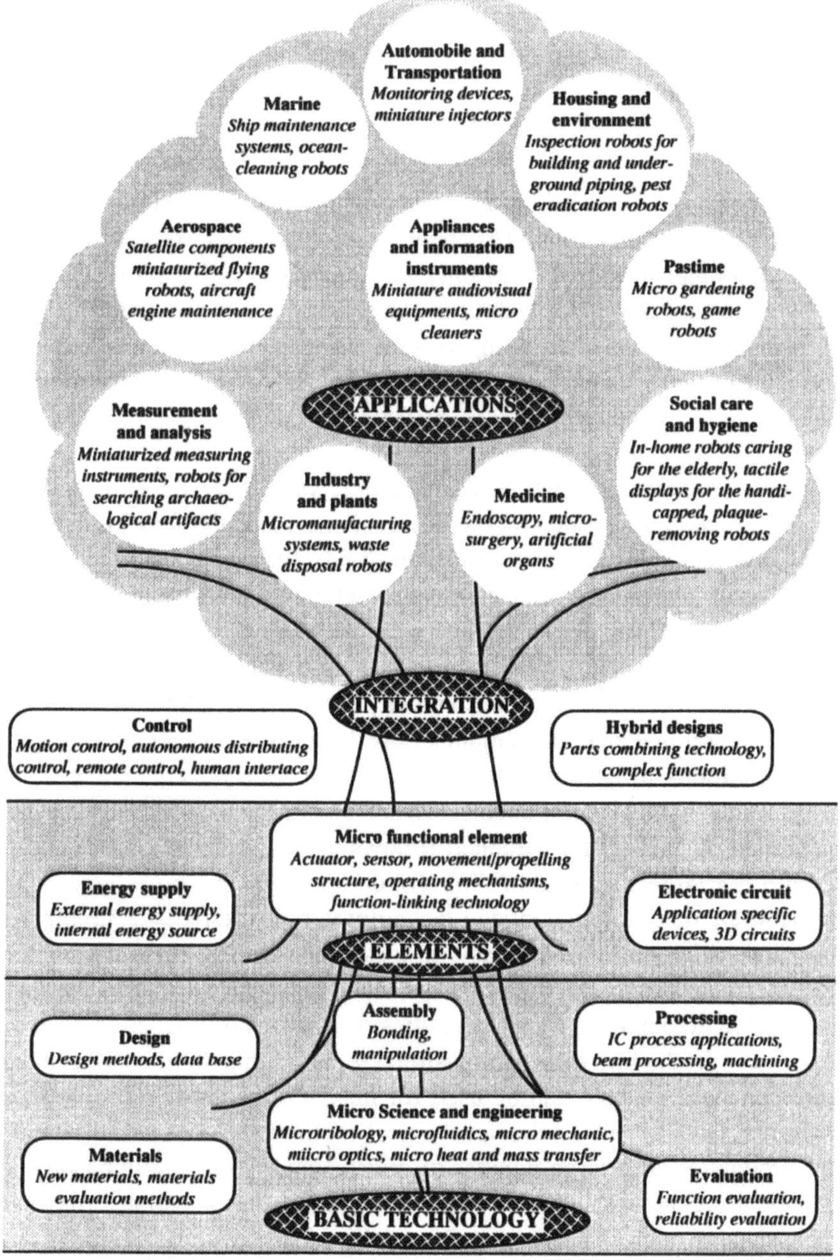

Fig. 3.12: A general structure of the MST-world. Courtesy of the Micromachine Center, Tokyo

4 Key Processes to Produce Micromechanical Components

Micromechanics has become an accepted technology and many devices are finding industrial applications. This new technology was created during the past few years and its impact can only be anticipated. The subject is being taught by almost all important technical universities as an important discipline. Micromechanics is concerned with manufacturing methods to produce small parts in the µm-range using special MST techniques, such as lithography, thin layer techniques, dry and wet etching, micromolding, etc.

Since many fundamental MST techniques were originally developed for the processing of silicon semiconductor devices, the basically two-dimensional planar silicon technology is also used worldwide for micromanufacturing by the majority of researchers and companies who make these small devices. Often the process is supplemented by micromechanical methods to realize a third structural dimension. The silicon technology takes advantage of the excellent electrical, chemical and mechanical properties of silicon and its compounds. Conventional planar techniques are based on this technology; they allow large circuits to be integrated onto one chip. Very stable three-dimensional micromechanical components can be constructed using an expanded silicon technology. The LIGA method follows a fundamentally different approach, with it a large spectrum of different components can be produced. It is a significant alternative to the silicon technology, but often also a supplement. It uses X-ray lithography as its basic process for forming three-dimensional plastic structures, which can then be used to make metal structures by applying microgalvanic deposits. Metal structures, formed as molds, are a prerequisite for mass-producing three-dimensional plastic microparts. The plastic structures themselves may also be used for mass-producing certain metal microstructures. Both of these micromechanical methods allow the individual or parallel production of a large number of microcomponents, applying physical and chemical techniques instead of mechanical. On one substrate, for example, thousands of microstructures can be produced in one step.

In the following sections, both methods will be described in detail. Their application will be shown with various examples of already realized components.

4.1 Silicon-based Micromechanics

Since silicon and its compounds are ideal materials for semiconductor manufacturing, researchers started to use them also for micromechanics. However, because only two-dimensional components could be made, other technologies had to be developed for three-dimensional structuring; they will be introduced below.

4.1.1 Special Silicon Properties

Only a few materials have at the same time good electric and mechanical properties; many microcomponents, however, require both. Silicon fulfills these requirements like no other material, almost ideally, and is very suitable for the manufacture of MST components. The properties of silicon have already been extensively researched, and for decades, silicon has been used for mass-production of semiconductor circuits and has made it possible to realize very complex, powerful products such as the 64 Megabit memory and processors with over 3 million semiconductors.

With the physical properties of silicon it is possible to realize a wide variety of sensor elements based on thermal, magnetic or optic effects. Also the mechanical properties of single-crystal silicon allow the making of a strong, miniaturized precision structure. A perfect crystal structure assures a reliable reproduction of exact dimensions, which often is impossible when using precision machining methods. Single crystals can be grown in rod form by well controlled methods. The rods are then cut into wafers, polished and processed micromechanically. The elasticity of silicon is comparable to that of steel, but due to silicon's low density of 2.3 kg/dm^3, it is stiffer than steel.

Some silicon compounds, such as silicon nitride and especially silicon dioxide, have physical and chemical properties that are excellent for surface micromachining (Section 4.1.2.5). The importance of these materials will become clear during the following discussions. Figure 4.1 gives an overview of types of microsensors that can be made by the silicon technology.

Silicon structures are also used in microactuators. However, MST motion principles often require other materials (Chapter 5).

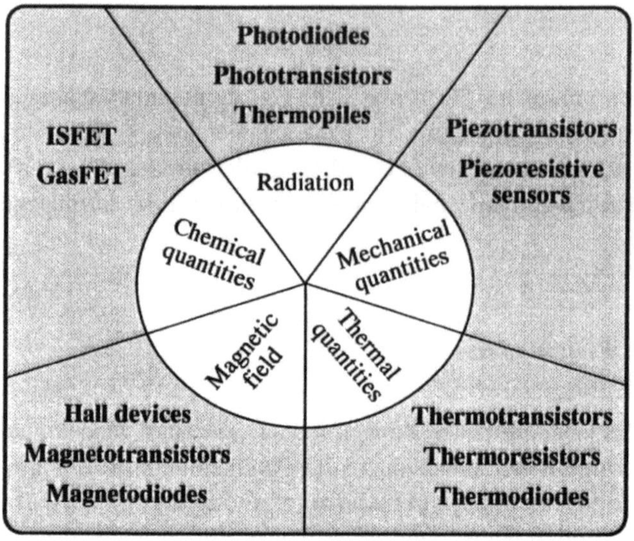

Fig. 4.1: Sensors based on silicon components. According to [Haup91]

Since the same silicon manufacturing methods can be used for electronic and mechanical components, silicon sensors can be made, which have their electronic circuit and microactuator integrated together to form a complete system on one silicon substrate. This monolithic design makes it possible to construct small, lightweight devices at low cost and high reliability. The integration of a variety of functional elements on silicon can also be realized by using advanced hybrid techniques. Whether the microsystem is constructed in a monolithic or hybrid manner depends on the availability of microsystem technology and its production costs.

4.1.2 Silicon Fabrication Techniques

The layer techniques described in Chapter 3 and the micromechanical structuring techniques, like doping, lithography and etching, allow the production of both tiny electronic circuits and mechanical structures from a single semiconductor material. Making a three-dimensional microstructure from silicon, however, requires a multi-step process. Below, the fundamental steps of the silicon micromechanics will be discussed in a readily comprehensible manner.

4.1.2.1 Doping

Doping is probably the best known semiconductor technique. Here, doping atoms are introduced to a silicon substrate in a defined way so that n or p conducting layers are formed, which play an important role in semiconductor components. Besides determining the electric properties, doping can improve properties, such as wear and corrosion. Some doping atoms, like boron or phosphorus can create an etching barrier in a silicon substrate, allowing the manufacture of a precise microstructure when the doping atoms are exactly placed in the crystal lattice (Section 4.1.2.5). Two doping methods, diffusion and ion implantation, are used in MST.

- In the diffusion method, silicon wafers are put into a furnace, where the diffusion process takes place. The doping atoms either come from the surrounding gas or from a thin, preapplied surface layer. They diffuse into the silicon crystal and replace regular lattice atoms. The main difficulty is the determination of the absolute concentration of the doping atoms in the silicon. In addition, it is only possible to create a doping profile on the surface of a wafer, which limits the applicability of this technique.

- Ion implantation involves shooting charged doping ions, which are externally accelerated in a vacuum, into the silicon wafer. The ions can penetrate up to a few micrometers below the surface. The silicon disc is irradiated uniformly by scanning it with a thin, focussed ion beam. By measuring the beam current, the process can be simply controlled. Thereby, the doping concentration gets an improved homogenity, its parameters can be easily adjusted, and the doping profile under the wafer's surface can be controlled more exactly. Also, so called "buried" maxima of doping material are possible using ion implantation. Ion implantation, however, is one of the most expensive technological processes, after lithography, due to the complex systems involved.

4.1.2.2 Lithography

Lithography is used for preparing the substrate of a wafer for the subsequent processing stages. In order to be able to etch the desired pattern of the workpiece selectively, a photosensitive layer is applied to the wafer surface. This photosensitive resist will be lithographically structured so that specified areas of the substrate remain covered and protected. This technique is the most critical and also the most expensive single micromechanical process and plays a key role in mass-production of microcomponents.

The approximately 1 μm thick photosensitive resist layer is either applied directly to the silicon wafer or on another deposited layer. This resist layer can then be masked and structured by using different sources of irradiation, such as light, X-rays, electron or ion beams. Ultraviolet light with a wavelength between 250 nm and 450 nm is mainly used for the silicon process. The mask is made of a light-transparent substrate, usually glass of a few mm thickness and a light-blocking layer deposited onto it. This so-called absorber layer, in general made of chromium, masks up the resist which is to remain unexposed. It is only a few 100 nm thick and is deposited onto the glass substrate, using a sputtering technique.

In practice, three basic lithographic methods are used, Fig. 4.2:

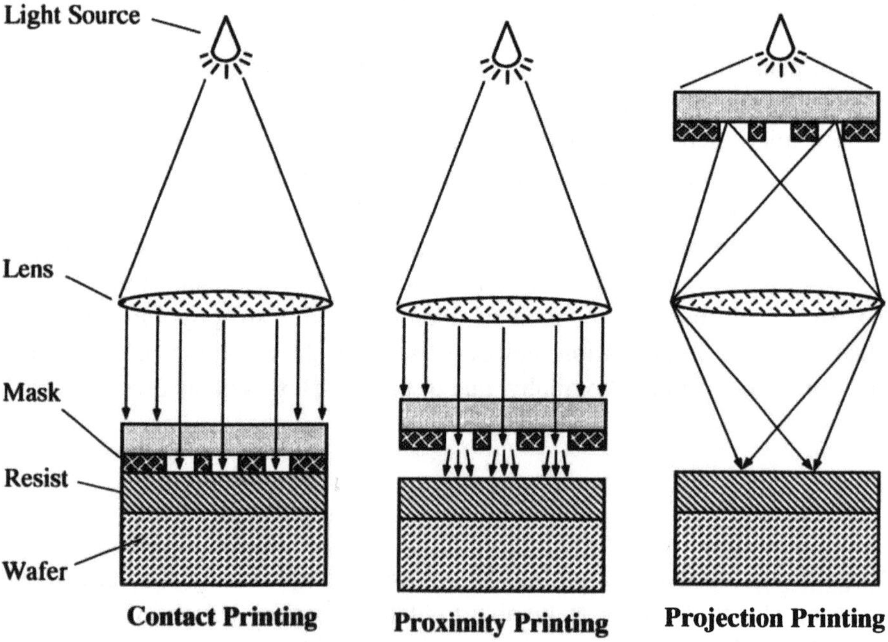

Fig. 4.2: The basic lithographic exposure methods. According to [Mokw93]

• Contact lithography means pressing the mask directly onto the substrate. A very precise structural resolution of better than 1 μm can be reached. However, contact lithography is not suitable for mass-production, since the mask is exposed to a high mechanical load during exposure and it can be easily damaged by dust particles deposited on it from the surrounding.

• In proximity lithography, there is an air gap of 20 to 50 μm between the mask and the wafer. This reduces the wear on the mask, but due to the Fresnel diffraction of light, the resolution is limited to 2 μm.

• The best possible structural resolution of about 0.5 μm is obtained by using optical projection instead of simple contact copies. The mask is projected onto the wafer using a reducing high resolution lens system; reductions of 5:1 to 10:1 are typical. Since often it is technically impossible to simultaneously expose the entire substrate surface, and since the exposing process is precisely carried out in a single step at a time using the step-and-repeat method, the equipment and processing costs are high.

Exposing the resistive layer causes molecule chains to break or to cross-link, depending on the type of resist used. Either the exposed part, which is the positive resist, or the unexposed part, which is the negative resist, are removed in a following step using a special developer solution, Fig. 4.3.

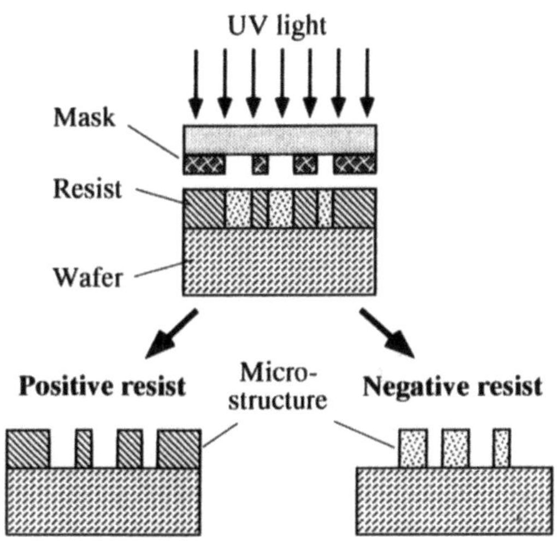

Fig. 4.3: Various development steps

During this step, the mask pattern is copied onto the resist, which was applied to a substrate in a preceding step. The aspect ratio which can be obtained by the UV lithography is about 1 or 2, but it can be raised to 6 or 8 by using special techniques such as reapplying a photoresist layer several times.

To obtain improved microstructures by lithography, a new lithographic process is being investigated, the so-called gray-tone lithography [Quen94], [Weng94]. Here, a special raster-screened photomask is used during the exposure process. The transparency of the mask changes along its length, either continuously or in discrete steps. This is reached by either changing the thickness of the chrome layer or by putting tiny, exactly placed holes in this layer. Thus, different areas of the resist are exposed to different light intensities. With this process various profiles of the remaining resist layer can be obtained after development. In a subsequent dry etching step (Section 4.1.2.3.2), which etches the whole uncovered surface, the resist profile is transferred to the underlying silicon substrate, Fig. 4.4.

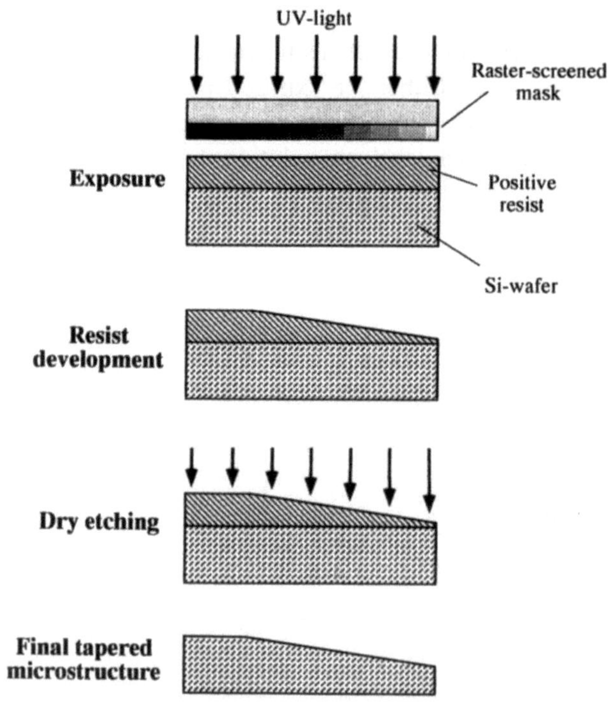

Fig. 4.4: Principle of gray-tone lithography. According to [Weng94]

A large variety of structural geometries are possible with this technique as compared to conventional lithography. Figure 4.5 shows a microstructure made by the use of gray-scale lithography [Quen94].

Fig. 4.5: A microstructure made using gray-tone lithography. Courtesy of the Fraunhofer-Institute for Silicon Technology, Itzehoe

The X-ray, electron and ion beam lithography will render a higher resolution than the other processes mentioned before, but there is a trade-off between process complexity and production cost. Electro-optical (writing) methods produce the structured layer point by point with a focussed electron or an ion beam. No lithography mask is needed but the processes are very time-consuming and not suitable for mass-production. Therefore, these methods are only used to produce special masks or to repair them. X-ray lithography is becoming more and more important, it is in principle a further improvement of the light lithography. Due to the very small wavelengths (under 1 nm), structures can be made with dimensions in the submicrometer range (up to about 0.2 μm). This technique will be described in more detail in Section 4.2 in connection with the LIGA method.

Many etching methods are available for transferring lithographically made mask patterns onto a substrate. There is a distinction between chemical wet and dry etching, which will be discussed below.

4.1.2.3 Etching Techniques

By etching, the exposed areas of the photoresist from the wafer and the deposited layers respectively (positive resist) are removed. The etching methods used for MST must be able to remove material exactly so that tiny µm-sized structures like grooves, bridges, membranes or beams are obtained. If an etching method only dissolves a specific material and it does not react with other materials, it is considered to be selective. Another characteristic property of etching is that it may produce different results in different directions of the material to be processed. An etching process is isotropic if the etching speed is the same in every direction of the material, otherwise it is anisotropic. Selectivity and directional properties play a decisive role in microstructuring.

4.1.2.3.1 Chemical Wet Etching

Wet etching is commonly used in microelectronics and can produce three-dimensional and surface structures needed in MST. The etching solution, which may be acidic or alkaline, dissolves the material to be removed. Etching occurs by dipping the substrate into an etching bath or spraying it with the etching solution. Depending on the structure of the material or the etching solution, the wet etching process can either be isotropic or anisotropic.

- **Isotropic etching**

If the material is amorphous or polycrystalline, the chemical wet etching process is always isotropic. The etching of a polysilicon substrate causes the formation of cavities with rounded off edges. At the same time, the resist is undercut, Fig. 4.6. In micromechanics, however, round edges are usually undesired unless they are needed to obtain good fatigue strength in moving parts.

This fact limits the application of this method to deep forming, especially when the undercut area exceeds the accuracy of the dimensions of the desired part. Therefore, the isotropic wet etching is not suitable for a cavity width smaller than about 2–3 µm. Despite of these problems, isotropic etching is useful for many applications in surface micromechanics, where individual layers are se-

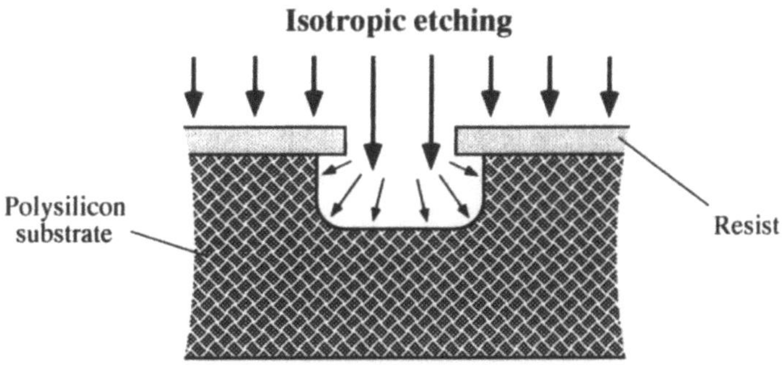

Fig. 4.6: Isotropic wet chemical etching

lectively etched in a multi-layered surface of a substrate. More details on surface micromechanics can be found in Section 4.1.2.5.

- **Anisotropic etching**

Single-crystal silicon, as opposed to polycrystalline silicon, can be anisotropically etched, which means the single crystal wafer can be precisely structured in three dimensions. Single crystal silicon and diamond have the same atomic lattice structure. The characteristic atoms are located at the corners of the cubical lattice stucture and in the center of the cube's faces of the crystal. The cube's edges form a cartesian coordinate system, Fig. 4.7.

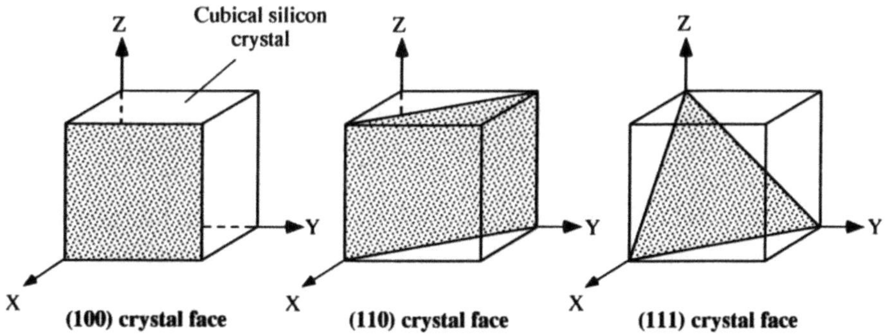

Fig. 4.7: Arrangement of the crystal faces in silicon lattice structure

As opposed to polysilicon, the etching speed is dependent on the crystal's orientation in single-crystal silicon. Many etching solutions have a distinct minimal etching speed in direction of the (111) crystal face, which practically brings the etching process to a halt in this direction. This allows the formation of geometrically precise microstructures with sharp corners and edges that can be defined in a single crystal substrate. If a silicon wafer is cut in such a way that the lateral etching rate is much lower than the vertical one, the resist mask will not be undercut. Thereby vertical and V-like cavities are formed with the (111) crystal face bordering them. The walls of the structure are very smooth since the etch rate along the (111) crystal face is more than a hundred times higher than the one perpendicular to it, Fig. 4.8. The designations (100), (110) and (111) indicate along which crystal face the substrate wafer is cut, i.e. which crystal face is parallel to the substrate's surface.

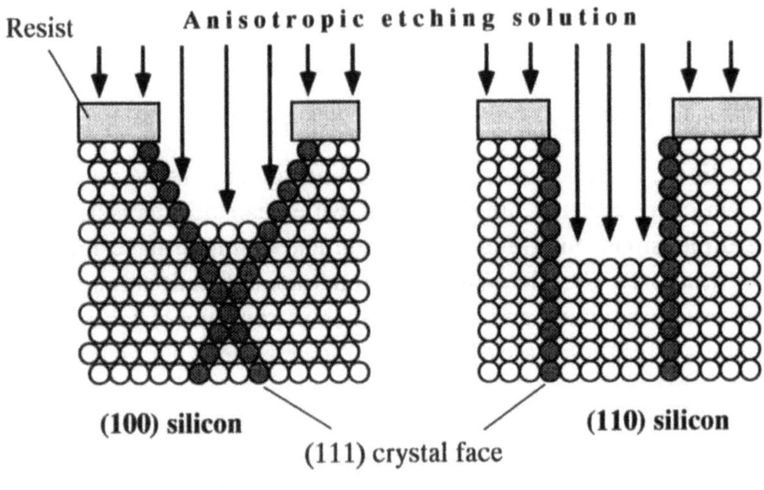

Fig. 4.8: Anisotropic etching in single-crystal silicon. According to [Menz93]

The most often used anisotropic etching solutions are alkaline like potassium hydroxide, sodium hydroxide or EDP (ethylene diamine pyrocatechol). Silicon dioxide or silicon nitride are the main resist materials. Typical structures that can be produced by several crystal orientations are shown in Figure 4.9. The attainable dimensions are limited by the used photolithographic method. They are usually small enough to allow the making of compact cavity patterns in

a

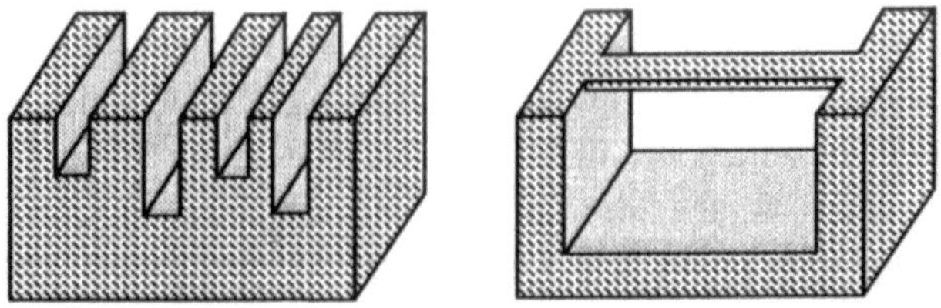

b

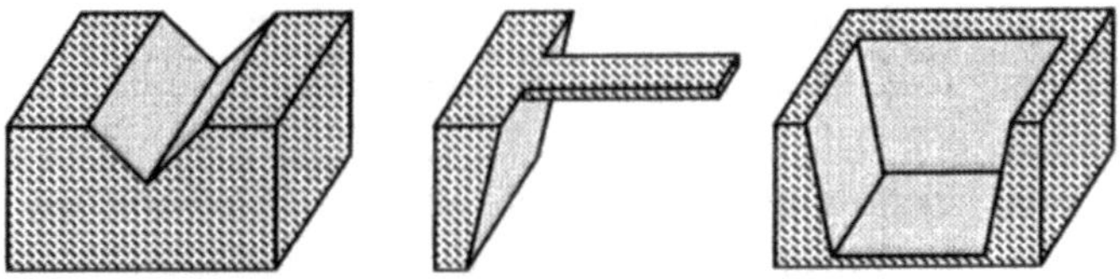

Fig. 4.9: Basic structures that can be produced in silicon
a) (110) silicon; b) (100) silicon

the substrate. Therefore, the wet chemical anisotropic etching process has become a key technology of micromechanics. Many different micromechanical silicon components have been realized with this method.

4.1.2.3.2 Dry Etching

Dry etching involves the exposure of the substrate to an ionized gas. Etching occurs through chemical or physical interaction between the ions in the gas and the atoms of the substrate. Dry etching techniques can be used to structure almost any material suited for MST. It is also used for materials on which wet etching is too difficult, such as platinum or tin dioxide. The most often applied techniques can be divided into three groups:

- physical sputter etching or ion beam etching
- chemical plasma etching
- combined physical/chemical etching

Sputter etching is similar to the sand blasting technique at the atomic level. Instead of sand particles, chemically neutral positive argon, helium or xenon ions are used. The etching takes place in a vacuum reactor which contains two flat electrodes separated by a few centimeters. The electrodes are surrounded by the plasma in which the inert ions are generated and accelerated by an electric field. They "crash" into the substrate which is fixed to the cathode. The impacting ions physically loosen particles from the bombarded substrate surface, Fig. 4.10. The other physical process, ion beam etching, also uses the concept of electrostatically accelerating chemically inert ions. It has technical advantages over sputtering, but is much more complicated and expensive. In chemical plasma etching, reactive gases are used which decompose in the plasma, creating active radicals. These particles diffuse into the surface of the substrate and initiate the etching process, Fig. 4.10. Chlorine or fluorine gases are applied in connection with either single crystal or polycrystalline silicon, silicon dioxide, silicon nitride or metals.

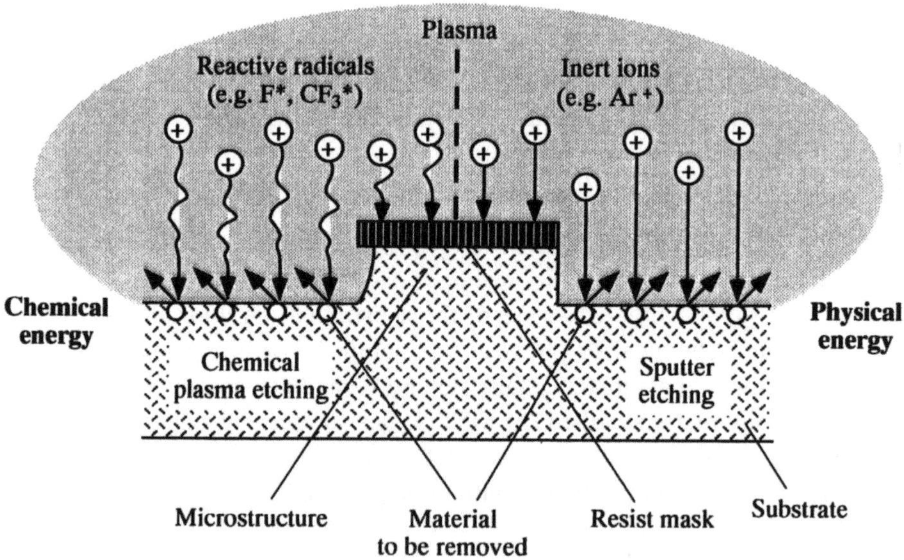

Fig. 4.10: Basic dry etching methods. According to [Krull93]

All physical methods are anisotropic, allowing a precise structure of smooth walls; selectivity is not guaranteed, which means that all the used materials will be etched at about the same speed. Plasma etching is completely isotropic as opposed to sputter etching; it has excellent selective properties. The etching rates of the physical method only reach a few 10 nm/min, but the rates of the chemical method reach up to 100 nm/min [Menz93]. An alternative are the combined dry etching methods which use both the physical and the chemical energy of reactive particles, combining high selectivity and anisotropy. The reactive ion etching (RIE), as opposed to plasma etching, accelerates reactive ions and radicals in an electric field in direction of the surface to be etched. Inert gases can be added to reinforce the physical etching process. The etching rate ranges from 20 to 200 nm/min. Another combined method is reactive ion beam etching. It requires more equipment to carry out but it has the fastest etching rate of up to 500 nm/min [Menz93].

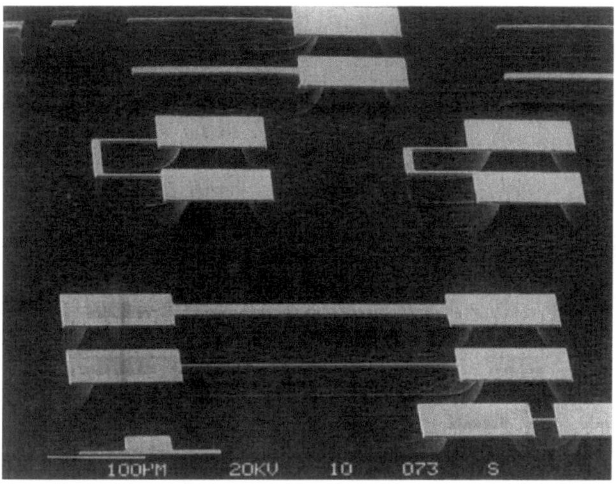

Fig. 4.11: Aluminum microstructures made by the plasma etching process. Courtesy of the University of Neuchâtel (Institute of Microtechnology)

An example of a micromechanical structure produced by dry etching can be seen in Figure 4.11. This figure shows a freely standing aluminum structure on a silicon substrate.

4.1.2.4 Lift-Off Technique

The lift-off technique is a different structuring technique often used for making metallic pads on a chip. Here, a resist layer is applied to the substrate and is structured by a lithography process. The developed resist is an inverse projection of the desired metal structure, Fig. 4.12.

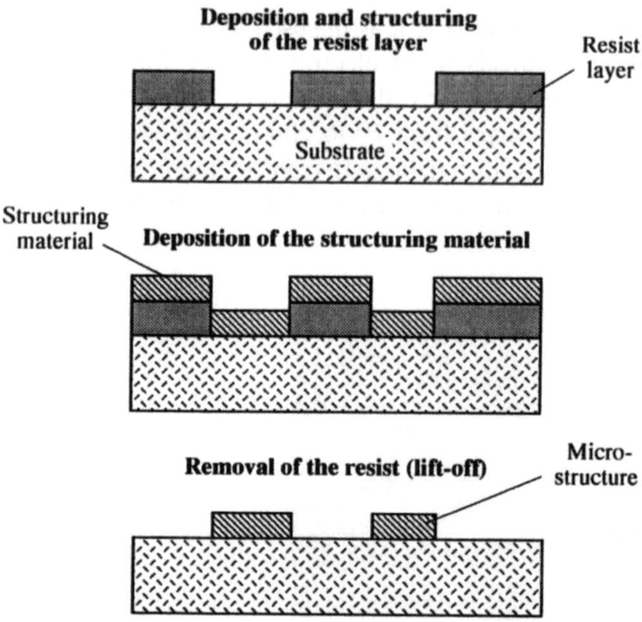

Fig. 4.12: Lift-off technique used for structuring

After the structuring material has been deposited by sputtering, the resist layer is removed by a lift-off process with a suitable solvent. Thereby, the deposited structure material layer above the covered substrate areas is lifted off with the resist. The desired structure remains on the substrate. In the lift-off technique, usually a photosensitive resist is used and aceton serves as the solvent. This technique is especially suited for making electrically conductive layers from hard-to-etch metals, such as platinum. It is also used to construct *interdigital* capacitors, which are important for biochemical microsensors, Fig. 6.27. An interdigital structure consists of many planar metal electrodes integrated into a substrate; it is similar to a multi-tooth comb structure. A typical method for producing such a structure by the lift-off technique is shown in Figure 4.13. This figure is self-explanatory and the reader can understand it by following the manufacturing steps in their shown sequence.

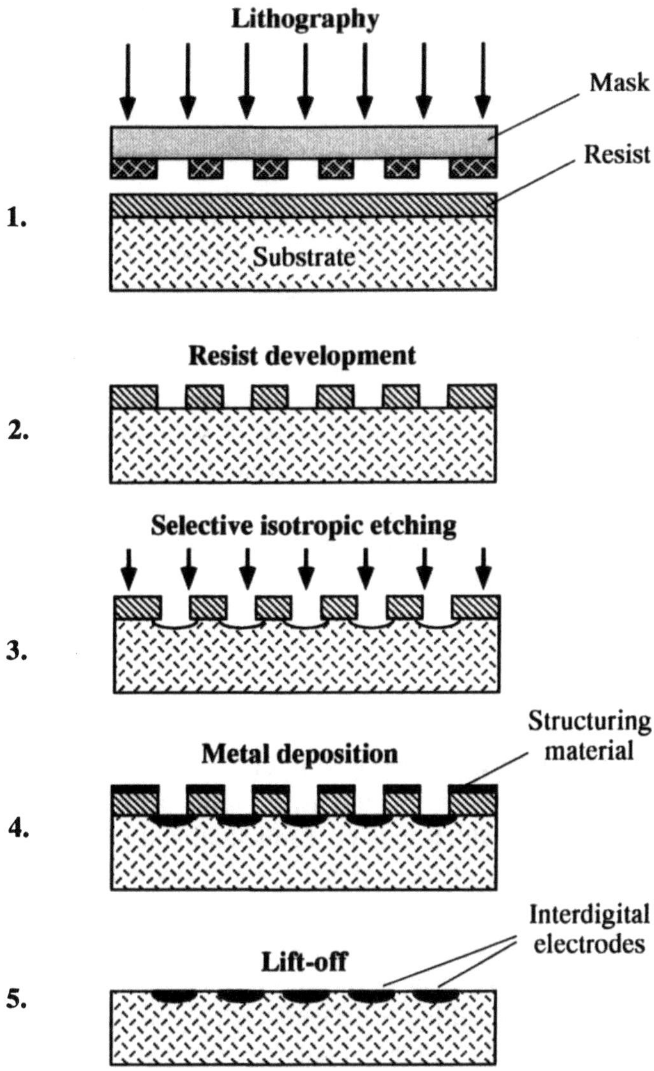

Fig. 4.13: Lift-off technique for making interdigital capacitor structures

The depth structuring step 3 in Figure 4.13 only plays a minor role and can be omitted. This step is carried out when the flat substrate surface is to be further processed or when the electrodes have to adhere very tightly to the workpiece. Metal is usually deposited by the chemical vapour deposition (CVD) technique; mainly platinum, titanium or gold are deposited.

4.1.2.5 Surface and Bulk Micromachining

Microcomponents are either made from single-crystal or polycrystalline silicon by using surface and bulk micromachining.

- Surface micromachining involves planar structuring of the surface of a substrate. Many layers of different materials are successively deposited on top of each other, whereby they are structured and also etched away. This technique can be used to make complicated planar structures with a maximum thickness of a few micrometers.

- Bulk or substrate micromachining is the structuring in three dimensions. The entire substrate, which can also be structured from the back, may be used. Anisotropic wet chemical etching techniques are applied to process single-crystal silicon. The three-dimensional structuring is an alternative or supplement to surface micromachining.

4.1.2.5.1 Bulk Micromachining

The bulk micromachining technique was developed in the 60's and allows the structuring of silicon in three dimensions. A high aspect ratio can be reached for micromechanical components which can be formed directly from the silicon wafer. The crystal orientation of the wafer plays a decisive role (Section 4.1.2.3.1). The silicon wafer is preprocessed with optical lithography and the exposed resist material is removed. By using anisotropic etching solutions selectively on the resist, deep grooves can be made on a substrate. The remaining resist acts as a mask. The resulting form depends only on the crystal orientation of the substrate, i.e. along which crystal face the wafer was cut, Fig. 4.7. Different constructions, such as bridges, beams, membranes etc., can be made by this technique, Fig. 4.14.

Many wafers produced this way can be connected by using bonding or other interconnection technologies to form complex three-dimensional structures. The possibilities are still limited, however, since the lattice structure of the silicon crystal is not variable. Simple circular, cylindrical cavities or columns cannot be realized with this method.

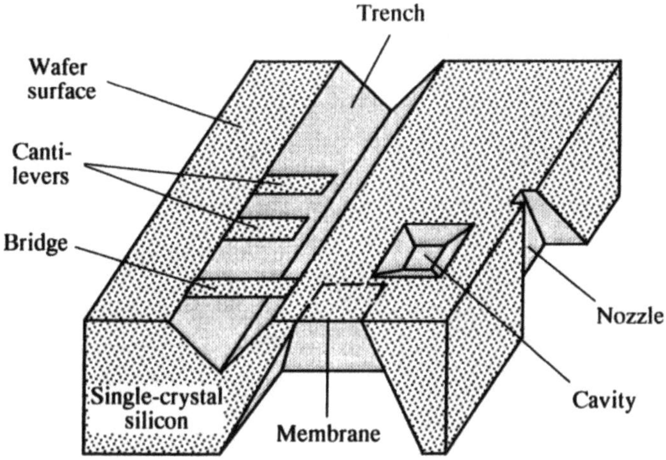

Fig. 4.14: Various bulk micromachining structures. According to [Howe90]

In order to be able to structure the substrate exactly, additional techniques are necessary to interrupt the etching process at the right time. Otherwise the substrate will simply be etched through. Etch-stop techniques are applied here to form very thin membranes or grooves with flat bottoms, which are otherwise difficult to make. A detailed description of the etch-stop techniques can be found in [Peet93]. Two basic methods will be described below.

- **Boron implantation (p^+ etch-stop)**

The p^+ etch-stop technique has many applications. It is based on doping the silicon substrate with germanium, phosphorus or boron atoms. This way, the etch rate of various selective etch solutions can be drastically reduced. Boron has the strongest effect and is therefore almost always used. In order to form a thin membrane, the silicon substrate is doped with boron atoms on one side, using a boron concentration of about 10^{20} cm^{-3} or higher. Then the substrate is etched on the other side. As soon as the solution meets the boron-doped layer, the etching process is drastically slowed down. Thereby, a 2–3 μm thick membrane can be obtained depending on the doping depth. Figure 4.15 shows the manufacturing steps of a silicon membrane as a sandwich structure using the bulk micromachining technique in connection with doping.

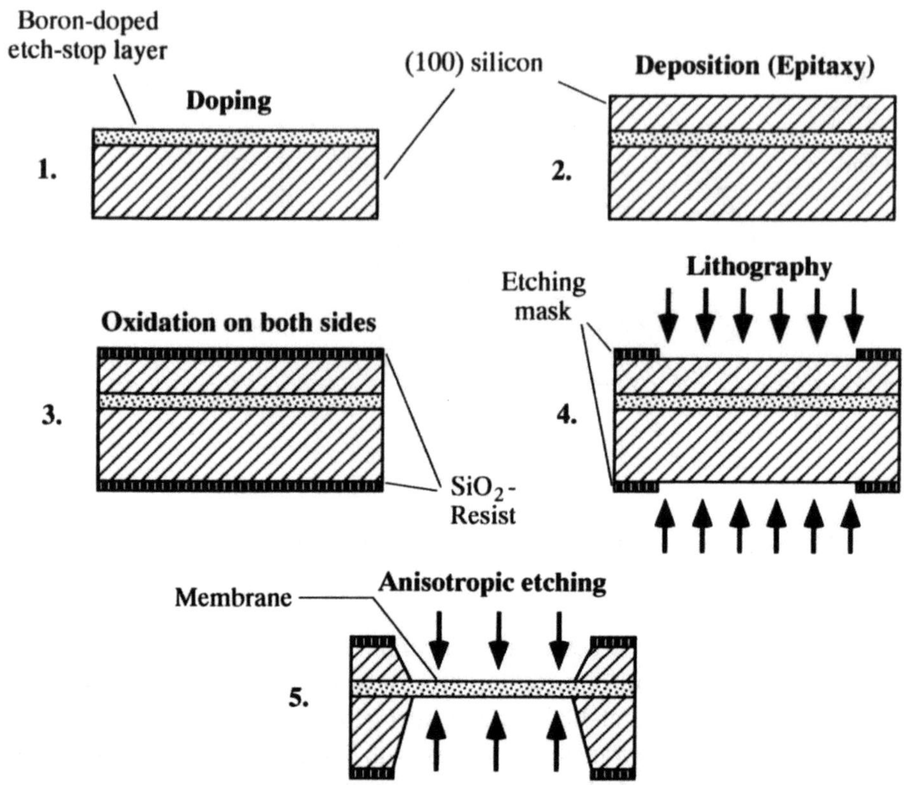

Fig. 4.15: The production of a membrane by the p+ etch-stop technique

First, the substrate surface is saturated with boron atoms until the desired concentration is reached. This forms an etch-stop layer (1). A second silicon layer is put on the etch-stop layer by using epitaxy (2) and then both sides of the workpiece are oxidized (3). The silicon dioxide layer can then be structured photolithographically (4), and after isotropic etching on both sides, the boron-doped layer remains as a membrane (5). Thin membranes make it difficult to get a uniform membrane thickness by doping with the diffusion method, because the membrane thickness depends on various parameters, such as the ion concentration, the wafer thickness etc. Ion implantation can be used instead, although the process duration is very long. For a boron implantation of 2 µm with a concentration of 10^{20} cm^{-3}, a 150 µA implanter needs up to an hour per wafer [Mokw93].

- **Electrochemical etch-stop technique**

An alternative is the so-called electrochemical etch-stop technique. Here, for certain etch solutions the etching process is interrupted when an electric voltage is applied to an np or pn silicon substrate. For example, a protective n-doped silicon layer is deposited onto a p-substrate. When the interruption voltage is applied, the etching process stops at the pn-transition, Fig. 4.16.

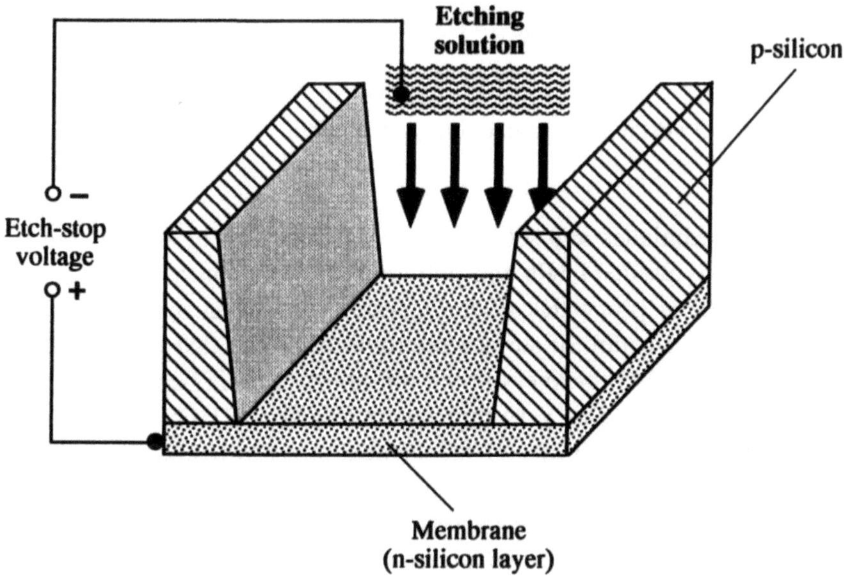

Fig. 4.16: Membrane produced by the electrochemical etch-stop method

The thickness of the n-layer, which remains as the desired membrane, can be determined during the deposition process.

4.1.2.5.2 Surface Micromachining

In addition to making cavities, microstructures can be built up step-by-step on a silicon surface. Surface micromachining originates from microelectronics and is therefore oriented towards processes and materials used for chip manufacture. Freely standing planar structures can be made such as cantilevers or membranes which hover just above the substrate surface, Fig. 4.17.

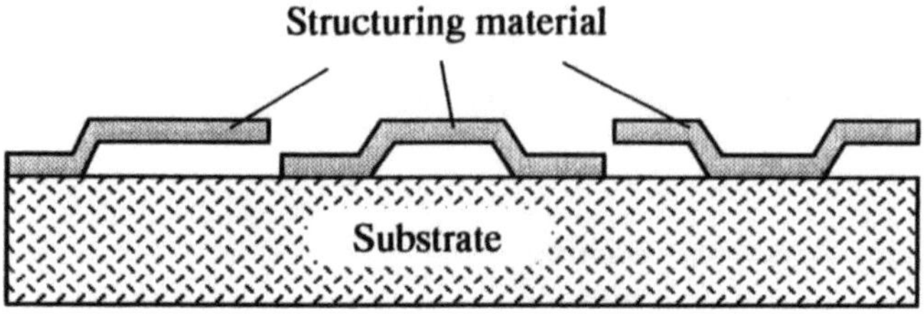

Fig. 4.17: Typical surface micromachining structures

The thin layer techniques and selective etching discussed before are the fundamental methods used. Metals and polycrystalline silicon are the base materials. The various properties of polycrystalline silicon are described in [Faro95]. A special feature of surface micromachining is that the microstructure can be made with a thickness in the nm range. This is because the structure material is deposited as very thin layers (Section 3.1.1). One disadvantage, however, is the possibility of mechanical stresses occurring inside the thin layer, which limits the lateral structuring dimensions. A schematic representation of the process steps for making a free-standing cantilever is shown in Figure 4.18. This picture helps us to give a brief overview of the possibilities of surface micromachining.

The actual structuring process starts with the deposition of a silicon dioxide layer onto the substrate, e.g. by thermal oxidation. The thickness of this layer determines the distance of the microstructure to the substrate surface. This so-called sacrificial layer is formed by a lithographic method (1) and the selective removal of the surface material (2). Afterwards, a polysilicon structure layer is deposited (3) and structured by a second lithographic etching process (4) and (5). When this process is complete, the sacrificial material is etched away and the desired microstructure remains (6). In order to obtain a more complex, three-dimensional microstructure, the sacrificial layers and structure layers can be successively applied and formed using the appropriate lithographic and etching techniques, respectively.

The processing steps depicted in Figure 4.18 show why surface micromachining is often referred to as the sacrificial layer technique. As mentioned before, this technique can be used to make complex, three-dimensional microstructures with a height of about 20 μm from many thin layers. The structu-

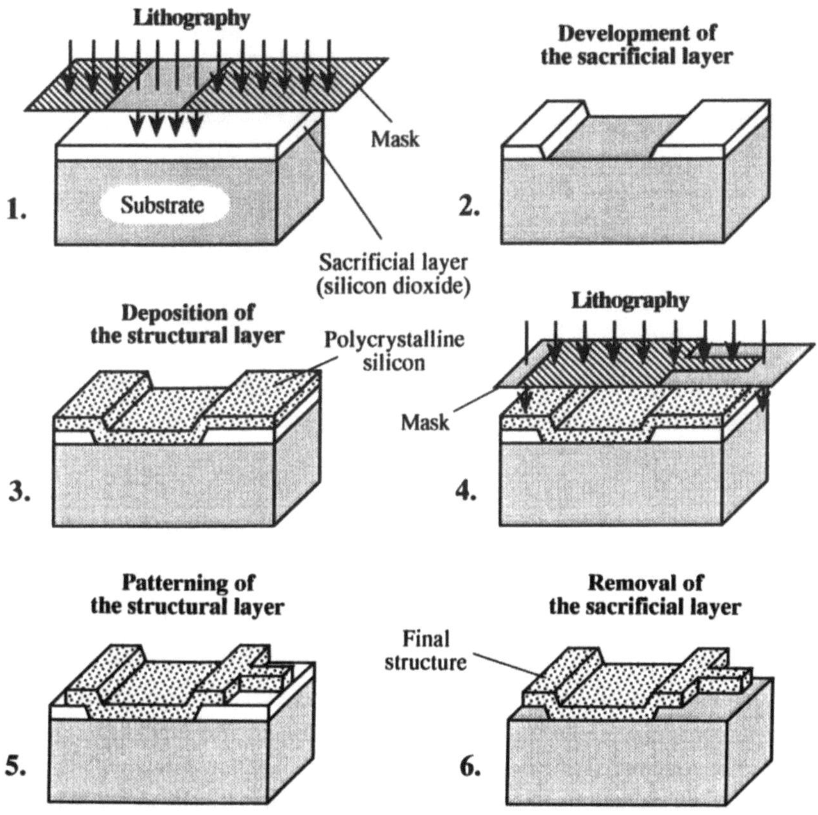

Fig. 4.18: Processing steps of surface micromachining. According to [Stix92]

ring may start with an already processed component as a substrate, but a passivating layer such as silicon nitride must be applied to its surface first to protect this micromachined component. Figure 4.19 shows an application of surface micromachining to manufacture a comb-like, free-standing microstructure from polysilicon [Jaeck92]. Each beam is only 0.8 μm wide and 2 μm high. Such structures are important elements of microactuators.

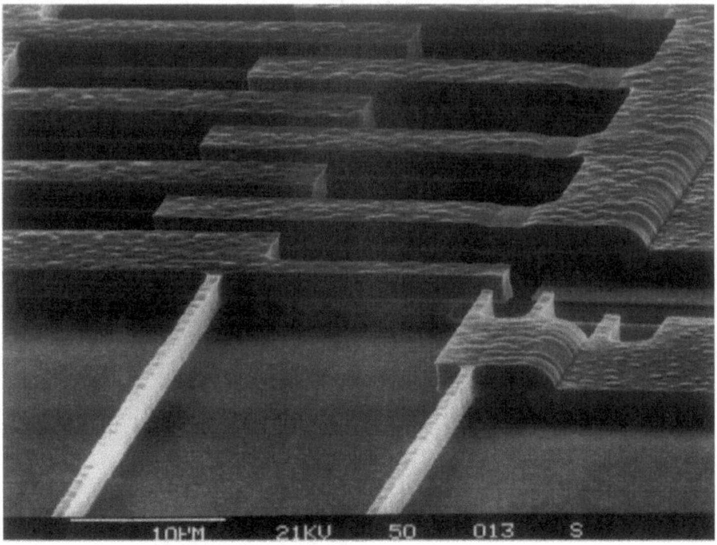

Fig. 4.19: Application example for surface micromachining. Courtesy of the University of Neuchâtel (Institute of Microtechnology)

4.1.3 Various Prototypes Manufactured by the Silicon Technology

Every successful scientific and technical idea has to prove itself as an industrial product at some point of time. Therefore, various micromechanical prototypes are being produced and then tested under realistic conditions to investigate their performance. Several prototypes will be introduced here; in the following chapters, many additional silicon sensors and actuators will be presented. Figure 4.20 shows a microgripper made of polysilicon by the surface micromachining method. The gripper is moved by electrostatic forces which are generated when opposing voltages are applied to two opposite comb-like elements. Such effectors can be used for microsurgical operations and for various microassembly tasks.

Another example is an electrostatic micromotor, Fig. 4.21. The motor diameter is about 100 μm and the distance between the rotor and the stator is about 1–2 μm. A rotational speed of about 15,000 rpm was reached with an operating voltage of 35 V. This is an interesting prototype for many possible MST applications.

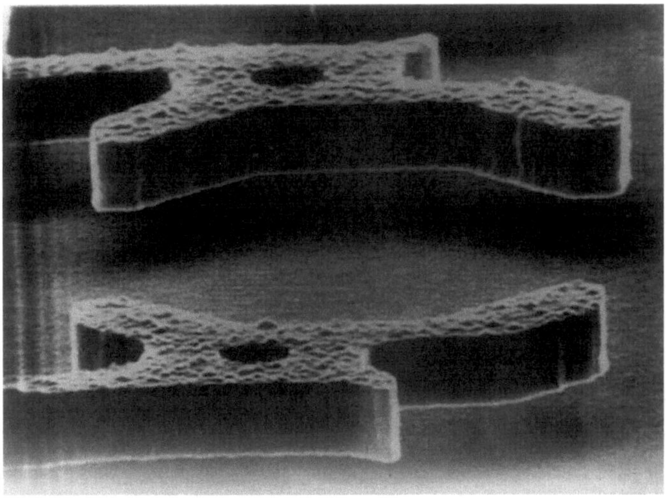

Fig. 4.20: Microgripper made by the surface micromachining technique. Courtesy of the Berkeley Sensor and Actuator Center, University of California, Berkeley.

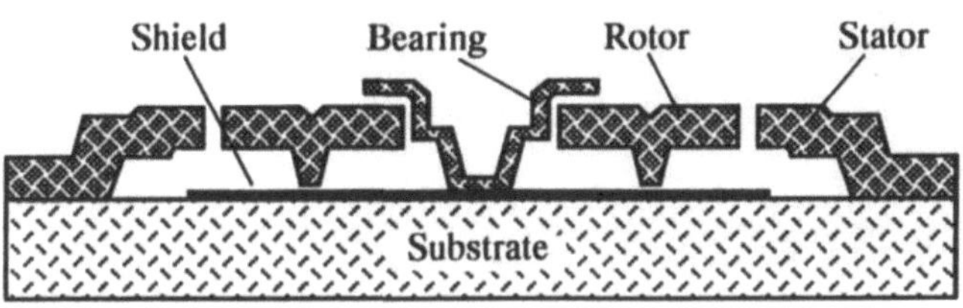

Fig. 4.21: Cross-section of an electrostatic micromotor made by the surface micromachining technique. According to [MIT92]

An optical microshutter, which also makes use of the comb actuator principle, is shown in Figure 4.22. This microstructure was made from a 2 μm thick polysilicon layer with the surface micromachining technique. The shutter has a dimension of 100 μm × 30 μm and can move along the wafer up to 8.3 μm under a voltage of 53 V. In Figure 4.23 the cross-section of a high precision optical fiber connection device is shown. The grooves were made anisotropically using the bulk micromachining technique on a single-crystal silicon wa-

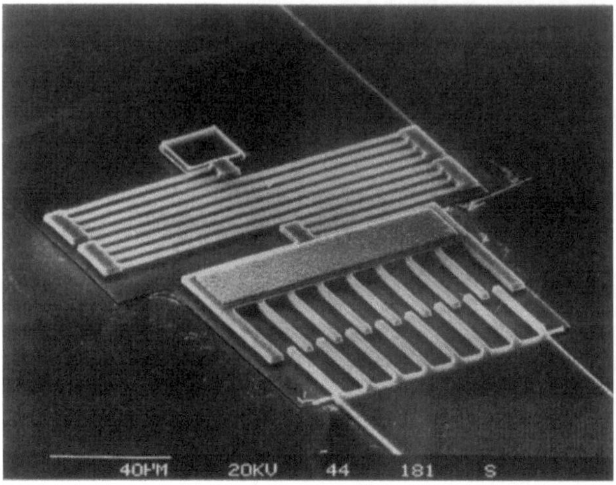

Fig. 4.22: A microshutter made by the surface micromachining technique. Courtesy of the University of Neuchâtel (Institute of Microtechnology)

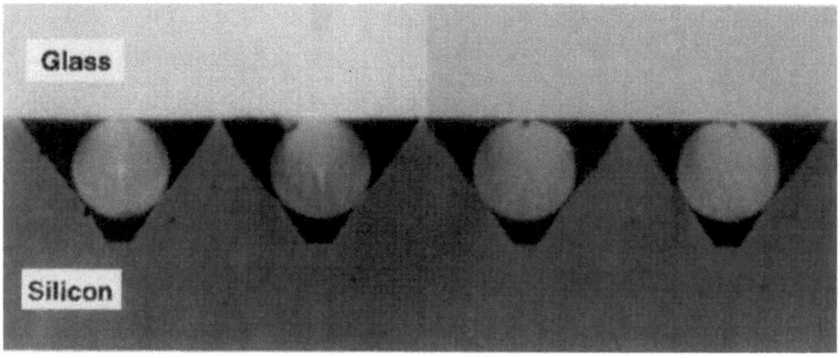

Fig. 4.23: Optical fiber aligment device made by the bulk micromachining technique. Courtesy of the Industrial Microelectronics Center, Kista

fer. The device is used for precision aligning optical fibers in a light coupler; with a good alignment the attenuation of the light is reduced. Glass is used for the casing of the alignment device.

Both structuring techniques introduced here, the bulk and surface micromachining methods, are used to make the majority of microstructures. The method chosen is determined by the problem to be solved and the production cost. When more complicated silicon-based components are to be made, often both techniques are applied.

4.2 LIGA Technology

Another fundamental technology of micromechanics is the LIGA process; LIGA is a German acronym for LIthographie, Galvanoformung, Abformung (lithography, galvanoforming, molding). It was developed at the Research Center Karlsruhe in the early 80's to make diffusion nozzles for uranium enrichment. The technology used X-ray lithography for mask exposure and galvanoforming for making the parts. Soon it was discovered that it could also be used for various other products with small dimensions. It was particularly useful for making plastic molds for very small parts.

Numerous investigations indicated that the LIGA process could be used for a broad range of MST products. It is an important alternative and in many cases a supplement to silicon-based micromechanics. With the LIGA process it is possible to make more complex microstructures and to work in the third dimension with structure heights of up to several hundred micrometers. In the lateral dimensions it is possible to form parts in the nanometer range of about 0.2 μm. Different plastics, metals, ceramics or combinations of them can be used in connection with the LIGA process in order to produce difficile structures such as actuators and sensors which are not possible to make with the silicon-based methods. The multitude of usable materials also allows electric, magnetic, optic, and insulating components to be integrated in one hybrid microsystem [Lehr94]. The disadvantage of this process is its high production costs. This can only be alleviated to some degree by mass-producing the microcomponents.

The single steps of the LIGA process, the X-ray lithography and the galvanic and plastic molding techniques are well mastered today, but this must still be optimized. Figure 4.24 shows a schematic diagram of the process sequence. In the first step, a radiation-sensitive plastic resist, several hundred micrometers thick, is applied to a conductive substrate and is exposed to synchrotron radiation through a mask (1). The irradiated areas of the resist are chemically removed, and the unirradiated portion remains (2). The gaps in the resist are then galvanically filled with metal (3). The rest of the unirradiated plastic is removed and a metallic microstructure remains (4). This can either be used as a mold for the mass production of plastic parts by micromolding techniques (5) or as a "master" for casting plastic molds for galvanic repro-

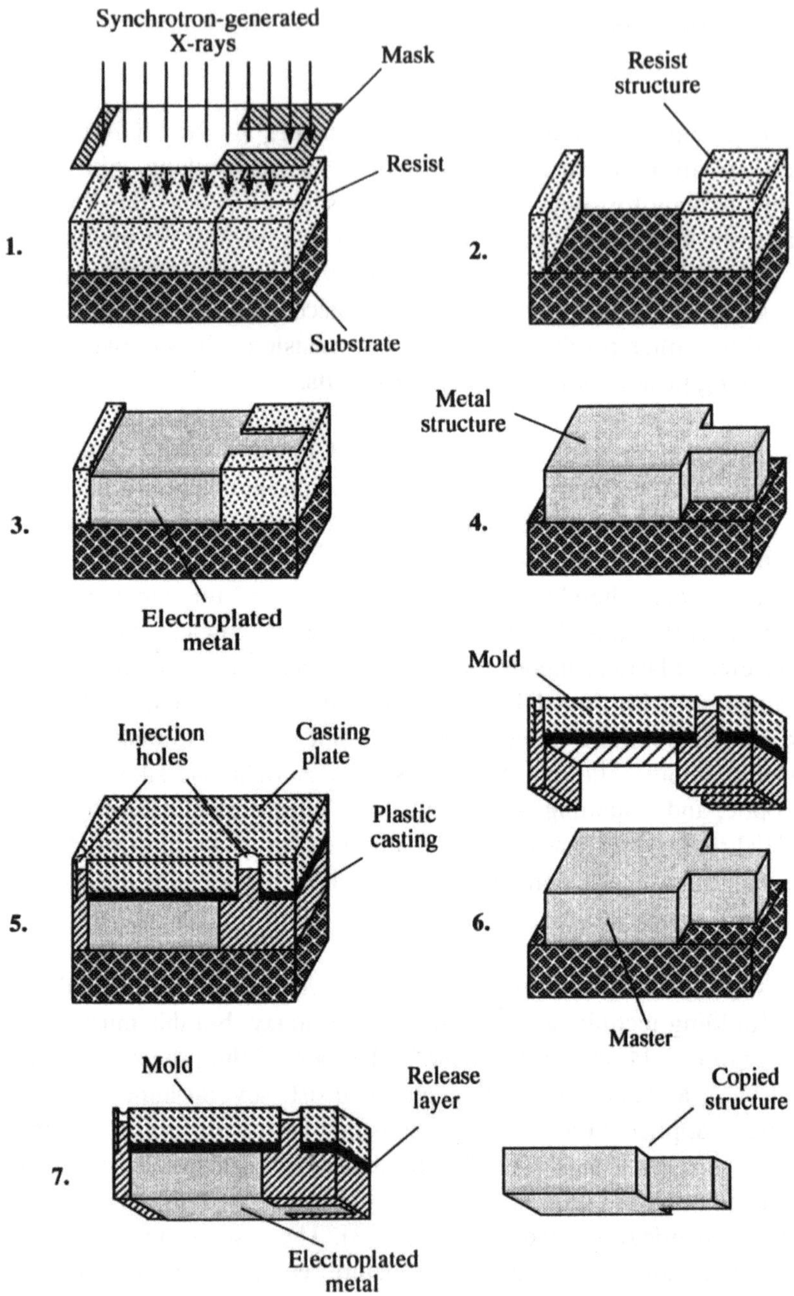

Fig. 4.24: Steps of the LIGA process. According to [Stix92]

duction of metal parts. The latter process is depicted in the steps (5), (6) and (7); here, the metal parts are produced similarly as shown in (3) and (4) using the "lost mold" technique. In the following sections the individual steps to the LIGA process will be described.

4.2.1 Mask Fabrication

An important prerequisite for producing very precise microstructures with the LIGA process is the use of a high-contrast X-ray mask. An X-ray mask usually consists of a material which can absorb X-rays very well, e.g. gold or tantalum. The material is deposited onto a thin foil of X-ray transparent material of a lower atomic number, e.g. titanium or beryllium. The significant steps for producing an X-ray mask are shown in Figure 4.25.

First, a titanium layer about 2 μm thick or a beryllium layer about 20–60 μm thick is sputtered onto a mask base made of a special alloy (1). On top of this layer, a plastic resist layer, usually polymethylmethacrylate (PMMA), is applied (2). The PMMA, better known as plexiglass, is an amorphic thermoplast. It is also used in the actual LIGA microstructuring process, as we will see later. Characteristic of this material is that it can be handled very well. The PMMA layer is about 20 μm thick and is structured by X-ray lithography (Section 4.2.2) (3). The gaps formed in the PMMA resist are galvanically filled with an X-ray absorber, e.g. made of gold (Section 4.2.3) (4). For a thick microstructure, an equally thick plastic resist must be produced, which implies the use of a strong radiation dosis and a thick absorber layer. To make a structure of about 400 μm thickness with an X-ray wavelength of 0.225 nm, the gold layer must be about 10 μm thick [Bach91].

In the last step, the plastic layer is removed and also the alloy of the desired exposure area of the mask is selectively etched away (5). Thus, a metal frame remains along the edges of the mask which supports the mask foil. Such an absorber mask is usually about 20 μm thick, allowing structures of about 650 μm height to be made. The production cost of such a mask can be up to several thousands of dollars due to the high precision required. This makes it the most expensive step of the LIGA process.

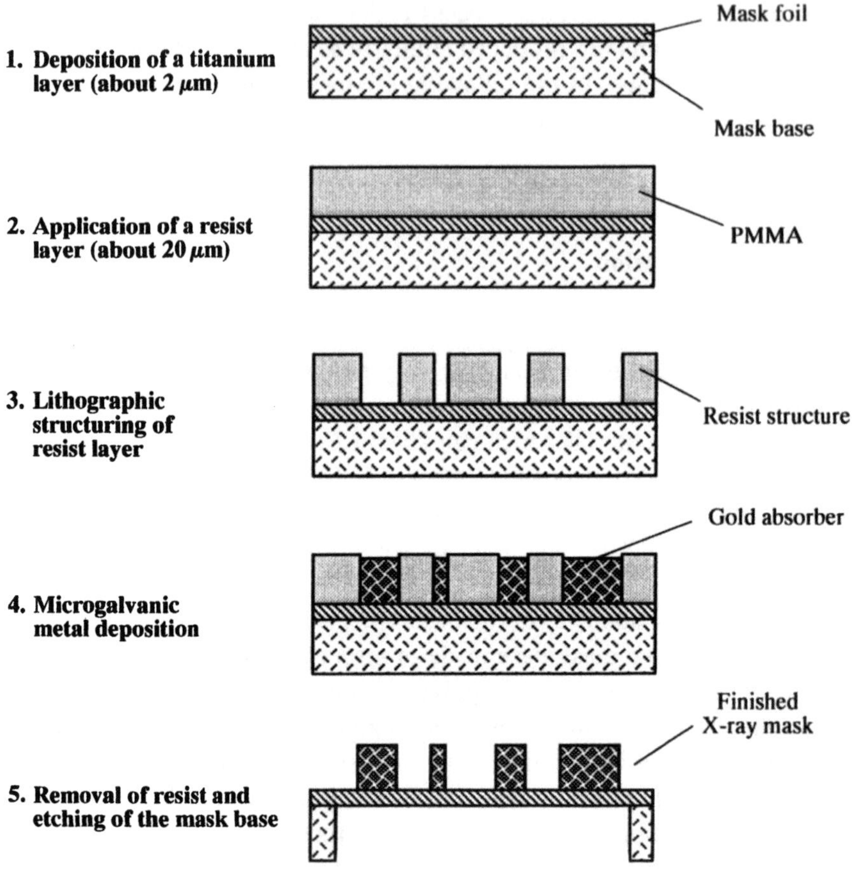

Fig. 4.25: Production of a mask for X-ray lithography. According to [Menz93]

4.2.2 X-Ray Lithography

The most important step in the LIGA process is the X-ray lithography using a synchrotron irradiation source. The radiation is produced by magnetic acceleration of electrons. The wavelengths of 0.2–0.6 nm are extremely short, which causes a high energy density and high parallelism. Due to the deep penetration capability, the X-ray lithography is well-suited for making three-dimensional structures of up to a millimeter height. The parts may have lateral dimensions in the micrometer or submicrometer range. This range of dimensions can only be achieved by using a high energy as rendered by a synchrotron radiation source.

Similar to UV lithography, a mask image is projected onto a substrate made of a irradiation-sensitive resist, usually PMMA. The synchrotron radiation beam is highly parallel, with a spread of about 0.2 mrad; thus only a very small area of the substrate can be exposed at one time. Therefore, the absorber mask and the substrate are moved evenly under the X-ray beam by a scanning mechanism, Fig. 4.26. The part structure is projected on the substrate using the shadow projection of the mask; it has a proximity distance of a few micrometers. Inside the non-masked plastic resist, chemical changes occur in the material during irradiation, e.g. in a positive resist molecule chains are broken. The irradiated resist areas are then dissolved, leaving behind a three-dimensional microstructure which is the spacial image of the flat X-ray mask.

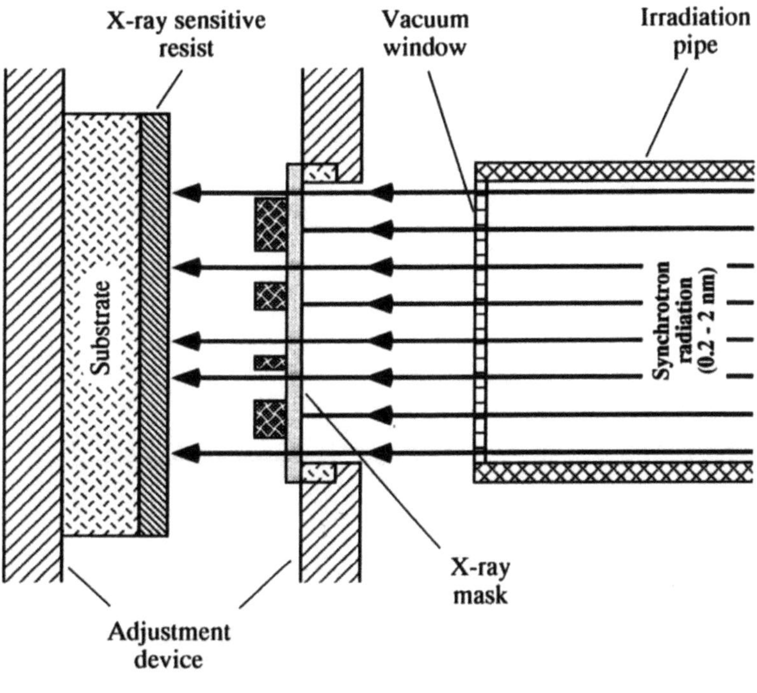

Fig. 4.26: Substrate exposure by X-ray lithography. According to [Menz93]

The PMMA resist, although used most often, also has its disadvantages. It is relatively insensitive to irradiation, which means that it has to be radiated for a long time. For example, to obtain a structure height of 350 μm, the PMMA resist must be exposed for about 8 hours to a high-energy dose of radiation. PMMA is also susceptible to stress corrosion cracking. This might be a problem when producing very fine microstructures. Other improved resist materials are not available yet, however, they are still searched for. Better resist materials would greatly contribute to an economic utilization of the LIGA process.

4.2.3 Microgalvanics

The primary plastic structure obtained after the X-ray lithography step is only an intermediate product. The goal is to create a metallic microstructure. Thus, the next important step of the LIGA process is the galvanic deposition of a metal in gaps of the plastic structure, Fig. 4.27.

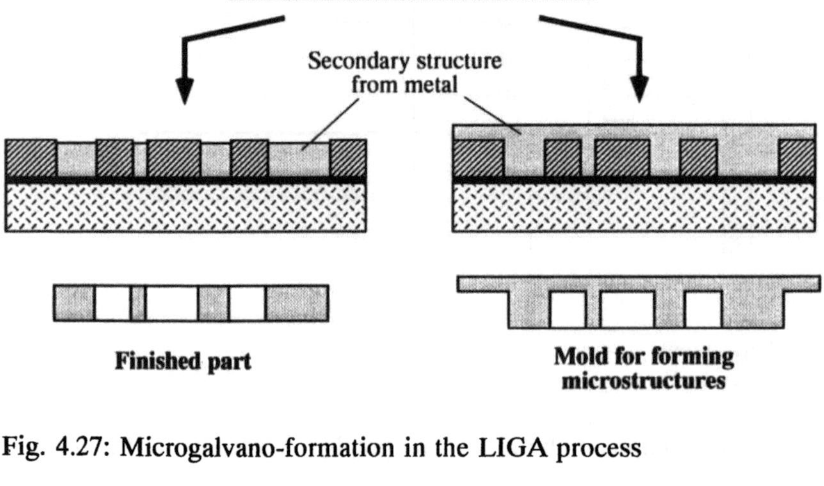

Fig. 4.27: Microgalvano-formation in the LIGA process

The substrate material (the base plate) must be electrically conductive in order to start the galvanic process. Usually, a metal plate serves as the substrate which has an additional layer of oxidized titanium on it to make the surface more adherent. At the moment, microstructures of nickel, copper and gold, as well as nickel-cobalt and nickel-iron alloys can be made by the LIGA process. Deep cavities and narrow channels in the submicrometer range put strong demands on the galvanic process. E.g. the exact composition of the electrolyte is of extreme importance. Technically, the microgalvanic process can best be carried out by the use of nickel-based electrolytes; for this reason nickel is the most often used material. The purity of the electrolytic solution must also be guaranteed for the galvanic deposition. Present research on microgalvanics is mostly concentrated on obtaining a reliable process control and on making new structuring materials available.

After removing the plastic and substrate, a metal complement to the primary structure remains which could already be used as the desired product. However, generally the main goal of this step is to produce a metal mold for mass production of high-precision plastic parts using micromolding techniques. In this case, the metal is deposited in such a way on the resist structure that it covers the resist completely in order to generate the mold with its base plate, Fig. 4.27, right.

4.2.4 Plastic Molding

The key to the inexpensive mass production of microcomponents using the LIGA process is plastic molding. The metal molding tools made for this purpose by microgalvanic techniques differ from conventional, mechanically produced tools in their small lateral dimensions and the variety of structures that can be produced in the μm range. Using the various micromolding processes, such as casting or coining techniques, any number of plastic parts can be produced as copies of the primary form.

Techniques such as injection and reaction injection molding have rendered good results in the mass production of plastic components. The basic method of reaction and injection molding for making micro-plastic parts is schematically shown in Figure 4.28. Both methods use a vacuum process in which the mold is hermetically sealed and preheated liquid material is injected into

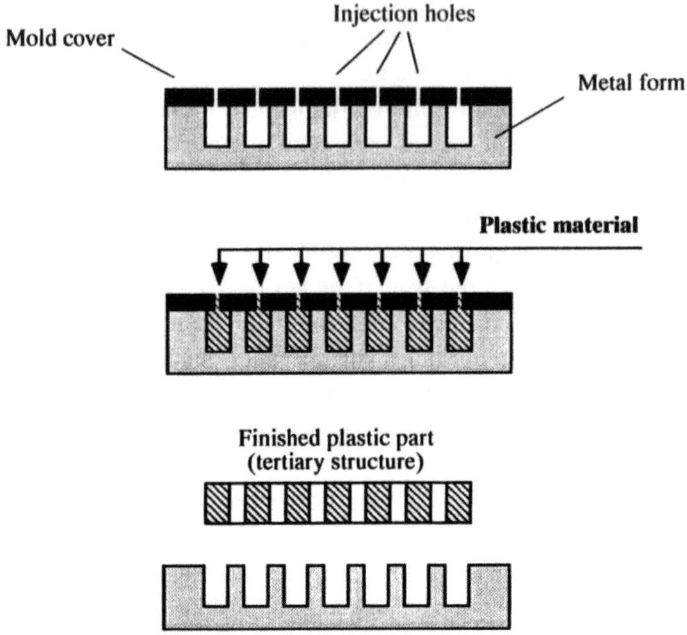

Fig. 4.28: Part molding by injection techniques

the mold to form the desired plastic part. In reaction molding, a chemically reactive resin is the base material, it is injected into a metal form and polymerizes to become a stable part. Typical materials are reaction resins based on methacrylatene, silicone and caprolactame. In injection molding, a non-viscous thermoplast is heated and melted and then injected into the form mold. The material hardens as it is carefully cooled down. Thermoplasts are used for this molding technique, like polyoxymethylene or PMMA. Due to the plastic's low viscosity and the corresponding low injection pressure the reaction molding method is used more often.

In general, the end product are the plastic structures that result from the molding process. Figure 4.29 shows two cylindrical, 150 μm thick lenses made of PMMA which were produced for optical applications. It is also possible to use metals and metal alloys and ceramics for mass producing microstructures. In this case, the tertiary plastic structure is used as a mold in a second galvanic process (Fig. 4.24, Step 7) or for a casting by the lost mold

Fig. 4.29: Cylindrical lenses made of PMMA. Courtesy of the Institut für Mikrotechnik GmbH, Mainz

technique (for ceramic parts). For the last mentioned technique ceramic powder is prepared as a castable suspension and then injected into a mold made of plastics. The raw ceramic part is then made denser by drying it at a temperature of about 90°C; the plastic mold is melted out (lost mold). The remaining part is sintered to obtain the desired ceramic microcomponent. It is a negative replication of the tertiary plastic structure. If, however, the tertiary plastic parts are used for galvanically formed metal components, a mold cover made from conductive metal must be used for reaction molding. In this case, the mold cover serves also as a base plate for the following galvanic process, Fig. 4.27. The plastic remains connected to the mold cover and serves as the mold for the galvanic process. As an alternative to the metal base plate, a conductive plastic base can be used. In this case, a base plate is firmly pressed over the entire surface of the mold filled with reaction resin. Once the resin has hardened, the base plate is connected to the mold material. A plastic structure results which can be galvanically filled with metal.

The coining technique makes use of a thermoplastic transformation process. A thermoplast is warmed to its working temperature and is pressed into a mold under vacuum. After it cools, the plastic can be removed from the mold. This method has a much lower production cost than the others, but results in a much lower accuracy of the final structure. For this reason this technique is hardly used in MST.

4.2.5 Sacrificial LIGA Technique

By using the sacrificial layer technique (Section 4.1.2.5), LIGA structures separated from the substrate can be realized. For this purpose, a sacrificial titanium layer is applied to the substrate and structured to the desired shape before the beginning of actual LIGA process. After processing the LIGA microstructure, the titanium layer underneath is selectively etched away by using a wet etching technique.

The sacrificial layer technique allows the formation of microcomponents for making movable microactuators and sensors. If an electrically insulating substance is used as the substrate, additional electric functions can be realized in the microsystem by depositing structured metal layers onto this substrate.

The required steps of the sacrificial LIGA technique (SLIGA) for making movable parts are shown in Figure 4.30. In the first step, different metal layers several 100 nm thick are applied to the silicon wafer covered by an insulating layer (1). The metal layers are predominantly made of chromium and silver, they also serve as electric connectors and as starting layers for the galvanic process and as the adhesive layers for the substrate. These metal layers are structured by photolithography and dry etching, forming a conducting strip (2). Then, a 3 to 5 µm thick titanium sacrificial layer is deposited on the work piece and is structured by an additional lithographic etching step (3).

The substrate is now ready for the actual LIGA process. First, a resist layer of several hundred micrometers thickness is applied to the substrate by direct polymerization. The layer is exposed to synchrotron radiation via an X-ray mask (4). The exposed resist areas are then removed (5) and the metallic secondary structure, usually nickel, is deposited by means of galvanic molding (6). Finally, the unexposed resist is removed and the titanium layer is selectively etched away (7). Thus, a freely standing microstructure results which can be electrically connected to a microelectronic circuit.

Highly integrated one-chip microsystems can be produced by the LIGA process when using prestructured silicon wafers. By adjusting the X-rays, mechanical microstructures are formed on the wafer after the galvanic molding process. They can then be used as actuator or sensor components.

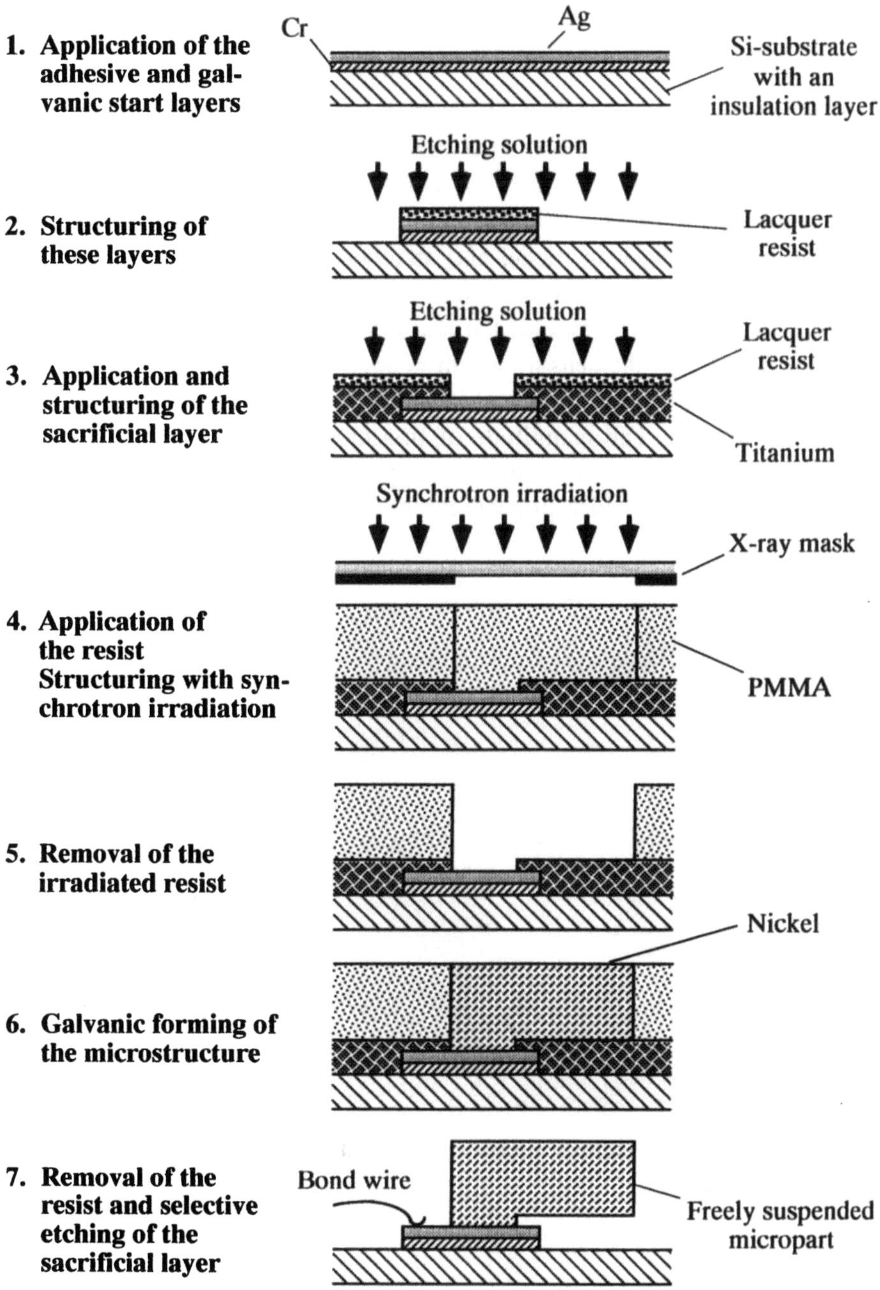

Fig. 4.30: Steps of the sacrificial LIGA technique. According to [Mohr91]

4.2.6 Prototypes of LIGA Components

A few examples of microcomponents made by the LIGA process were given in Chapter 2. Now we will introduce various characteristic microstructures and components made by this process. The first movable structure produced by the LIGA process was a displacement sensor of an accelerometer. It is an oscillating mass made of nickel and attached to a cantilever type spring, Figure 4.31. It is located between two electrodes which are connected to a substrate. The gap on each side of the mass is 3 μm; the height of the structure is 100 μm. When the sensor system is accelerated, the gap changes between the mass and the electrodes which in turn changes the capacitance of the capacitor system. The magnitude of this change is an indication of the magnitude of acceleration. The sensor is equipped with a mechanical overload stop (below right), which protects the structure from possible damage.

Fig. 4.31: Acceleration sensor components. Courtesy of the Karlsruhe Research Center, IMT

The SLIGA technique also allows the manufacture of movable components which are neither connected to the substrate nor to any other structural component. For example, it is possible to make microturbines or micromotors with freely suspended rotors which are contained by an axle for rotation. Figure 4.32 shows an impeller of a microturbine compared to the size of an ant's head. The nickel impeller has a diameter of 260 μm and is 150 μm high.

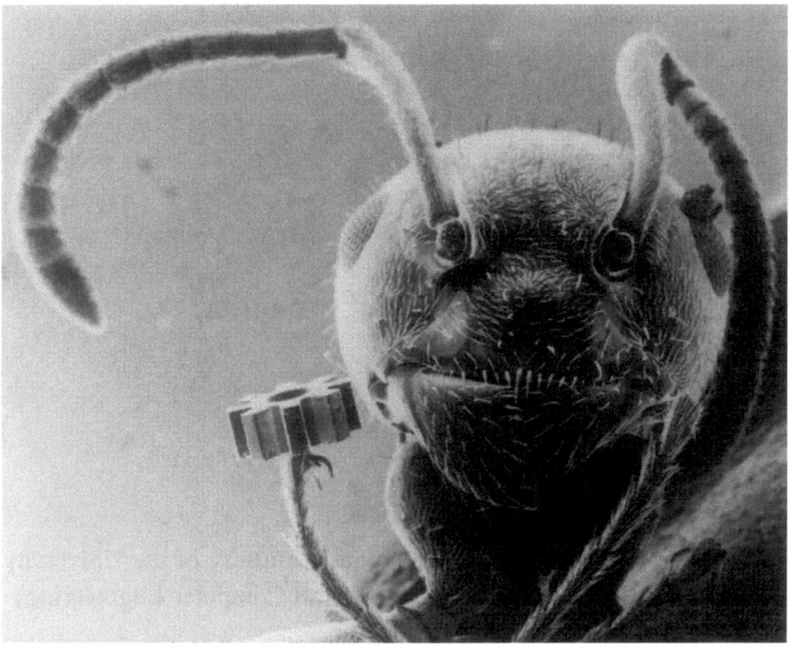

Fig. 4.32: Microturbine impeller made by the LIGA technique. Courtesy of the Karlsruhe Research Center, IMT

A magnetic micromotor with a gear train transmission is shown in Figure 4.33. It is made of nickel by the SLIGA technique and has a height of 100 μm; the clearance between the axle and the rotor is only 500 nm.

The LIGA technique, like the silicon technology, can also be used to realize electrostatic comb actuators, Fig. 4.19. Here, the effect that two charged capacitor plates attract or reject each other, depending on the charge, is utilized (Section 5.2). In Figure 4.34, a comb-like microstructure made from nickel by the LIGA technique is depicted. The lower comb is connected to a substrate, the upper comb is etched to be free floating and can move into the lower comb structure when a voltage is applied. The movable comb structure has a micro-spring system that exerts the necessary restoring force. The comb

Fig. 4.33: Micromotor using the SLIGA technique. Courtesy of the University of Wisconsin-Madison (Department of Electrical and Computer Engineering)

Fig. 4.34: LIGA microstructure with a comb drive. Courtesy of the Karlsruhe Research Center, IMT

prongs are 200 μm long and 50 μm wide. The gaps between the prongs are 5 μm and the entire microstructure is 70 μm high.

The following example represents a more complex component having a hybrid construction. It is a micropump which was developed within the context of the microanalysis system presented in Section 2.3 for dosing and transporting of small amounts of liquids and gases. The pump is capable of transporting microscopic amounts of matter to a sensor for measurement and then it removes it again. The schematic design of the pump is depicted in Figure 4.35.

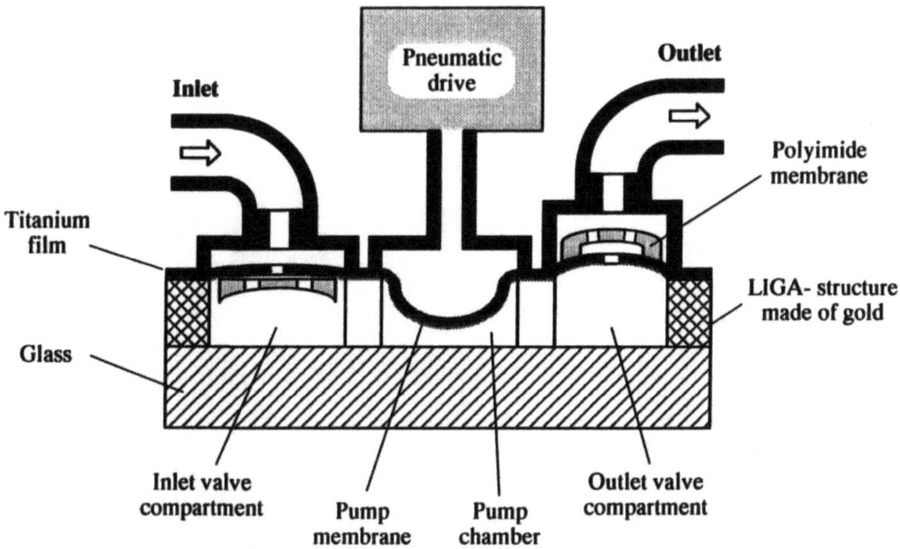

Fig. 4.35: Schematic sketch of a micropump. According to [Schom93]

The pump casing consists of a 100 μm high LIGA structure made of gold, a glass plate and a thin titanium film of about 2.7 μm thickness. Inside the pump there is a 5 mm diameter chamber and two valve compartments, each with a diameter of 1 mm. The valves themselves are made from a very durable elastic polyimide membrane. The pump has a high compression ratio, which is obtained with the help of an external pneumatic actuator. This actuator pushes the pump membrane (the middle part of the titanium film) down to near the floor of the chamber. A pressure increase causes the polyimide membrane of the outlet valve to curve away from the titanium membrane since it is more elastic than the titanium membrane. The valve opens and the medium inside is pushed out. At the same time, the inlet valve is closed since the polyimide membrane contacts the titanium membrane. If the pump membrane is moved upwards, the pressure and valve states reverse and the medium flows into the pump. The device can transport 70 µl/min of water or 86 µl/min of air, respectively.

The LIGA technique can also be used to realize components for microfluidic controllers which are mostly used in biomedicine. A bistable switch for microfluid flow control was presented in literature [Menz93a], Fig. 4.36.

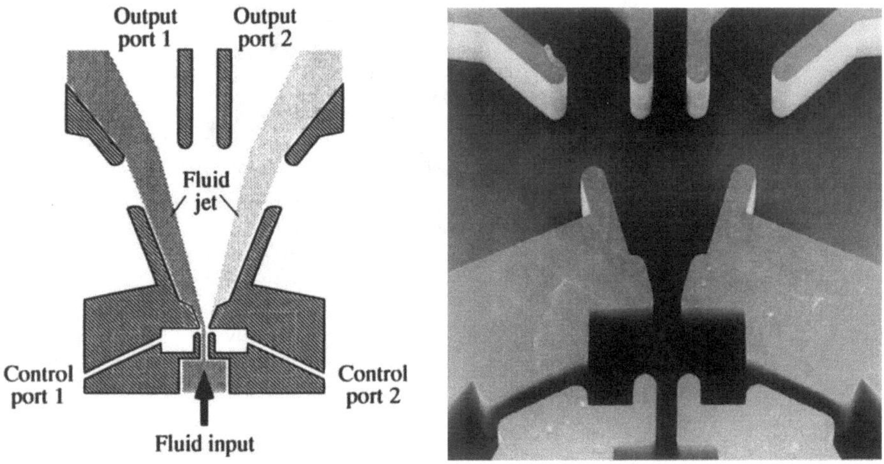

a b

Fig. 4.36: A bistable fluidic switch made by the LIGA technique
a) schematic design; b) a 500 µm high prototype made of PMMA with a nozzle width of 30 µm. Courtesy of the Karlsruhe Research Center, IMT

A fluid stream can be switched back and forth in a flip-flop manner by pressure impulses from a control channel. During operation, the fluid stream adheres to one of the two switch walls due to the Coanda effect. It is stable as long as there are only small disturbances. This condition denotes a state 1 and in our figure the fluid flows through Output 1. When a control signal is applied to Control port 1 in form of a pressure increase, the fluid jet flips over to Output 2, changing the state of the switch to state 2. The switch does not need any movable parts.

LIGA structures have very smooth walls and are suitable for microoptics. Various optical components have already been realized and some are in the state of development [FZK93]. An infrared band pass and a high pass optical filter were described in [Bach91a] which were made for the space observatory of the European Space Agency. The high pass filter was made by the LIGA technique by reaction molding using form molds made of nickel, Fig. 4.37. The nickel mold, which was a secondary structure in the LIGA process sequence, consists of a honeycomb-like screen with a total of 74,400 holes, each having a diameter of 80 μm. The gaps between the individual columns in the tertiary PMMA structure are 8 μm wide. The height of the filter is 170 μm. This PMMA structure is used to galvanically produce the copper high pass filter foils which are copies of a secondary nickel structure.

a b

Fig. 4.37: LIGA fabrication of high pass optical filters
a) a nickel mold; b) the tertiary microstructure made of PMMA. Courtesy of the Karlsruhe Research Center, IMT

The LIGA steps described before can be used to make high, perpendicular structures. It is also possible to make microstructures with slanted walls. The substrate and the mask just have to be tilted to a certain angle during the radiation period. A side wall roughness of 50 nm was obtained, which is comparable to structures produced with vertical walls [El-Kh93]. Complicated, non-linear microstructure profiles can be produced by irradiating twice with exactly dosed X-rays, taking advantage of the damage done to the unirradiated mask-covered part of the resist, usually undesired. During radiation, rays are bent at the mask boundaries and affect sections of the masked resist. The resist is a few hundred μm thick and only gets slightly damaged in the upper area. After briefly radiating the substrate a second time without the mask, the damaged resist gets an additional dose of irradiation necessary to remove the upper parts of the resist during development. After development, complicated curves appear across the resist thickness.

The examples given here show the application potential of the LIGA technique for various branches of industry. Opportunities are e.g. in the automobile, medical equipment or environmental protection equipment industries. The advancement of this technology is hampered, however, due to the fact that synchrotrons are extremely expensive. In order to create a powerful microstructuring technique, low cost irradiation sources for industrial use must be made available. Also, other necessary process equipment must be made smaller and more cost affordable in order to be in reach for small and medium-sized companies. In addition, further development and improvement of the existing interconnection technologies are necessary to be able to produce LIGA-based hybrid microsystems. If these requirements are fulfilled, the LIGA process has a good potential of being the basis for mass producing of microparts. Due to a high number of units their costs can be dropped and the LIGA process would be competitive to the silicon technologies.

5 Microactuators: Principles and Examples

5.1 Introduction

The MST applications introduced in Chapter 2 suggest the use of new micro-actuator systems which allow motions to be realized with micrometer accuracy. Conventional motion concepts or manufacturing methods are no longer able to fulfill the demands concerning miniaturization and all questions connected with it. Microsystems, and in particular future microrobots, require the development of new advanced actuators with very small dimensions, simple mechanical construction and high reliability.

It is rather difficult to determine exactly what the name "microactuator" implies. In the literature, the term "microactuators" is used for devices ranging in size from micrometers to several decimeters, proving the typical classification difficulties of this young scientific field. This book considers a microactuator as a device of a few micrometers to a few centimeters in size having a functional principle applicable in the microworld. Microactuators also include those actuators which are manufactured by using micromechanics (Chapter 4). The authors are aware of the imperfection of the above definition. However, it is pointless to make the classification stricter as it is unknown which constructions and actuator principles will be useful in MST and which of them will be successful in the long-run.

In addition to the miniaturization, mechanical microdevices having components such as pumps, valves, robot grippers, linear and rotational positioning elements, simple cantilever actuators and complex artificial muscle systems, must be functional to provide a microsystem with task-dependent capabilities. Micropumps and -valves for treating liquids and gasses at the microlevel can be applied in medicine, where implantable, highly accurate microsystems are needed for the dosing of medication or for chemical and biotechnological analysis where minute volumes of liquid must be transported and measured. They can also be applied for technical devices such as ink jet printers. Micro-actuators using the cantilever principle can be applied for various applications to generate minute motions. In optics such microactuators can serve as electronically tunable mirrors, in fluid dynamics as valves, and in microrobotics as grippers. Micromotors also have a great potential for MST applications. The development of micromotors is not very advanced yet, since the fundamental principles for a good realization have not yet been developed. In present micromotors, a rotor is moved by electrostatic, electromagnetic or piezoelectric forces which are applied inside or outside of a stator. The development of micromotors is going through repeated cycles of designing, manufacturing, testing and modification; optimal dimensions, elementary design principles and materials to be used for each specification are subject of investigations. Also performance parameters such as torques and rotational speed do not yet reach the requirements of most applications. The flexibility, versatility, high strength and other properties of muscles encourage researchers to take a closer look at natural actuator principles [Pelr92], [Ecke92], and they are trying to imitate muscle structures. An artificial muscle is usually made up of a series of microactuators put together to form a more powerful large integrated bundle, in analogy to a natural muscle.

It is not clear today how much influence microactuators will have in the future, but they are expected to have an especially large impact on medical and environmental devices, various mechanical designs and automotive components where actuators of small dimensions are needed in large quantities.

Usually, microactuators are complex system components or independent subsystems. Regarding the system integration, an actuator connects the information processing part of the system's controller with the process to be influenced [Jano93] and is therefore indispensable for a complete microsystem. A simple schematic representation of an integrated actuator is shown in Figure 5.1. Here, the control signal generated with the help of sensor information is transformed into a motion by means of the actuator.

Presently, there are very few functional microactuators available compared with microsensors. Most microactuators are still in the development stage, having significant problems in regard to design concepts, control principles, accuracy, resistance to environmental influences, etc. Another practical difficulty originates from friction which often has disastrous consequences for the microactuator system. This area is therefore regarded as a challenge for research and industry; the above listed problems are specific to MST and must be approached from several different points of view in order to be able to offer system-compatible actuators in the near future. Characteristic of a

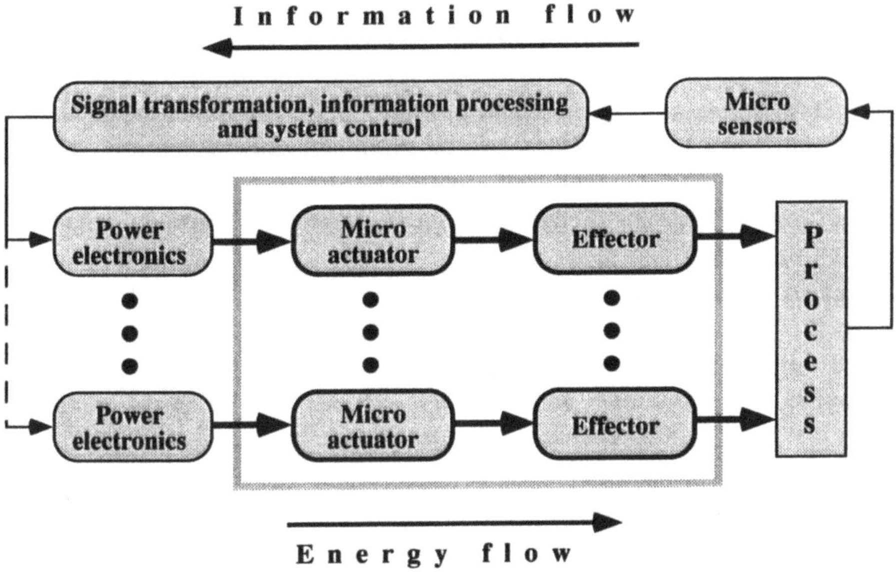

Fig. 5.1: Integration of a microactuator into a microsystem. According to [Kall94]

microactuator system is that the manufacture of the device and of the corresponding signal processing components are often based on the same silicon technology; thus a monolithic system construction saves space and cost. On the other hand, in specific applications it may be necessary to combine different kinds of system components (not necessarily microtechnological ones) to one complex microsystem. There are also many other materials and/or effects (Chapter 3) which may lead to the conception of a microactuator system, but cannot be realized with the present silicon technology. In this case, a hybrid system construction has to be selected. The LIGA method e.g. offers a way to structure a system from a variety of materials. In any case, the manufacturing methods for MST must be cost-effective, in particular when it comes to mass production. This is in contrast to the expensive miniaturized microactuators made by conventional precision machining.

By the use of appropriate energy transformation techniques microactuators must be able to generate exactly specified forces and torques or positional changes. The function of a microactuator can be based on conventional force-producing principles, as well as on new principles especially tailored to MST and suited for the micro world. MST researchers have come up with various new force-producing principles. An overview of all force-producing principles used in MST is presented in Figure 5.2. In addition to the electrostatic, piezoelectric and electromagnetic microactuators which are currently being investigated, magneto- and electrostrictive materials and shape memory alloys (SMA) are becoming of great interest. Electrorheological, hydraulic, pneumatic and thermomechanical microactuators are also being investigated to find their possible use [Dario92], [Fati93], [Remb95]. These microactuators open up a world of new possibilities since they are based on entirely different functional principles than conventional macroactuators. As compared to conventional actuators, microactuators will be using mainly direct drives without mechanical transmission elements.

The piezoelectric effect, for example, is based on atomic interactions in ceramic materials to which a voltage is applied. The electric field causes expansion in ceramics, which can be exactly controlled, and thus allows a movement in the nanometer range. Magnetostrictive alloy actuators, when exposed to a magnetic field, cause an expansion or contraction of the alloy. The movements that can be produced here are in a similar range as those of piezoelectric actuators. Of special interest are the shape memory alloys. After being plastically deformed, they return to their original shape when heated (thermal shape memory). This effect is a result of a thermoelastic, martensitic transformation in the alloy. Electrorheologic actuators have caught the attention of many researchers; interesting is that electrorheologic liquids change their viscosity under the influence of an electric field and switch from a liquid to a plastic state. This unusual property can be used for many applications such as clutches, valves or vibration absorbers. The above mentioned materials are often characterized as being "smart", since they display both sensor and actuator properties at the same time. These actuator principles have the great advantage that a given input current results in a predictable displacement, so that the actuator system can operate without force or position sensors [Jano92], [Jano93].

The development of new actuator concepts is expected to profit from research on biological motion principles, although their technical realization is still in the distant future. An attempt e.g. is being made to design muscles with the help of microactuators connected in parallel to increase the performance (motion and strength) of the actuator [Pelr90], [Fuji93]. Thereby macroscopic

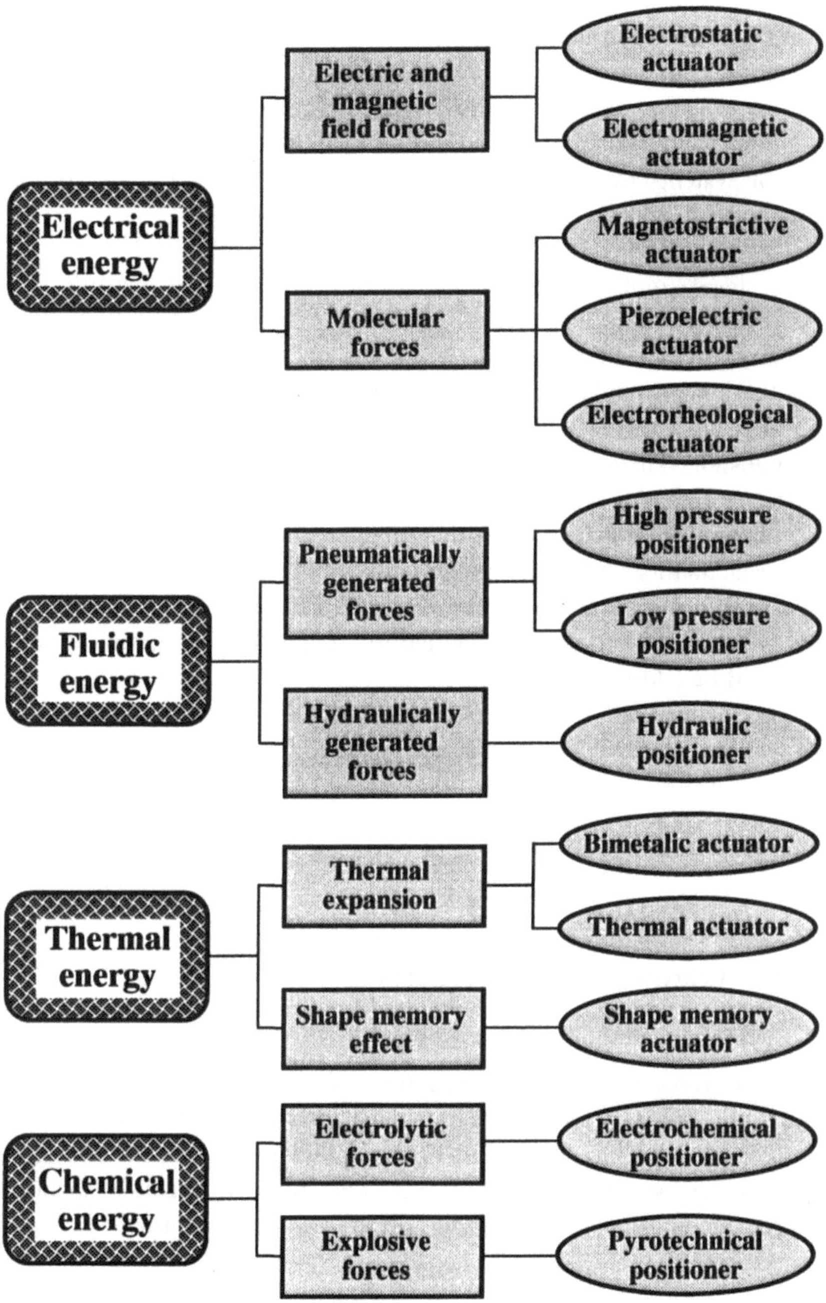

Fig. 5.2: Classification of actuators. According to [Schrö93]

effects can be reached through the coordination of a group of elementary actuators since the motions and strengths of each individual actuator are additive. In order to carry out more complex motions, each microactuator must only be equipped with a series of individual sensors and a controller. Such microactuator systems are robust to malfunctions of individual elements, can easily be adapted to specific problems and can be extended.

There is no "ideal" motion principle that can be always used. Despite this, researchers often try to find a "cure-all" by using one microactuator principle for any kind of micromachines. As we will see soon, the selection of the most suitable actuation principle depends on many factors, the most important being the necessary forces and the amount of motion needed by the application in question. The advantages and disadvantages of the different energy transformation principles with regard to their application to micromachines and microrobots will be discussed in this chapter, illustrated by examples of various microactuators and their components which have already been realized. We will also give an overview of the current development state and the future potential of various microactuators. Typical applications of microactuators will be addressed, too.

5.2 Electrostatic Microactuators

5.2.1 Motion Principle and Its Properties

An electrostatic force is created by applying a voltage across two capacitor plates which are separated by an insulator, e.g. an air gap. Thereby, opposite charges are formed on the plates and this potential results in an attractive force between the two electrodes, Fig. 5.3. This force F is called an electrostatic force and is calculated according to the following equation:

$$F = \frac{\varepsilon \cdot S \cdot V^2}{2 \cdot d^2}$$

Here:
- F – electrostatic force [N]
- ε – dielectric constant [$C^2N^{-1}m^{-2}$]
- S – electrode surface area [m^2]
- V – applied voltage [V]
- d – electrode distance [m]

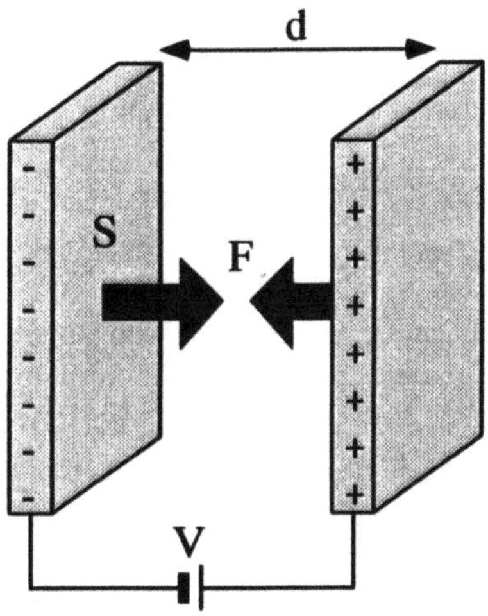

Fig. 5.3: Generation of an electrostatic force

By varying the voltage (usually between 40 V and 200 V) which is applied to the specially designed electrodes, a deflection can be obtained. The deflection is in the range of a few micrometers. The smaller the distance between the two electrodes the greater is the minimal electrical field strength. This results in an increased field energy density and consequently in a greater actuator force. The design methods described in Chapter 4 allow very narrow gaps to be manufactured between the electrodes, and the silicon technology allows a monolithic design of electrodes on a chip.

Electrostatic actuators can easily be miniaturized since the electrostatic force between two capacitor plates depends on the applied voltage, the gap distance and the plate area, but not on the plate thickness or volume. That is why the electrostatic force is also known as the surface force; this is in contrast to the magnetic force which depends both on the thickness and on the area of an element and which is known as the body force. The fact that the electrostatic force only depends on the opposing surfaces and the distance between them, makes it well suited for the micro world. The use of light materials such as aluminum as an active element is especially advantageous for applications in which weight is important. In this regard, electrostatic actuators differ from many other actuator types. E.g. magnetic actuators are made of iron or cobalt alloys, SMA actuators usually of nickel-titanium alloys and piezoactuators of solid barium titanate bars.

Like all other actuator types, electrostatic microactuators have their disadvantages, too. Even when the gap is very narrow, the applied voltage must be high. For a gap of 1 μm the voltage necessary for an actuator pressure of 1 kg/cm^2 is about 150 V. Another disadvantage is that under certain circumstances the actuator can fail catastrophically due to an electrostatic collapse. Even microscopically small surface defects can cause such a collapse; therefore these devices need very smooth surfaces compared to other actuator technologies. It must also be kept in mind that electric fields interact with many materials, making a suitable insulating layer necessary. Also, electric fields attract all kinds of dust which can be detrimental to the functioning of unprotected electric circuits.

Linear and rotational drives can both be realized by using the electrostatic principle, Fig. 5.4. Both actuator designs rely on the fact that the movable electrodes tend to reach a state of maximum capacitance in a certain position. In the comb arrangement in Figure 5.4a, a movable comblike element is inserted in a mating fixed comblike element, realizing a small vertical linear motion. The movable comb usually has a spring back mechanism; its elasticity can supply the pull back force. The possible motion distance depends on the pull back force of the spring, the depth of the comb elements and the applied voltage, respectively. The arrangement in Figure 5.4b shows how a motion horizontal to the comb structure can be obtained. As we will see later, this arrangement is most suitable for rotational microdrives.

The functional principle of an electrostatic micromotor is shown in Figure 5.4b. If two oppositely charged comblike electrodes are facing each other, they are not only attracted in the normal direction but also parallel to their surfaces due to a tangential force until the plates are in an exact alignment. In order to generate a torque or a rotation the rotor and stator must be equipped with partially offset electrodes; this causes a tangential force in every position of the rotor. Because the generated force is proportional to the

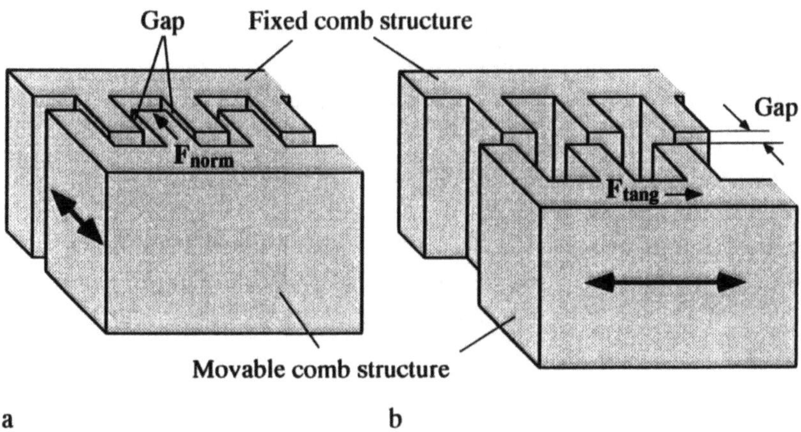

Fig. 5.4: Two types of motion which can be obtained by electrostatic forces. According to [Mohr91]

number of electrodes, often many electrodes are used in parallel in a small amount of space. During the past few years, a variety of rotating electrostatic micromotors had been developed. In general, their torque was very small and they had a short lifetime due to friction caused by the mechanical contact of the stationary and mobile elements. New concepts are now being investigated for electrostatic micromotors, so that the rotor can be levitated without friction [Kumar92], [Ziad94], [Bleu92].

Another way to design electrostatic actuators is to use a membrane structure. Here, flat electrodes face each other; they approach each other when a voltage is applied. The force, which reacts only vertically to the electrode surfaces, increases quadratically with decreasing distance between them. This motion principle is used to make valves, pumps and artificial muscles.

5.2.2 Concepts and Prototypes of Electrostatic Microactuators

After the electrostatic effect and its most significant properties for MST have been discussed, some interesting R&D work to develop electrostatic microactuators will be discussed. Many prototypes of different kinds of microactuators have been made and investigated. Since the investigations of MST-specific actuation principles have been started, many new and original ideas and principles were introduced. As the examples presented here were selected based on their simplicity and generality they are easy to comprehend. They can be applied in their simple form or serve as components of a more complex actuator system. In the following discussions important actuator designs will be explained in detail; we will be trying to make these discussions comprehensible. The chosen order of presentation is coincidental, both here and later on in the book, and does not suggest a priority by which the actuators are selected.

Electrostatic microshutter

Electrostatic actuators using membranes will be presented first. In metrology and in microoptics, so-called microshutters have become of great interest. One of such electrostatic microshutters was proposed in [Lin93]. The principle of this shutter is based on the electrostatic deflection of a movable electrode (microshutter), made of aluminum, gold or doped polysilicon. During an operation, the shutter moves against a fixed silicon electrode (substrate) which was produced by anisotropic wet etching in (110) silicon. In Figure 5.5 this principle of such a shutter is shown.

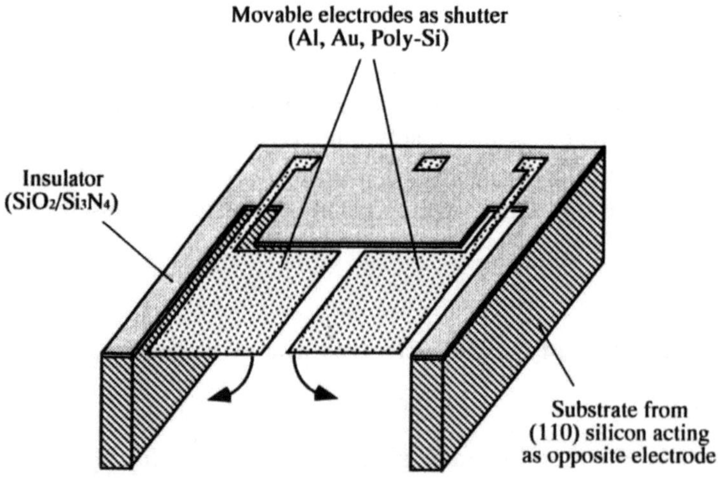

Fig. 5.5: Electrostatic microshutter. According to [Lin93]

When no voltage is applied, the movable electrodes (microshutters) are in their horizontal home position, and the shutter is completely closed. If a voltage is applied between the microshutters and the substrate, the shutters open. Each shutter has a narrow torsional beam attached to it about which it rotates. The torsional beam is insulated from the silicon substrate by an oxide layer. When the shutter is completely open (90°), a maximum amount of light can pass through. When the voltage is removed, the shutters return to their original position due to the restoring spring action of the torsional beams.

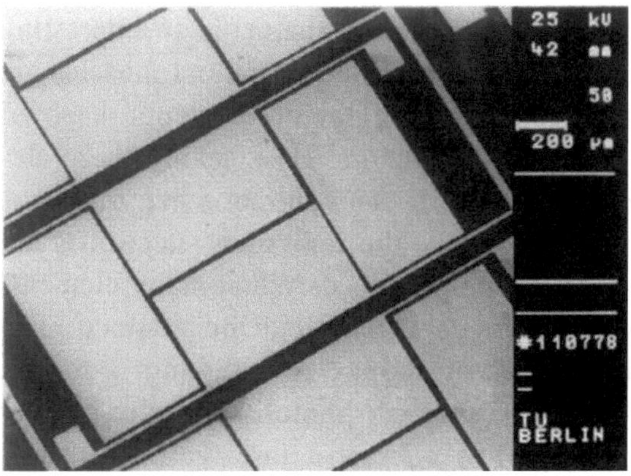

a

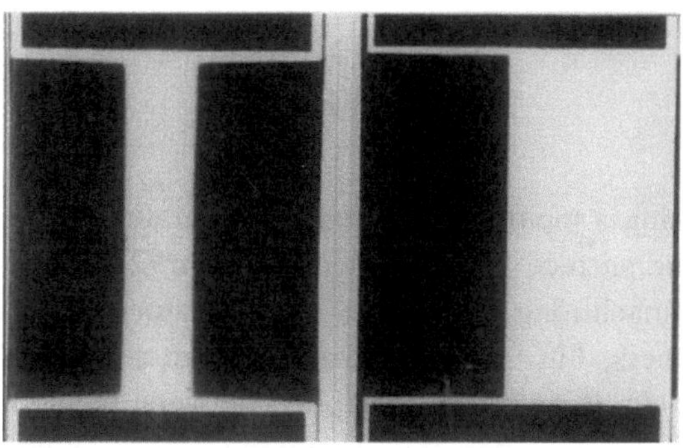

b

Fig. 5.6: Microshutter made of Al
a) shutter in its closed position; b) a double-sided 30° opening (left) and a single-sided 90° opening (right). Courtesy of Technical University of Berlin

The dimensions of each of the movable electrodes are 0.8 mm × 0.5 mm, Fig. 5.6a. With a beam of 450 μm length, 10 μm width and 1 μm thickness, a voltage of 20.2 V can deflect the shutter to its maximum position of 90°, Fig. 5.6b. The microshutter has a low mass, a relatively low drive voltage, a high cut-off frequency and a long operating time. All main components of the structure are monolithically integrated onto one chip which eliminates tedious adjustment and mounting processes.

Microactuator pile

A microactuator pile (distributed electrostatic microactuator) is an interesting actuator principle [Pelr92], [Yamag93]. With this concept natural muscles are imitated. The device consists of several small actuator elements, each of which has two bow-shaped electrodes and insulators. By stacking these elements together an artificial muscle is formed; it can contract when a voltage is applied across the electrodes, Fig. 5.7. Since the individual electrodes are close together, relatively large forces can be produced. When many elements are stacked in series, relatively large displacements can be attained. The prototype of this actuator has dimensions of about 5 mm × 4 mm × 5 mm. The microactuator elements are produced through photolithography, sputter technique and anisotropic wet etching. The microstructure realized by this design achieved a maximum displacement of 28 μm with a voltage of 160 V; the force exerted was approximately 6 μN.

Two-chamber actuator

A unique electrostatic actuator using a membrane to generate a motion in the direction normal to the membrane surface was reported in [Gabr92]; it was produced with the surface micromachining technique. The actuator consists of two concentric air-tight chambers, Fig. 5.8. The chambers are connected with each other via several channels. If a voltage is applied between the substrate and the membrane of the outer chamber, the membrane arches toward the substrate and presses the air out of the chamber. The air is pushed through the connecting channels into the inner chamber and then the inner membrane arches outward. By having a large outer chamber and small inner chamber, a small electrostatic motion applied to the outer membrane can be

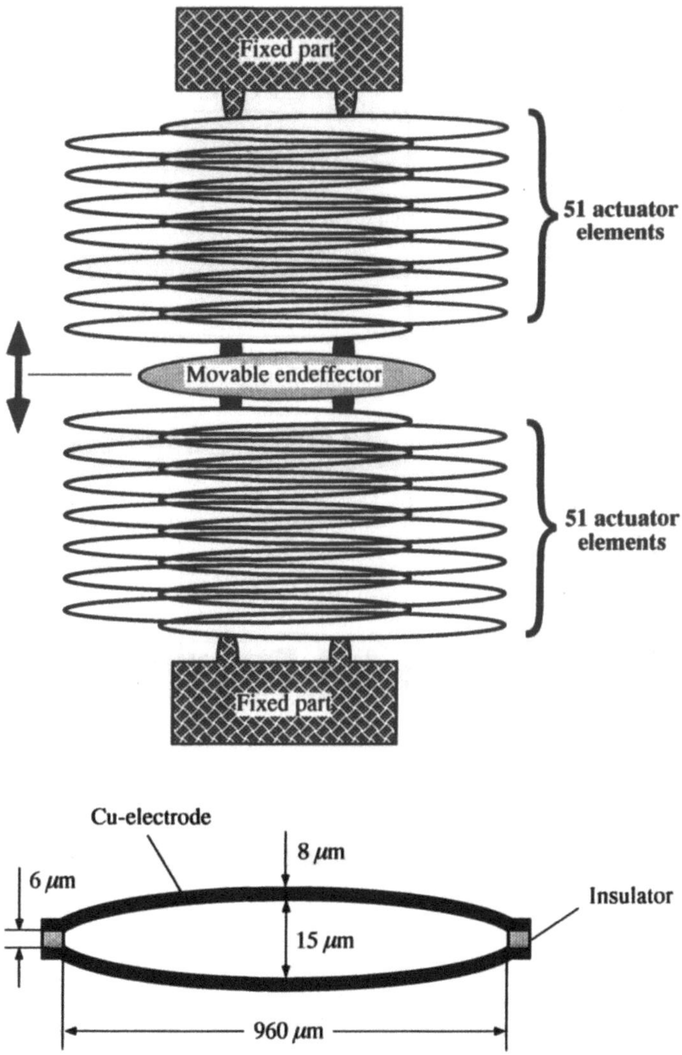

Fig. 5.7: An electrostatic pile actuator and its single actuator element. According to [Esashi93]

transformed into a larger motion of the inner membrane. Several prototypes of this actuator were made from polysilicon. The inner-chamber of which had a radius of 100–250 μm and outer-chamber a radius 200–750 μm. They were operated by 50 V, attaining motions from 1 μm to 4 μm.

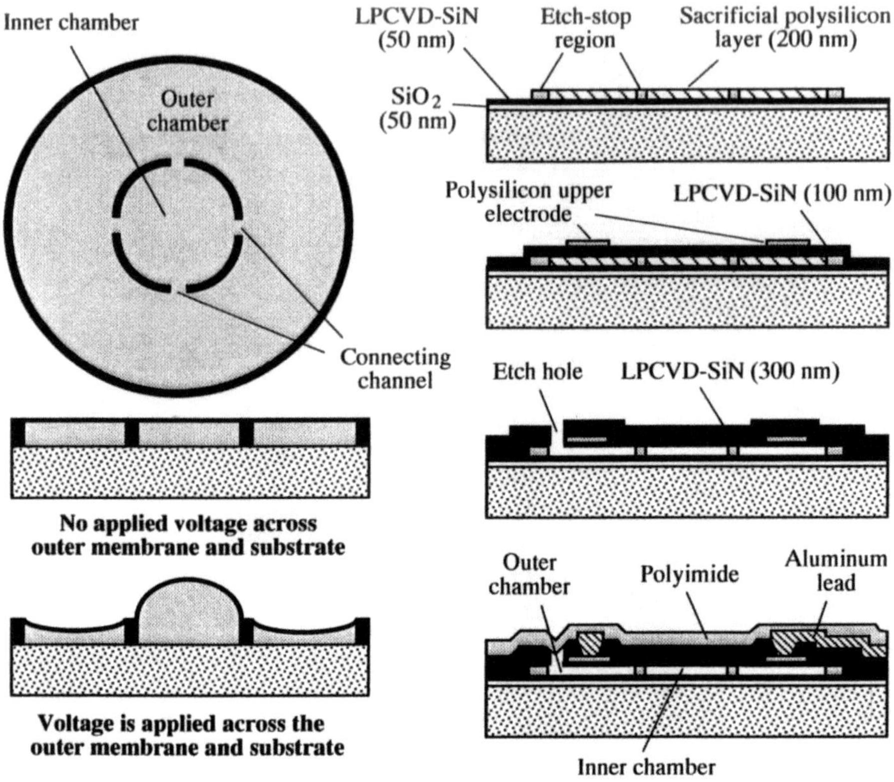

Fig. 5.8: Schematic representation of the two-chamber actuator and its production method. According to [Gabr92]

Electrostatic micropump

The electrostatic membrane principle is well-suited for designing micropumps. Figure 5.9 shows a sketch of an electrostatic micromembrane pump. The device consisting of 4 silicon chips was produced by the bulk micromachining technique, [Rich92], [Zeng92]. The two upper chips form the drive consisting of the membrane and the electrode; the latter is part of the outer frame. The identical lower chips form the inlet and outlet valves. If a voltage is applied between the membrane and the electrode, the membrane arches toward the electrode thereby generating a partial vacuum in the chamber. This causes the inlet valve to open and liquid to be drawn into the pump chamber. By removing the voltage, the liquid is pushed through the outlet valve. Since the drive unit and the pump chamber are separated, the liquid is not affected by the electric field. This is important when the liquid contains ions, e.g. as in salt solutions or medicines.

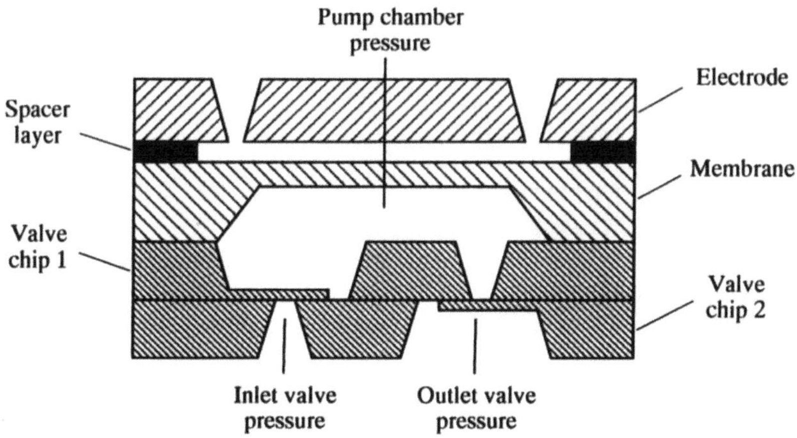

Fig. 5.9: Schematic representation of a micropump. According to [Rich92]

The built pump had an outer dimension of 7 mm × 7 mm × 2 mm and the valves a cross-sectional area of 0.16 mm^2. The membrane had an area of 16 mm^2 and a thickness of 50 μm. The distance between the membrane and the electrode was 6.3 μm. For a voltage of 170 V and a frequency of 25 Hz, the device is able to pump 70 μl/min; it could also be operated at frequencis of up to 100 Hz. Since the moved volume for one cycle is about 10–50 nl, the desired liquid flow can be very precisely metered. This is of great importance for medical applications, e.g. for implanted dosing systems.

Electrostatic foil actuator

An electrostatic microactuator using a travelling foil bend was proposed in [Sato92]. The actuator consists of two silicon electrodes forming a chamber. Between the electrodes a metallic foil made of steel or an Fe-Ni alloy is placed, Fig. 5.10. The foil is formed into an S-bend which can travel from one

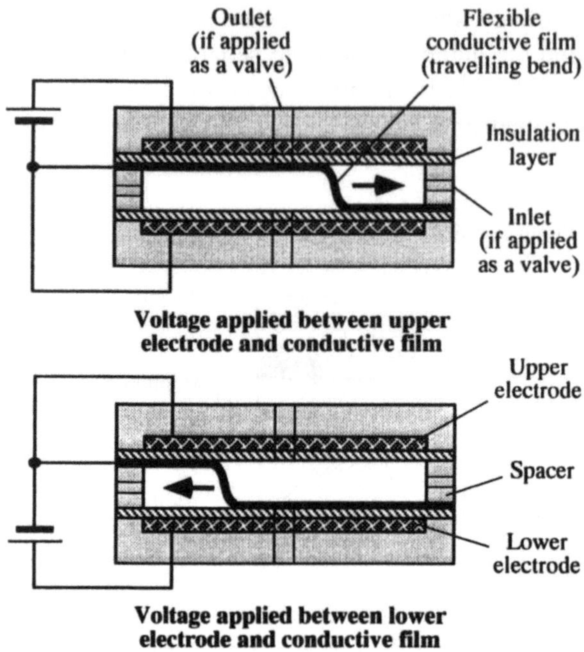

Fig. 5.10: Schematic representation of a foil-like actuator. According to [Sato92]

end of the chamber to the other. For operation, a voltage is applied either between the upper electrode and foil or the lower electrode and foil. The foil is pulled towards the respective electrode by an electrostatic force, with the S-like bend travelling through the chamber, Fig. 5.10. A larger displacement can be achieved by increasing the volume between the electrodes. The mechanism can also be used as a valve to control the flow of gas. The gas inlets open into the space between the electrodes, and the two outlets are formed by recesses in the electrodes. As the foil bend travels along the inside of the valve it opens up a path for the gas on one side of the chamber and closes it on the other side. A miniaturized prototype was made by using this concept with a 5 µm thick and 12 mm wide FeNi foil. The free space was 2.5 mm wide and the working area for the foil was about 40 mm. When using an insulating oxide layer of 0.5 µm on the silicon electrodes the actuator can be operated with a voltage of 100–150 V. The foil can be moved at 4 m/s with a voltage of 150 V.

Electrostatic microvalve

Now an electrostatic microvalve with an elastic membrane-like SiO_2 strip as valve cover will be introduced [Haji94]. Usually, when a SiO_2 layer is produced to form a membrane, internal stresses are created which can lead to the deformation of the layer. In general this is avoided in the manufacture of microelectronic components. However, in this device it is used as the actuator principle. The device consists of a base made from silicon into which a valve opening is etched, Fig. 5.11. A SiO_2 membrane is attached to its top.

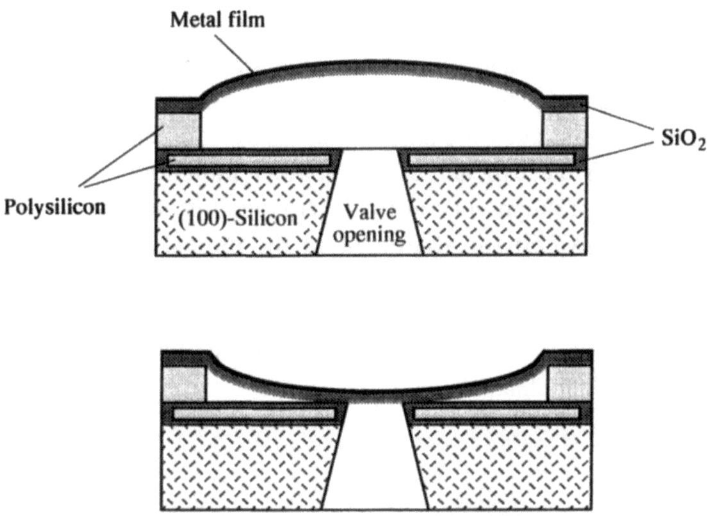

Fig. 5.11: Working principle of an electrostatic microvalve. According to [Haji94]

Due to inner stresses, the membrane bends up and opens and clears the substrate, thus gas or liquid can flow, Fig. 5.11 top. The SiO_2 membrane is coated with a thin chrome layer, which acts as the movable electrode. If a voltage is applied across the polysilicon electrodes, the membrane is pulled down and the flow stops, Fig. 5.11, below. Several prototypes of this microvalve have been produced from a (100) silicon substrate using a combination of the surface micromachining and bulk micromachining techniques. The valve openings of the prototypes ranged from $10 \times 10\ \mu m^2$ up to $100 \times 100\ \mu m^2$, the lengths of the membrane strips ranged from 80 μm up to 1000 μm and the width from 20 μm to 120 μm. In absence of additional gas pressure the "off" voltage was held at 68 V and the "on" voltage at 120 V. Such a microvalve can be mass-produced at low cost, it has a long service life and a negligible dead volume. Due to its mechanical nature, it is immune to electromagnetic fields and can endure mechanical shocks, since the movable part has a very low mass. Several valves can be fabricated on one substrate by a batch process to simultaneously control different gas flows.

Electrostatic micropositioner

Optical communication networks and optical computers need flexible optical coupling and interfacing techniques. Here, very high precision is required, to align connecting optical fibers; even the smallest deviation can lead to excessive light attenuation. Also the expensive optical fiber mountings make an economical solution to this problem very difficult. An electrostatic micropositioner was presented in [Kiku93] for alignment of the optical fibers in the coupling, Fig. 5.12. The whole device consists of two parts: the micropositioner and a stationary part with integrated optical channels (the latter part is not shown in the picture). It is the task of the micropositioner to align the fibers with their corresponding channels.

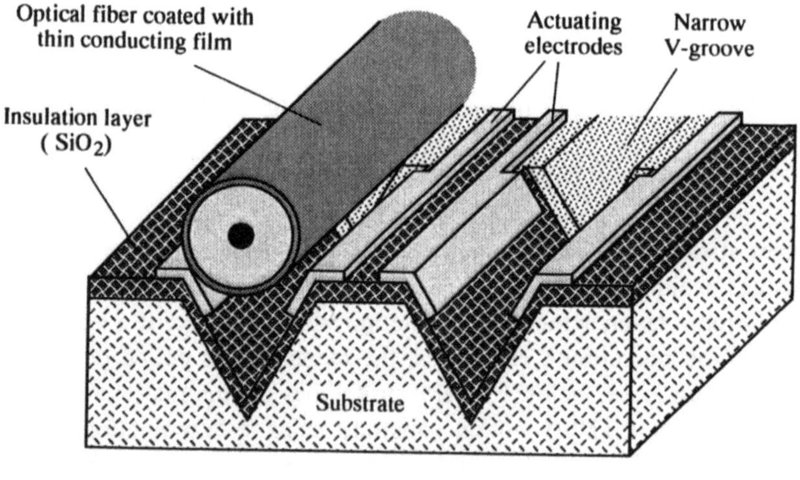

Fig. 5.12: Electrostatic micropositioner. According to [Kiku93]

The micropositioner has a base containing several V-shaped grooves into which the optical fibers to be aligned are placed. The groove itself has a wide part and a narrow part. In the wide part there are alignment electrodes on either side. The optical fibers are coated with a thin metal layer so that they can be actuated electrostatically. If an electric voltage is applied across one of the electrodes, the fiber is attracted by this electrode. Therefore, by controlling the voltage of both electrodes, the fiber can be exactly aligned.

The metal layer of the optical fiber is applied by a sputtering process, and the groove is produced by the bulk micromachining technique on a (001) silicon substrate. The smallest distance between the optical fiber and the electrode is 15–30 μm. The electrodes were produced by depositing a thin aluminum layer with the lift-off technique.

Electrostatic micromirror

Figure 5.13 shows a sketch of an electrically driven micromirror proposed in [Jaeck93]. The device consists of a micromirror which can be tilted about a torsion beam with respect to a base made from silicon. Two positioning electrodes, one for addressing and one for landing, are mounted underneath the mirror. If a voltage of approximately 31 V is applied to the address electrode, the mirror rotates 7.6° and touches the landing electrode which is grounded. In order to bring the mirror back to its original position the voltage is decrea-

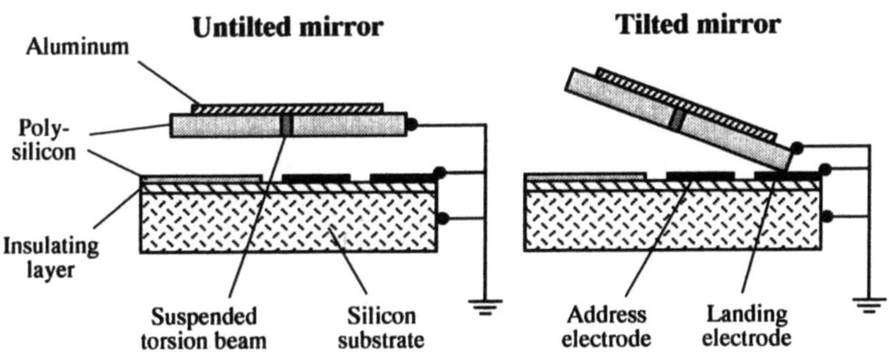

Fig. 5.13: Electrostatic micromirror. According to [Jaeck93]

sed to 16 V. The aluminum mirror has a reflectivity of 83% and very short response times; its dimensions are 30 μm × 30 μm, and it is suspended on two 10 μm long and 0.6 μm wide torsion beams, Fig. 5.14. The micromirror can be used in different types of optical measuring and display systems. The goal of similar projects in the USA is to produce a stable, high-resolution projection of computer and video images [Nach94].

Fig. 5.14: SEM image of a mirror prototype. Courtesy of the University of Neuchâtel (Institute of Microtechnology)

Electrostatic frictionless microactuator

High friction has been a typical problem of microcomponents. A frictionless electrostatic microactuator was described in [Tiro93] and [Tiro93a]. Its movable electrode levitates with the help of two relatively long and flexible "arms". The microactuator is of a hybrid silicon structure made up of two parts on a glass wafer. One part is bonded to the glass base and the other part is movable and levitates with the help of two elastic cantilevers as mentioned before. The counter electrode (not shown in the picture) is located next to the movable part which drives it by using electrostatic forces.

The bulk micromachining technique was used for the production of the functional parts including the elastic beams using (110) silicon. The prototype of the actuator has 1800 μm long, 450 μm high and about 10 μm thick beams,

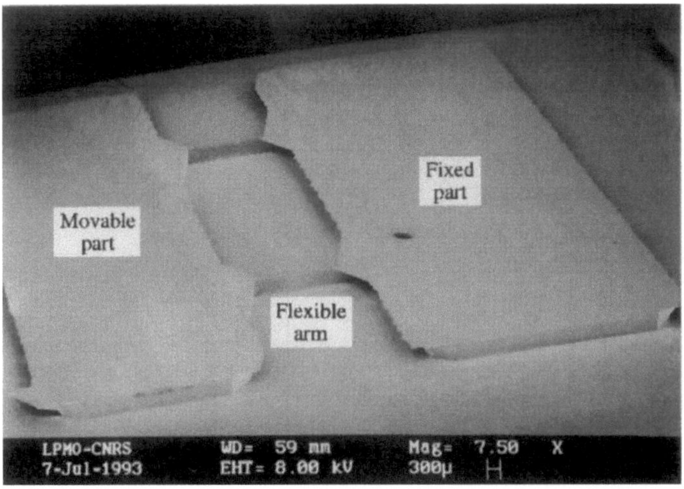

Fig. 5.15: Prototype of the cantilever-based electrostatic microactuator. Courtesy of the LPMO/CNRS and CETEHOR, Besançon

Fig. 5.15. The actuator was operated and tested with a frequency of about 120 Hz and a voltage of 50 V. Movements in the 100 µm range were attained; after 8 hours of operation, which corresponds to about 3 million work cycles, no damage could be found in the microstructure. Possible applications for this device include acceleration sensors and actuators for microscopy using the scanning tunneling principle. Strictly speaking, this microactuator cannot be classified as a membrane actuator, but we still present it, due to our simplified classification of the electrostatic microactuators.

Two dimensional positioner

Linear comblike microactuators will be introduced now. Often, precise two-dimensional positioning is an important requirement, e.g. for the production of semiconductors, optoelectronic elements, magnetic high-density memory units and especially for atomic force microscopy (Section 2.5). For such applications, an xy-nanopositioning system was made by the surface micromachining techniques [SAML93], [Brug93], [Inde95], Fig. 5.16. The device consists of two actuators holding a probe pin in their center; each actuator consists of a pair of combs, one is a fixed electrode and the other a movable one. The mechanism is mounted on a substrate. The large opening in the substrate

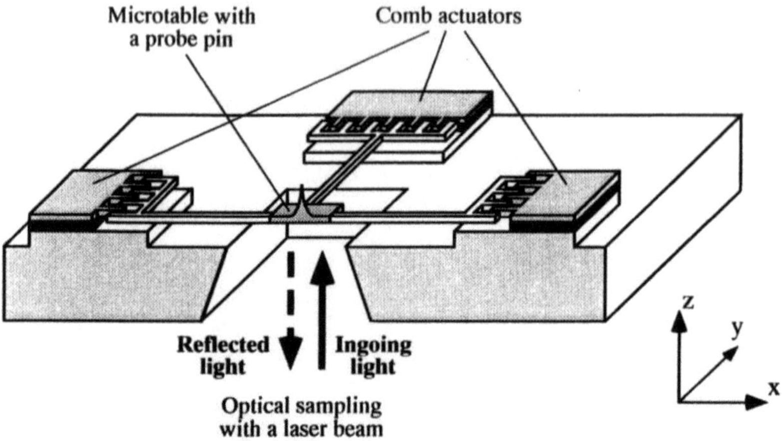

Fig. 5.16: Sketch of the AFM positioning system. According to [SAML93]

underneath the movable microtable is for the optical measurement of the probe movement in the z-direction; it is detected with a laser beam from the back of the device by an interferometric method. The resolution of the measurements in the z-direction is in the subnanometer range. The movable electrodes that have already been presented in Figure 4.19 are mounted on long beams, so that they move parallel to the electrode fingers. The nanopositioning system (Fig. 5.17a) is produced from two oxidized silicon wafers; anisotropic wet etching and reactive ion etching processes are used. It is also possible to produce the device from single crystal silicon. To position the probe in the x-y plane, two of the four comblike actuators are used. To perform measurements in AFM as described in Chapter 2, a probe is mounted on the microtable (Fig. 5.17b). The dimensions of the microtable are 8 μm × 8 μm, and each beam is 270 μm long, 0.6 μm wide and 2 μm high. The probe is 8 μm high and has a maximum radius of 40 nm. When a voltage of 40 V is applied, the microtable is pulled 5 μm in the direction of the corresponding actuator.

Oscillator drive motor

The electrostatic comb actuator principle was used for operating a motor with an oscillator drive. The driving mechanism is depicted in Figure 5.18. The motor's rotor is driven by a friction rod which "pushes" the rotor along

a

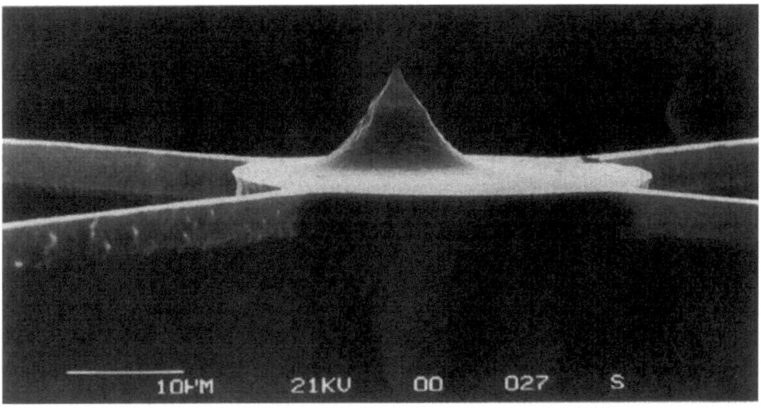

b

Fig. 5.17: The nanopositioning system
a) overall view of the system; b) microtable with a probe. Courtesy of the University of Neuchâtel (Institute of Microtechnology)

its circumference in a oscillating mode causing it to rotate. The rod is connected to two comblike oscillating actuators which are operated with frequencies between 10 kHz and 20 kHz. The oscillating linear rod movements are transformed into a continuous rotor rotation. A special feature of this micromotor is that the rotor can be turned in either direction, Fig. 5.18b.

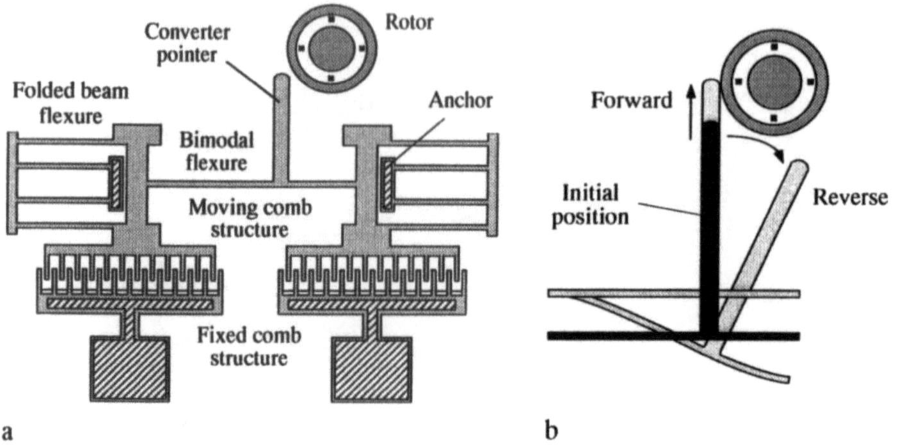

Fig. 5.18: Schematic diagram of the vibration micromotor
a) top view; b) two driving modes. According to [Lee92]

The surface micromachining technique was used, as it is normally the case for comb structures. Two prototypes with rotor diameters of 60 µm and 100 µm were produced. The microstructure is only 2.4 µm thick and its distance to the substrate is also 2.4 µm; the rotor has four 2 µm × 2 µm notches which allow contact with the surface of the substrate. The flexible connector is 200 µm long, the converter pointer is 200 µm long and 10 µm wide. So far, the produced prototypes had a limited service life of a few minutes in air and a few hours in vacuum. However, oscillator drive motors are very interesting for future applications, and a newer prototype was operated with a rotational speed of 60,000 rpm for a duration of 4 hours. It can be expected that longer service lives will be reached soon.

Linear step motor

An electrostatically driven bidirectional linear step motor which utilizes the principle of variable capacitance was fabricated by the LIGA process. The forces produced by a linear actuator are proportional to the height of the microstructure (Fig. 5.4b). In other words, the higher the actuator in the z-direction the more force it can produce. The LIGA process is well-suited to make actuators with a height of up to 500 µm.

A schematic diagram of such an actuator is shown in Figure 5.19. The device consists of stationary and movable electrodes. When a voltage is applied across the electrodes, the movable part is shifted relatively to the fixed part. In order to increase the performance of this actuator, a multitude of toothlike capacitors having several thousand teeth are arranged in parallel to each other. The device shown in Fig. 5.19 has three stator groups mounted on a substrate, and movable electrodes (actuators) which are fastened to a flat spring structure. The actuator and stator teeth are offset in such a manner that a tangential force is created in at least two of the three electrode rows when a voltage is applied. A switching of the voltage between the electrode groups results in a shifting to left or right until the end position is reached. This is limited by the spring structure. The amount of shift obtained is determined by the comb and spring structure as well as the driving voltage.

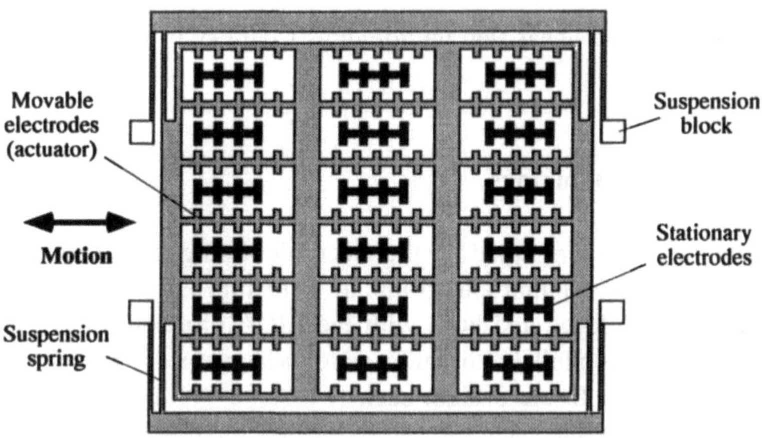

Fig. 5.19: Schematic diagram of an electrostatic linear step micromotor. According to [Kalb94]

Several prototypes of this microactuator were built on a Al_2O_3 substrate. A section of the linear actuator structure is depicted in Fig. 5.20. This prototype is 70 μm high, and there is a space of 4 μm between the electrodes. The entire structure is 4.5 mm × 5 mm and can be operated with a minimum voltage of 130 V. Other prototypes with a height of 150 μm and an air gap of 3 μm between the electrodes were tested with a voltage of 200 V; they reached a displacement of about 100 μm and a maximum force up to 50 mN.

Fig. 5.20: Prototype of the linear step micromotor. Courtesy of the Karlsruhe Research Center, IMT

Electrostatic rotational motors

Electrostatic rotational motors based on the electrode arrangement shown in Figure 5.4b form another important group of actuators. An electrostatic Wobble motor was developed and built with the LIGA process [Menz93a]. The motor consists of a comblike rotor and a stator with six poles; it is driven by a three-phase voltage. In the design, each two stator poles belonging to one phase are placed opposite each other to ensure continuous motion of the rotor. During operation, a radial force pulls the rotor to the stator until it touches the axle. The contact point travels around the axle and causes friction. The friction coefficient must be kept as small as possible, otherwise the rotation is slowed down and the durability of the motor is drastically reduced. By optimizing the construction, the motor can be made to only roll and not slip, which minimizes friction. Several prototypes of this micromotor were made from nickel using the extended LIGA process (Section 4.2.5).

One of these motors is depicted in Figure 5.21. This micromotor has 56 rotor teeth and an outer rotor radius of 267 μm, the height of the motor is 100 μm. The distance between the rotor and the axle (bearing clearance) is 4.8 μm. The motor is driven by a 3-phase voltage ranging from 60 to 100 V. The rota-

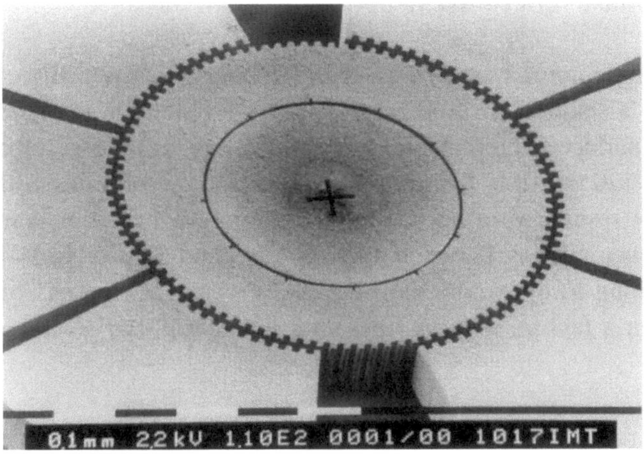

Fig. 5.21: Prototype of the electrostatic rotational micromotor. Courtesy of the Karlsruhe Research Center, IMT

tional speed can be varied from a single-step operation up to 3,400 revolutions per minute. Due to friction, the torque of the motor is only 10 pNm. This is too low for most practical applications. Because of this, an attempt is being made to lower the mass of the rotor to reduce friction.

Fig. 5.22: Electrostatic polysilicon micromotor. Courtesy of the LAAS/CNRS, Toulouse

Electrostatic three-phase micromotor

Another electrostatic micromotor was discussed in [Camon93]. It is a three-phase micromotor with a stator/rotor pole ratio of 3:1. The motor was made from polysilicon by the surface micromachining technique, Fig. 5.22. The rotor has a diameter of about 120 μm, it is 1,5 μm thick and has a gap of 4 μm with the stator. Practical test results were not reported. The motor behavior was also simulated and torques ranging between 10^{-16} Nm/V^2, and $5 \cdot 10^{-16}$ Nm/V^2 were calculated. According to the simulation, the motor can reach speeds of up to 70,000 rpm within a few μs with an operating voltage of 100 V and a frequency of 450 Hz.

Electrostatic needle-bearing micromotor

An electrostatic micromotor was made from nickel with the LIGA process [Hira93]. The special feature of this motor was the use of a needle structure as a mechanical bearing between the substrate and rotor. The motor consists of a fixed shaft, a freely rotating rotor and 8 stator electrodes, Fig. 5.23. The operating voltage is successively switched from one pole to the next so that

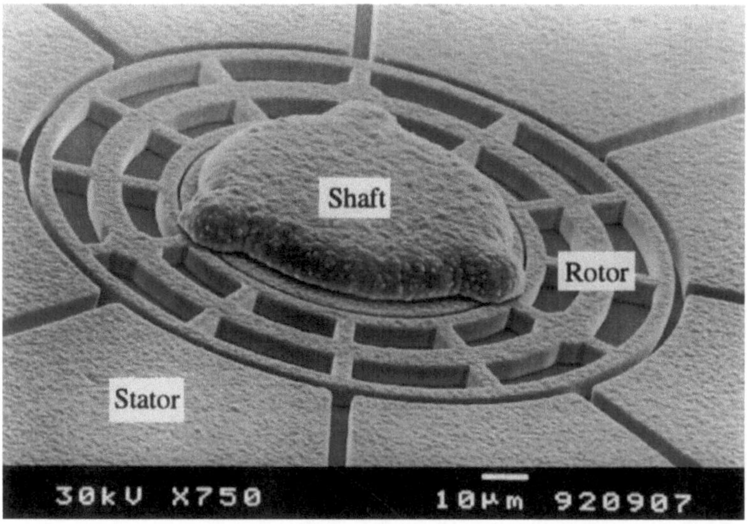

Fig. 5.23: Electrostatic micromotor with a needle-like bearing. Courtesy of IBM Research, Tokyo Research Laboratory and the University of Tokyo

the rotor always follows the "active" pole. The gap between the shaft and the rotor is 2 μm wide, which causes a wobbling motion. The rotor itself consists of 3 interconnected rings. The outer ring serves as a capacitor plate and is 3 μm away from the stator. There are several needle bearings underneath the middle ring, supporting the rotor during the eccentric wobble motion.

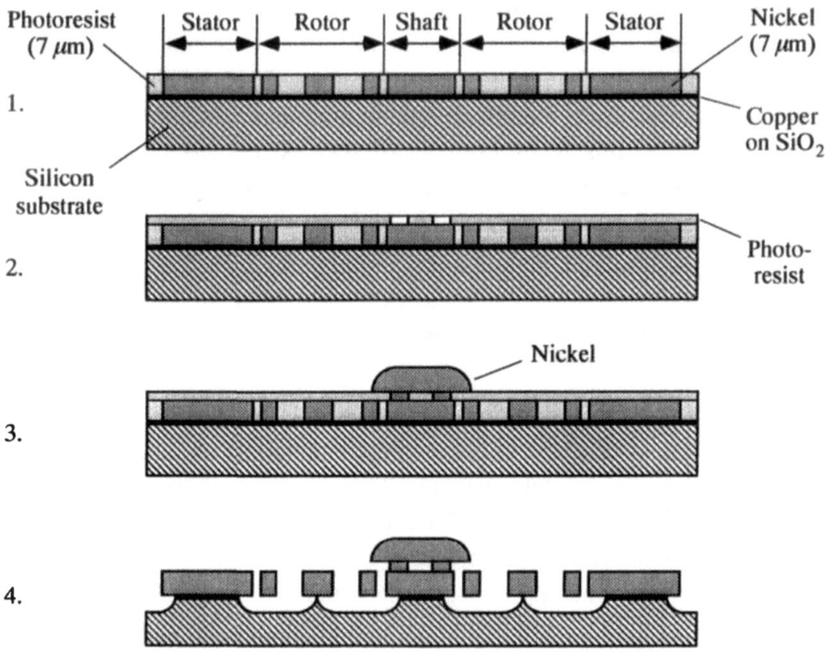

Fig. 5.24: Fabrication of the electrostatic micromotor. According to [Hira93]

The main steps to produce this motor are depicted in Fig. 5.24. After the deposition a silicon dioxide layer and a copper layer on the substrate, a photoresist with a thickness of 7 μm is coated and patterned by photolithography; then, nickel is electroplated (a). Another photoresist with a thickness of 1 μm is deposited and patterned (b). After that, nickel is electroplated to form the cap of the shaft (c). Then the resist, the copper layer and the insulation layer are partially removed. Finally, plasma etching of the silicon substrate is used to release the rotor (d). The diameter of the rotor is 140 μm, and that of the shaft is 80 μm. The motor reached a maximum speed of 10,000 rpm at 60 V.

5.3 Piezoelectric Microactuators

Actuators based on piezoelectric ceramics and magnetostrictive alloys are characterized by their precise movements in the nanometer range and very high reaction speeds, usually a few microseconds. Piezoelectric ceramic materials were discovered about 40 years ago, but they are just now gaining economic significance and have become an important component of piezoactuators that can be used in a variety of applications. Their working principle makes it possible to transform electrical signals ranging from 1 mV to 1000 V into motion. For MST, this technology is of interest because it allows the design of very small actuators that generate motion and exert very high forces. Therefore, many research activities focus on these materials.

5.3.1 Motion Principle and Its Properties

Piezoelectric materials can transform mechanical energy into electrical energy. When pressure is applied to piezoelectric material, an electric voltage is generated between the crystal surfaces. This is referred to as the direct piezoelectric effect and was described for the first time in 1880 by the Curie brothers. The inverse piezoelectric effect was discovered in the following year. An electric voltage was applied to an asymmetric crystal lattice, which caused the material to deform in a certain direction; the physical basis for this transformation will be described later. Because of the reciprocity of the piezoelectric effect, piezoelectric materials can be used for constructing "smart" actuators [Jano92], [Jano93]. Since cause (the voltage applied) and effect (the change in length) are strictly proportional, additional sensors are not necessary. Piezoactuators are also known for their quick reaction speed and the reproducability of the travelled distance. They are very efficient: About 50% of the applied electrical energy can be directly transformed into mechanical energy. Therefore, piezoactuators can exert very high forces, which is of great importance in MST. Other advantages of piezomaterials are their mechanical durability, they do not react to other electric components near them, and they are not sensitive to dust.

One basic problem concerning piezoelectric materials and other monolithic actuators is that the movement to be obtained is relatively small; it is in the range of several nm/V. A single piezoelement can expand by about 0.1–0.2% in one direction. Thus, it can be used in applications in which very small but precise displacements are needed, having high forces and short reaction times. The form and size of a ceramic component can be easily adapted to the given task. Multilayer stack and cantilever constructions are most often used.

When several piezoelements are placed on a stack, their displacements add. Thus, larger distances can be covered by using more than one piezoelement. Thin piezodiscs covered with flat metal electrodes are the principal machine elements for such actuators. They are useful in a variety of applications. Multi-layered ceramic actuator elements are manufactured by applying e.g. the *tape casting*-method. Electrodes are printed onto the raw ceramic similar to textile printing, and then the layers are stacked together and burned into a multi-layered structure. The actuators are about 10–200 μm thick, are operated by a voltage in a range of 50 and 300 V and can be mass-produced. The goal of current research is to reduce the operating voltage to a range of 5 to 15 V. However, the lifetime of these actuators may be limited by internal mechanical stresses caused by non-homogenities resulting from the electrodes, Fig. 5.25.

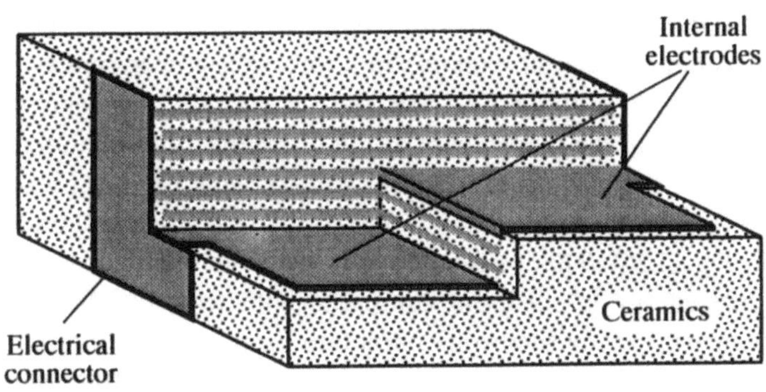

Fig. 5.25: Multi-layered piezoceramic structure. According to [Gibb94]

Another actuation principle is the cantilever. Piezoelectric cantilever structures which can be shaped like a tube or a plate make use of this principle. We will discuss the possibilities of tube-shaped piezoelectric microactuators in detail in Chapter 8 in which we describe a microrobot developed at the University of Karlsruhe [Magn94], [Magn95] and [Fati95]. Most often a plate-shaped cantilever type piezoactuator is a bimorphic element consisting of two piezoelectric ceramic plates, Fig. 5.26.

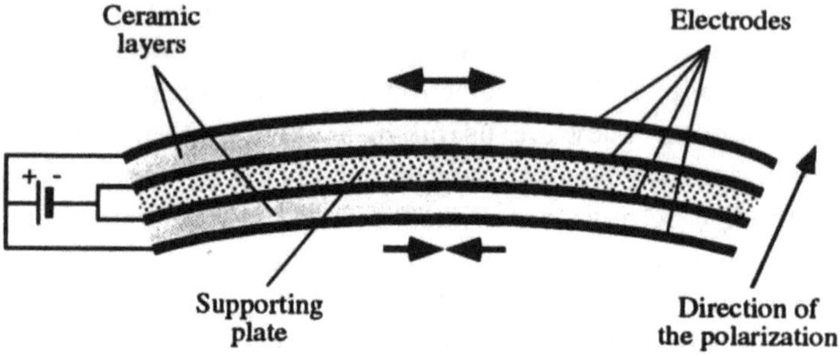

Fig. 5.26: Bimorphic piezoceramic structure. According to [Moil92]

The ceramic layers are mounted on both sides of a supporting plate. Each ceramic plate is covered by a conducting layer on both sides. The working principle of this actuator is based on the coordinated expansion/contraction of the piezolayers. The ceramic material expands in a direction perpendicular to the axis of its polarization. The applied electric field is parallel to or against the direction of polarization of the piezolayers. If the directions of the polarization and electric field are the same, the piezoceramic expands (upper piezolayer in Fig. 5.26); if they have opposite directions, the piezoceramic contracts (lower piezolayer in Fig. 5.26). This causes the actuator to be bent as shown. The electromechanical energy is not great compared with what can be achieved with stack elements, but the bimorphic structure is superior with regard to the displacement that can be obtained. However, the maximum force and working frequency are several times lower than that of the stack actuator.

In order to optimally select an actuator for an application, a series of parameters have to be considered; they depend on the situation, controlling distance, efficiency or stiffness. Table 5.1 shows an overview of the most important design principles of piezoelectric actuators and some typical parameters which describe their operating behavior.

Table 5.1: Piezoactuator designs and their characteristic values. According to [Jend94]

Standard shapes	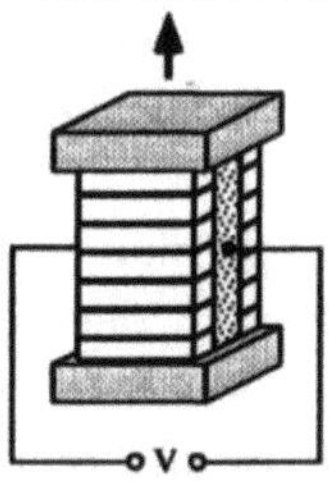 **Stack**	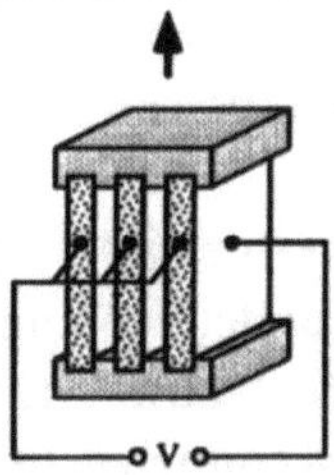 **Strips**	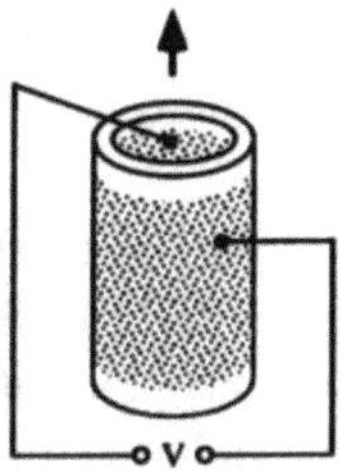 **Tubelet**	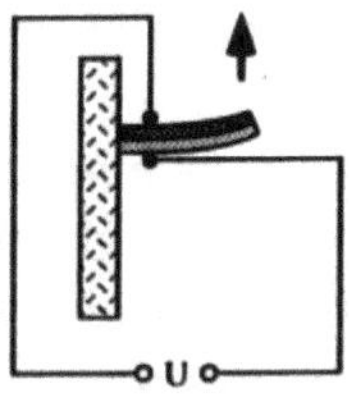 **Cantilever**
Typical displace-ments	20–200 μm	≤ 50 μm	≤ 50 μm	≤ 1000 μm
Typical forces	≤ 30000N	≤ 1000 N	≤ 1000 N	≤ 5 N
Typical supply voltages	60–1000 V	60–500 V	120–1000 V	60–400 V

The table shows that piezoactuators are usually operated with voltages of up to 1 kV. New production techniques allow the making of thinner ceramic layers (20–40 μm); they are covered with very thin electrode material. This makes it possible to reduce the required operating voltage to 100 V. The limitations of the individual materials must always be closely considered when using piezoactuators. The most important of these are the depolarization pressure and the coercive field strength. When these values are exceeded, the material will be depolarized, which is associated with a loss of the specific piezo properties.

For most applications in MST, piezomaterials are needed having a large piezoelectric coefficient, high electrical resistance, low internal voltage and high mechanical stability. Materials which can be easily formed micromechanically are also desired. E.g. many piezoelectric actuator elements can be directly manufactured with the LIGA technique. A few natural piezoelectric materials, such as quartz or turmaline, have been known for a long time, but their piezoelectric properties are not sufficient for technical applications. Since the forties, certain polycrystalline ceramics have been studied and they were found to have excellent properties. Among them Lead-Zirconate-Titanate (PZT) is the best, but it is hard to work with. ZnO also seems to be very suitable material for integrated piezoelectric microactuator systems.

The principle of the piezoelectric effect will be explained in the following sections to give the reader a basic understanding of it. A detailed description can be found in [Jend93]. The piezoelectric effect is obtained with a specific crystal structure which must be asymmetric; that means there is no symmetry center with regard to the positive and negative ions of an elementary cell. An external, mechanical compression causes the positive and negative charge centers of each elementary cell to separate, which leads to ionic polarization and an electric charge is appearing on the outer surfaces of the crystal. Figure 5.27 shows the lattice structure for barium titanate ($BaTiO_3$) which is the typical structure of an elementary piezoelectric cell.

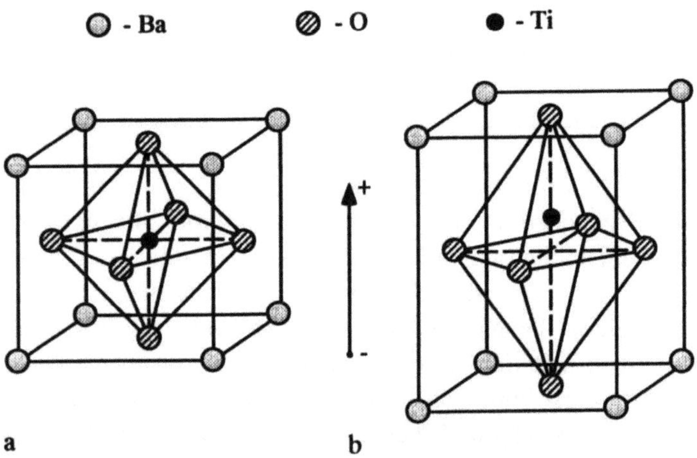

Fig. 5.27: Typical piezoelectric elementary cell ($BaTiO_3$)
a) above Curie temperature: cubical lattice; b) below Curie temperature: tetragonal lattice. According to [Jend93]

It also shows the changing of the lattice in relation to the Curie temperature. The Curie temperature is the temperature at which a material transforms from the paraelectric state to the ferroelectric state, which is between 120°C and 400°C, depending on the material. The Curie temperature defines the useful temperature range of a piezoceramic material. If a piezocrystal is warmer than the Curie temperature, its structure is made up of regular oxygen octahedrons with titanium ions in their center. The positive and negative charge centers compensate each other; there are no dipoles (paraelectric state). When the temperature of a piezocrystal sinks below the Curie temperature, the crystal lattice is deformed. This is because the positively and negatively charged ions attract/repel each other within the lattice, causing the charges to separate. This creates an electric dipole in each elementary cell. So called Weiss' domains, which are areas with unified dipole orientation, spontaneously appear and their orientation is distributed evenly within the material (spontaneous polarization). By applying a very strong electric field to piezoceramics during manufacture, all domains are oriented in a certain direction, causing the ceramic to become macroscopically polarized (ferroelectric state). Only then the actual piezoelectric effect can occur. The reversibility of this effect is advantageous. When an electric field is applied, the piezoceramics either expands or contracts, depending on the polarization (anisotropy). Developmental goals for piezoceramics include the improvement of the expansion capabilities, reducing the characteristic hysteresis curve, and increasing the Curie temperature and long-term stability.

Electrostriction is another effect that occurs in a polarized ferroelectric crystal. This causes the length of the material to change and can also occur in symmetric crystals, as opposed to the piezoelectric effect. However, only one effect can be dominant in a ceramic. The other effect is neglectible small. Although the electrical and mechanical properties are pretty much the same, electrostrictive materials do not offer a real alternative to piezoceramics since they have a greater hysteresis and are more temperature dependent.

The use of piezoelectric actuators in MST is of great importance. They are very versatile and can be employed practically in all industries. Piezoelectric microactuators can easily be integrated into microsystems. Another advantage is that they can be applied by using relatively simple control algorithms. The number of variants of piezoceramics that can be manufactured is also important. By exactly controlling the composition of the base material, the directions of expansion/contraction and physical properties of the actuator can be defined. There are many applications for piezoelectric microactuators. They can be used, for example, to accurately adjust mirrors in a CCD camera or in an electron microscope. They may also accurately align fiber optic cables, position fixtures in a precision milling machine or drive micromechanical ultrasonic-motors. Piezoactuators can be used to manufacture micropumps and valves for the exact dosing of substances in biology and medicine. They can also serve as shock wave generators to crush kidney stones or as microtools to perform minimal-invasive surgery.

Piezoactuators have found their entrance in the automobile industry. They can be used for active vibration control, noise suppression and for motors to operate windows and sun roofs. Since piezoelements have a short response time, they are ideal components for automotive valves which must quickly initiate or interrupt a certain car function, e.g. for electronically controlled inlet valves and injection nozzles. Presently, injection times in the range between 1.2 and 2 ms are possible, but for fuel reduction purposes about 0.2 ms would be necessary. Experiments are presently being done with piezoelements to integrate them in injection jets for regulating air pressure and volume. High-strength piezoelectric actuators are also of advantage to do noise and vibration suppression. Piezoelectric acceleration sensors can follow the motor movements and if necessary send an attenuation signal to a shock absorber. They could similarly be used for active wheel damping.

In order to make piezoactuators widely accepted as a design component, the displacement that can be obtained must be drastically improved through new design ideas; the price must be lowered, for example, by mass-production methods based on the LIGA process. Also development work is needed to conceive high performance control amplifiers which are indispensable for the dynamic control of piezoactuators. Should this be successful, piezoelectric microactuators will be a good alternative to electrostatic ones. By analyzing the following examples of MST-specific piezoactuators, the trend towards the use of piezoelectric actuators will become obvious.

5.3.2 Concepts and Prototypes of Piezoelectric Microactuators

Micromembrane pump

The development of a piezoelectrically driven pump was reported in [Rich92a]. The micromembrane pump consists of two glass plates and a silicon disc which is sandwiched between the plates, Fig. 5.28.

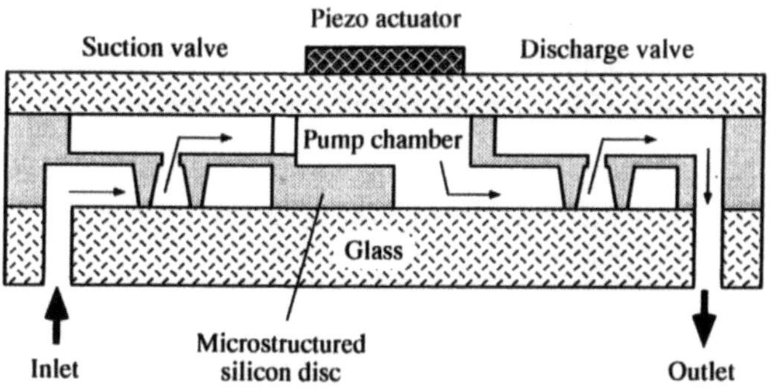

Fig. 5.28: Piezoelectric micromembrane pump. According to [Rich92a]

The silicon disc is structured by etching and contains a pump chamber as well as suction and discharge valves. The upper glass plate serves as a pressure-sensitive membrane. It can change the volume of the pump chamber with the help of a bonded piezodisc (the actuator). When a voltage is applied, the membrane buckles downward and the liquid is forced out through the discharge valve. When the voltage is removed, the membrane returns to its original position and the pump sucks in liquid through the suction valve. The diameter of a macro-prototype of this pump is 75 mm and has a thickness of 2 mm. The discharge pressure is 0.2 bar (max. 0.4 bar) and the pump has an operating voltage of 300 V at a flow rate of 0.6 ml/min. The micropump is able to dose very exactly with a displacement stroke of 0.2 μl, but only very clean liquids can be used since contaminants contained in the liquid may plug up the pump and impair its function.

Microvalve

A new design of a piezoelectric microvalve with pressure compensation was introduced in [Josw92]. A microvalve may have a problem when the external pressure is much higher than the internal pressure acting on the valve, thus the valve can no longer be opened after it has been closed. One solution is the use of a powerful microactuator which can overcome the external pressure. The schematic design and function principle of the piezoelectric microvalve is shown in Figure 5.29. The valve is constructed on a glass substrate. A valve membrane is made of silicon by etching techniques and is mounted on the glass substrate by an anodic bond. The actuator system consists of a glass membrane and a piezoceramic disc. The valve membrane and the actuator membrane are connected with prism-like connecting elements.

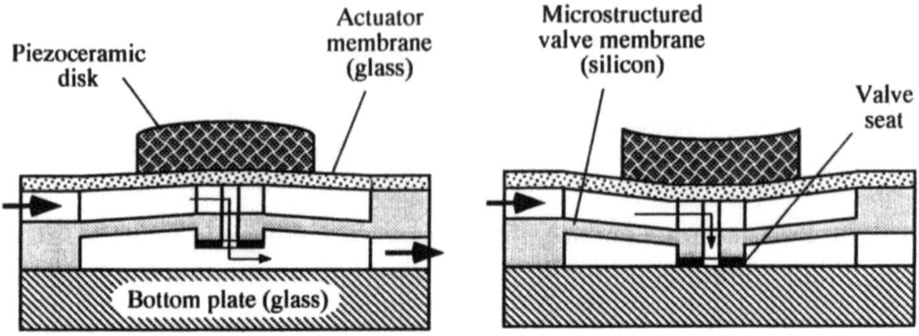

Fig. 5.29: The design and the functioning principle of a microvalve. According to [Josw92]

The microvalve opens and closes when the actuator and valve membranes simultaneously lift or release with the help of the piezoelement, Fig. 5.29. The valve prototype has a 130 μm thick actuator membrane with a diameter of 10 mm. The piezodisc acting as actuator is 300 μm thick. When a voltage of 50 V is applied, a gap of 4 μm can be reached. This microvalve can potentially be used for medical and automotive applications.

Chopstick gripper

A very important application area for piezoelectric actuators is the fine manipulation/adjustment of microobjects. Piezoactuators can precisely position these objects with an accuracy of about 10 nm. The development of a piezoelectric microhand for manipulating tiny objects has been pursued for the past several years [Arai92], [Arai93]. Two piezo-driven chopstick-like fingers, each having 6 degrees of freedom, are designed to manipulate microobjects like a human hand, Fig. 5.30. The two-finger design was specially conceived to work in the microworld, where gravity and moments of inertia play only a very minor role. Therefore, two fingers are sufficient for manipulating microobjects.

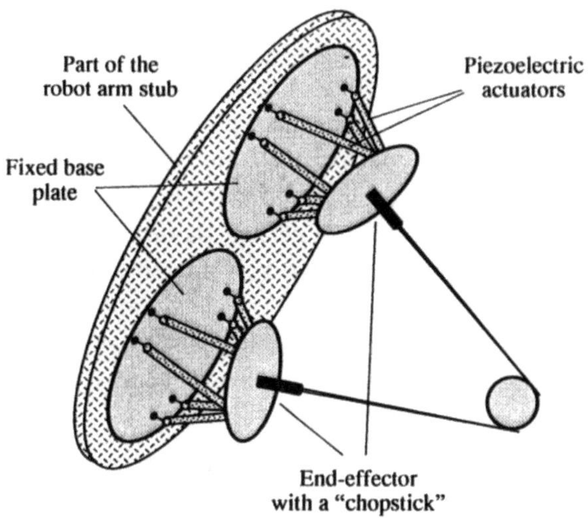

Fig. 5.30: Design of a piezoelectric chopstick hand. According to [Arai93]

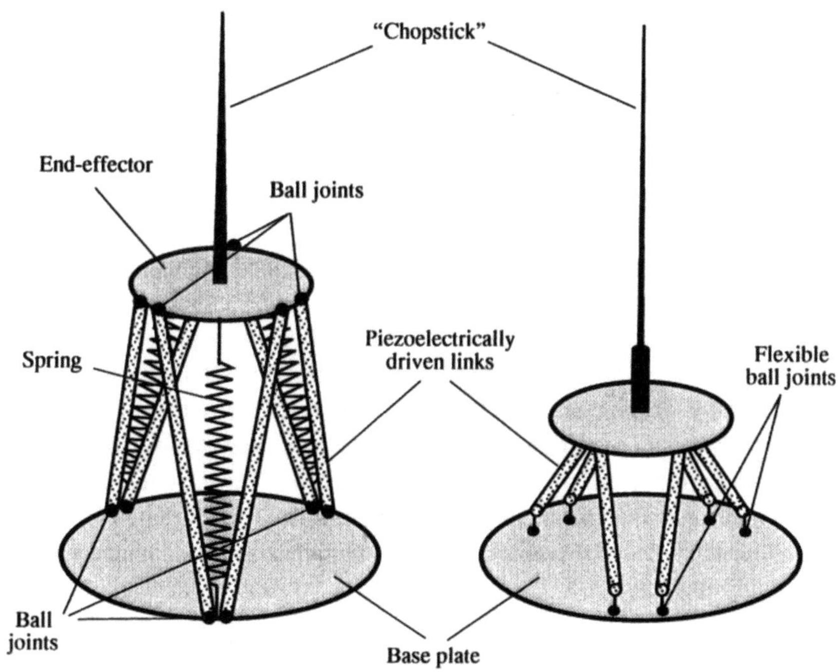

Fig. 5.31: Two prototypes of the chopstick fingers. According to [Arai93]

Two prototypes of the chopstick microfingers were developed and are shown in Figure 5.31. A parallel link mechanism was used to construct both prototypes. It is made up of 6 prismatic piezo connecting elements which are connected to the base plate and end effector. Springs are added to the first prototype to enable the fingers to move continuously and to maintain the hand's stability, Fig 5.31, left. The fingers are made of 50 mm long needles having a tip radius of 30 μm. The second prototype has flexible ball joints made of steel, Fig. 5.31, right. They increase the stiffness, thereby increasing the accuracy. An important design goal is to improve the mechanical structure in order to expand the work area of the micromanipulator and to improve the dynamic behavior of the actuator. The fingers were improved by replacing them with a glass pipette with a tip radius of under 1 μm.

Both finger prototypes are controlled by six piezoelements with a dimension of 2 mm × 3 mm × 8 mm. The diameter of the base plate is 56 mm, that of the end effector is 20 mm and the distance between the plates is 6.4 mm. The movement of the finger tip was 8 μm at a voltage of 150 V. In order to compensate for the hysteresis effect of the piezoelectric elements, an analog PI controller was added to each of the 6 connecting elements. This microhand is intended to be used for fine manipulations of biological cells, assembly of microsystems, and microsurgery.

Aligning device for optical fiber

Conventional aligning devices for optical waveguides usually have V-like grooves. Manufacturing these grooves to micrometer exact dimensions is very difficult. The accuracy has to be very high because optical waveguides can have excentric cores which must be matched. The attenuation achieved with conventional aligners is about 0.1 dB for a 1.55 μm DSF *(dispersion shifted fiber)*, which is significant for extremely long waveguides (e.g. in underwater applications). A piezoelectrically driven microaligner was developed, which can position glass fibers independently of each other. Figure 5.32 shows the construction of the aligning device which was developed to handle several waveguides.

The optical waveguides are placed in the parallel V-shaped grooves. The alignment is done by piezoelectric actuators which move the microarms up and down relative to their base. Each positioning unit consists of two piezo-

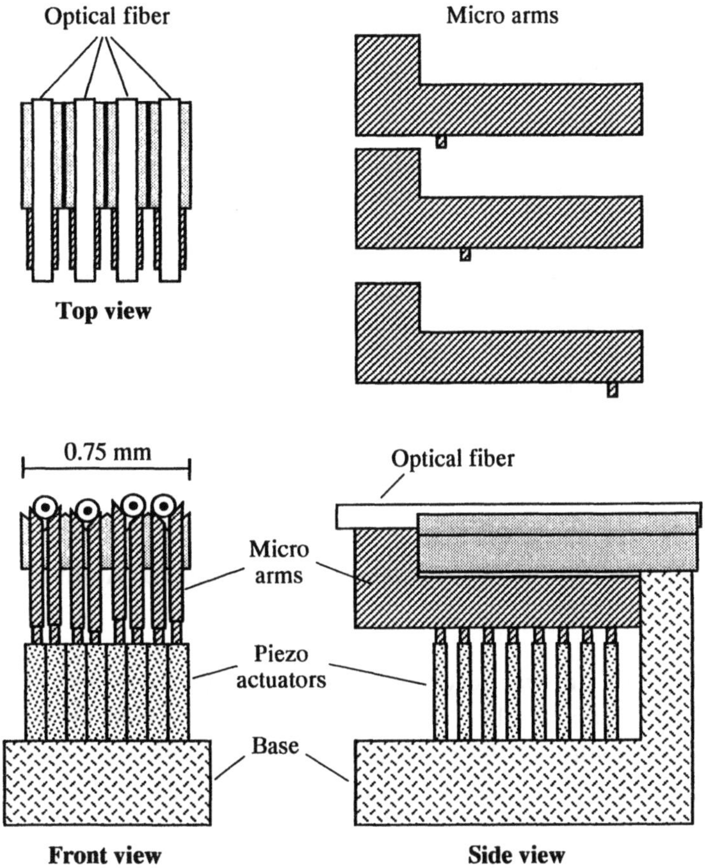

Fig. 5.32: The glass fiber alignment device. According to [Aosh92]

elements which are attached to the base and arms with tapered tips fastened to each piezoelement. Each arm can be individually controlled by applying voltage to its piezoelement. This causes it to move accurately in the micrometer range. The aligning process is illustrated in Figure 5.33. Each waveguide can be moved with a micrometer precision in two directions by controlling the piezoactuators; the overall alignment range is about 20 μm × 20 μm.

A prototype of this device was built for aligning several optical waveguides. $2 \times 3 \times 18\ mm^3$ large piezoactuators were used which can elongate 15 μm under a voltage of 100 V. During several alignment tests to position the glass

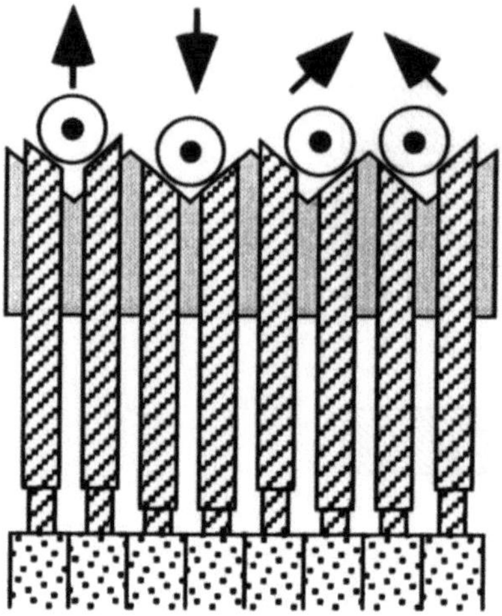

Fig. 5.33: The alignment of fibers by piezoelements. According to [Aosh92]

fibers, a resolution of 0.1 μm was reached. This microaligner was also successfully used as a connector for glass fiber bundles of a semiconductor laser device. By closed-loop control of the microaligner the attenuation across the interference could be reduced to 0.063 dB for a 1.55 μm DSF [Aosh92].

Cycloid micromotor

Ultrasonic motors are a new type of micromotors which can be used in microsystem applications. They are usually operated with a frequency of about 20–50 kHz. The rotational motion is obtained by the friction between the stator and the rotor. The advantages of ultrasound motors are their high torques at low rotational speed. The disadvantage, however, is the short lifetime due to the wear on the stator and rotor from friction. In Japan, a new concept was developed for these motors, which contains a piezoelectric drive. In the so-called cycloid motor, the lifetime is prolonged by reducing friction. This is realized by transmitting forces via a special geardrive. The inner part of the disc-like stator and the outer part of the rotor have gear teeth and engage with one another. The four piezoelectric drive elements are equally spaced at 90° along the outside of the stator, Fig. 5.34. If an alternating voltage is applied to oppositely located piezoelements with a phase difference of 90°, an oscillation is induced in the freely floating stator. The oscillation causes the rotor to move in the opposite direction due to the force transmission by the gears.

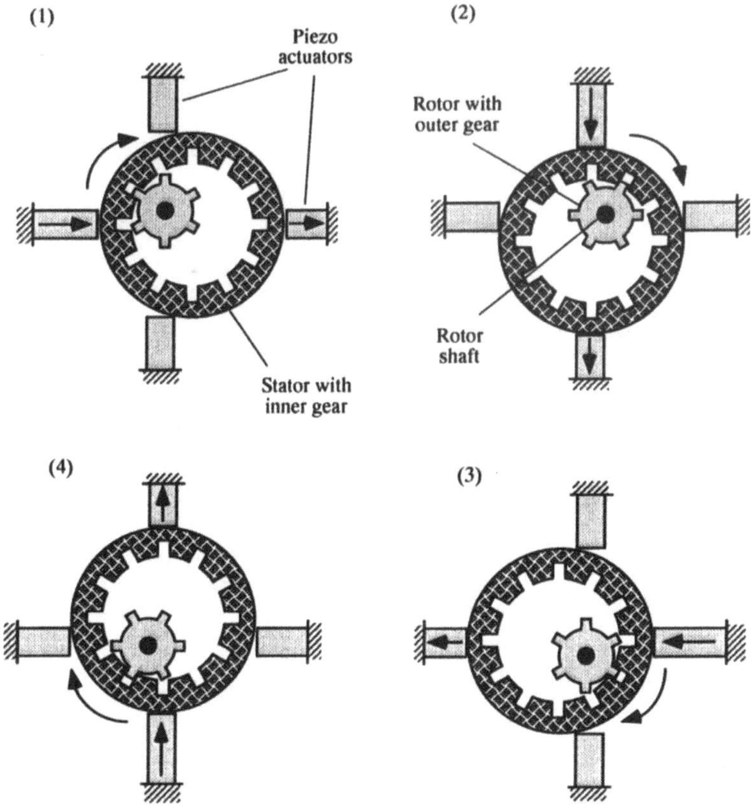

Fig. 5.34: Functional principle of a piezoelectric cycloid motor. According to [Haya91]

The first prototype of the cycloid motor had piezo actuators with a dimension of 5 × 5 × 20 mm³. The piezo elements have a maximum expansion of 16 µm under a voltage of 150 V. The motor's dimensions are mainly determined by the size of the piezoactuators. For further miniaturization of the motor, a newpiezoactuator design would be needed. However, this entails that the actuators can expand sufficently to guarantee the operation of the motor.

Laser scanner

The fast reaction speed and high positioning accuracy of piezoelectric actuators was utilized in a laser scanner, where a laser beam must be dynamically positioned in a two-dimensional optical system. A piezo-driven scanner can eliminate many problems usually encountered with conventional laser scanners, such as a very limited frequency range and a small deflection angle. The basic design of such a piezo-scanner is shown in Figure 5.35.

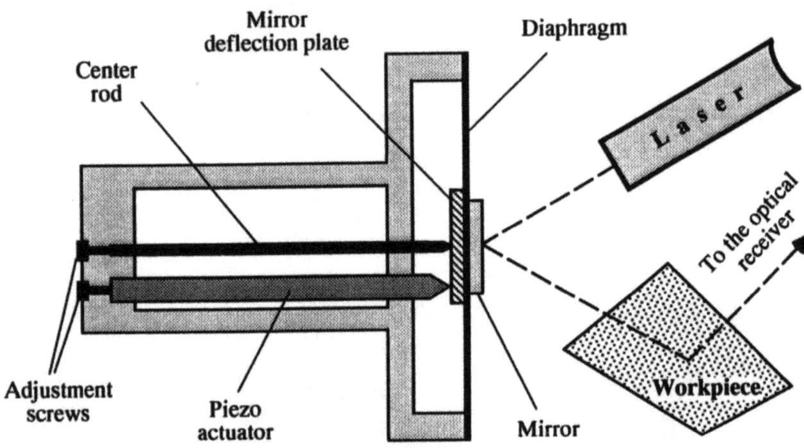

Fig. 5.35: Schematic design of the laser scanner. According to [Brand92]

The scanner consists of a deflection plate to which a mirror is attached, a metal membrane which serves as a spring element, a center rod and a piezo-actuator; the latter is installed perpendicular to the mirror. The mirror is tilted by energizing the piezoelement, making it expand and push against the deflection plate. By varying the distance between the center rod and the piezo-element, the deflection angle can be changed. The frequency that can be applied depends on the stiffness and internal stresses in the metal membrane. A prototype built of this piezo-scanner contained two piezoactuators to deflect the mirror in two dimensions. The obtained displacement of the piezo-elements was 60 μm, with a deflection angle of 8 mrad. The device could be operated at a frequency of 800 Hz. These characteristic values are more than sufficient for many optical applications.

5.4 Magnetostrictive Microactuators

Various solid state actuators have been investigated in the past few years; one example using piezoceramic was described in the previous section. While piezoactuators have served for a variety of applications, magnetostrictive actuators are still at the threshold of industrial exploration.

5.4.1 Motion Principle and Its Properties

In magnetostrictive materials electric energy is transformed into mechanical energy similar to piezoceramic materials. Magnetostriction means a dimension change of a ferromagnetic material by applying a magnetic field to it. In the normal state without an external magnetic field, the different polarizations in the Weiss' domains cancel each other. The magnetic field causes the Weiss' domains to align parallel to the external field direction. This causes the material to expand/contract in the direction of the magnetization (magnetostriction); the volumetric change can usually be neglected. In early experiments it was shown that magnetostictive properties of selected materials may induce a relative length change of about 0.1% at low temperatures [Clark92]. However, the Curie temperature (temperature at which material loses its ferromagnetic properties) of these metals was so low that there was no technical use for them.

Presently, the terbium-iron and terbium-disprosium-iron alloys are being actively investigated [Clark92], [Clae94]. They have better magnetostrictive properties and relative length changes of 0.15–0.2%, similar to that of piezoceramics. Especially the TbDyFe alloys, which are usually called Terfenol-D (Terbium, Ferrum, Naval Ordnance Laboratory, Dysprosium), have excellent magnetostrictive properties. Since these alloys have little or no hysteresis they only need simple control circuits. The materials are also very robust and insensitive to rough environmental conditions. Compared with the PZT ceramic materials, which make use of the inverse piezoelectric effect, the Curie temperature of Terfenol-D is high – about 380°C. Another advantage of Terfenol-D is the high energy density, about 20 times higher than that of PZT ceramics, which makes it specially suited for microactuators. Furthermore, magnetostrictive actuators are controlled by current, as opposed to piezoelectric actuators which are controlled by voltage. Therefore, voltages can be kept low, reducing the amount of necessary driving circuits. This is also less dangerous.

We have seen that the main disadvantage of piezoceramics is their small displacement. Unfortunately, this is also the "bottle neck" of magnetostrictive actuators. In static operation, Ohmic losses are a severe problem since the magnetization current must be applied permanently. As opposed to the piezoelectric effect, the magnetostrictive effect cannot be directly reversed.

At the moment it is very difficult to produce magnetostrictive components at a microscopic scale. Therefore, only few microactuators have been made of these materials. The scientists are sure that Terfenol-D has a great potential, since it can provide strong forces with its high energy density and a relatively small amount of material. Magnetostrictive actuators can be used for applications in which high forces, quick dynamic responses and short controlling distances with a very high positioning accuracy are required and in which the actuator is exposed to a high ambient temperature. They neither need movable electrodes nor a high electric voltage. These actuators are used as control elements for linear motors, as active vibration absorbers, as positioners, etc. They are being used for applications which previously were the exclusive domain of the piezoelectric actuators.

Terfenol-D actuators are usually made of a rod which is inserted into a magnetic coil. The coil must be exactly positioned to attain a good magnetic coupling. By carefully directing the magnetic flow, the flux losses can be kept at a minimum. Thereby, a high magnetic field strength can be generated in the Terfenol-D rod. To utilize the device more efficiently, a pretension obtained with a spring in direction of the axis of the rod can be of advantage. Its aim is to align precisely as many of the Weiss' domains as possible perpendicular to the length axis of the rod (zero position), which leads to a maximum change in length. When all Weiss' domains are aligned in the direction of the magnetic field, the Terfenol-D rod is magnetically saturated and no further expansion can be obtained. Since the tensile strength of Terfenol-D is very low, the actuator must be constructed in such a way that it is permanently under compression. Typical design principles of magnetostrictive rod-like actuators are shown in Figure 5.36. The figure suggests two possibilities: first, the rod can be magnetized with only one coil, Fig. 5.36, left, and second, an additional

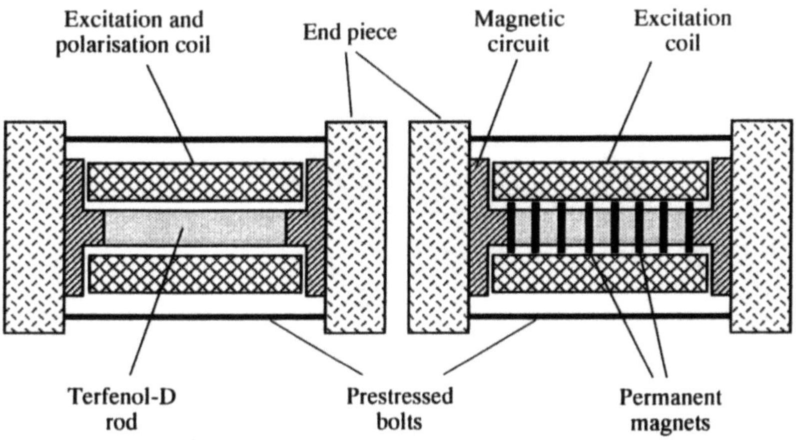

Fig. 5.36: Typical design of rod-like Terfenol-D actuators. According to [Lher92]

permanent magnet can be added to it, Fig. 5.36, right, which premagnitizes the rod for better alignment of the Weiss' domains. The second variant is more advantageous for practical applications since the premagnetization of the rod results in an almost linear relationship between the overimposed magnetic field and the resulting length changes.

At the moment only rod-like actuators are being investigated since other constructions are difficult to make. Magnetostrictive multilayered actuators are also becoming more and more of interest since they have been discovered as well suited converter elements for microsensors and actuators. Compared to conventional Terfenol-D rods, the devices which can be built are more lightweight and offer a higher movement potential at low cost. Bimorphic magnetostrictive actuators play an important role since a relative change of length in one component of the bimorphic structure can be transformed into large cantilever and membrane movements, Fig. 5.37. E.g. a silicon cantilever can be used as the elastic substrate layer in a bimorphic structure. Compared to piezoelectric layers, magnetostrictive materials have a relatively large change of length and need no direct electrical contacts.

So far, despite of intensive research and development work, magnetostrictive actuators have not been implemented successfully in many products. The

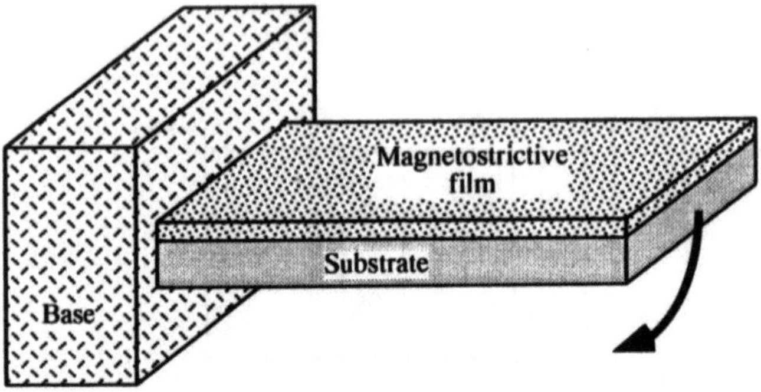

Fig. 5.37: Magnetostrictive cantilever

manufacturing methods for Terfenol-D are much too expensive. Improved mechanical material parameters to make other designs have not been found. Also good methods of integrating the microactuators into other systems and suitable control algorithms are missing.

In the following sections several magnetostrictive actuator ideas and prototypes will be introduced to give an overview of their application potential.

5.4.2 Concepts and Prototypes of Magnetostrictive Microactuators

The elastic wave motor

A very interesting development is the *Elastic Wave Motor* (EWM), which takes advantage of the properties of Terfenol-D [Kies88], [Roth92]. Here, electric energy is directly transformed into a continuous linear movement. A sketch of the function principle and a schematic design of the motor having a movable Terfenol-D rod and external coils (stator) is shown in Figure 5.38. In this device, the rod is placed into a guide tube, the end of which is attached to a rigid support. Several short coils are placed along the outer surface of the tube to produce the magnetic field. If the magnetic field is successively switched on and off from one end of the tube to the other in the coils, the Terfe-

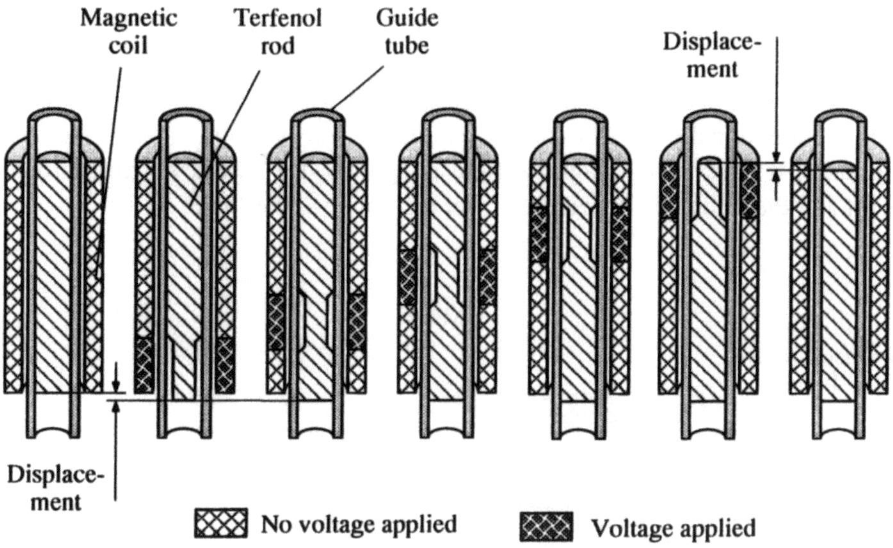

Fig.5.38: Schematic design and functional principle of the elastic wave motor. According to [Roth92]

nol-D rod moves within the tube in the opposite direction, as shown in Figure 5.38. The speed, force and position of the rod are controlled by the magnetic field. The design is successfully being used in the paper industry. It controls the paper thickness during manufacture and piling by moving a blade across the entire paper width [Roth92]. Conventional actuators need gears and are too inexact. The disadvantage of the Terfenol-D actuator is its high sensitivity to temperature and its high price.

Thin film magnetostrictive structures

The interest in thin film magnetostrictive material has increased steadily since its discovery. It can be useful for microactuator and sensor components. Compared to conventional Terfenol-D rods, the films can be better integrated into electromechanical devices and are made more cost-effective. Several interesting principles of magnetostictive microactuators using the film design are being investigated [Flik94].

- **Fluidic switch**

A TbDyFe film is used to make a bistable fluidic switch out of silicon. It consists of an external coil, a bimorphic cantilever arm, an inlet and two outlets, Fig. 5.39. If an external magnetic field is applied, the cantilever arm is bent down and the liquid jet is switched from the upper to the lower outlet.

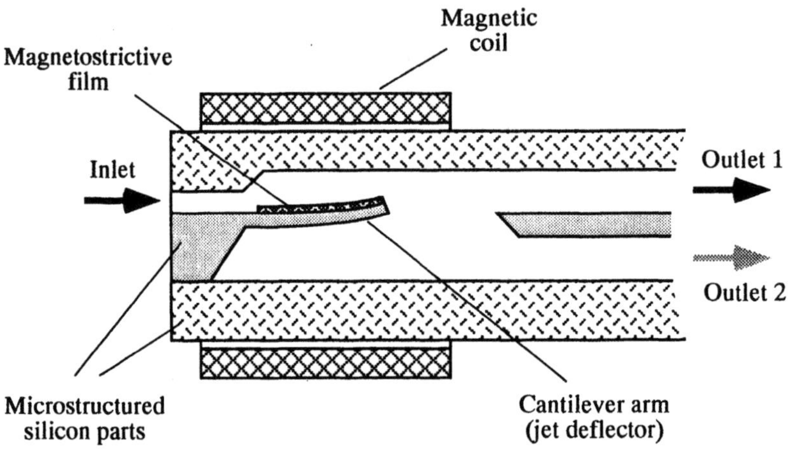

Fig. 5.39: Magnetostrictive fluidic switch. According to [Flik94]

The bulk micromachining technique is used to manufacture this switch. The cantilever arm is 2 mm long, 1 mm wide and has a layer of 20 μm (Si) and 5 μm (TbDyFe). The lift is 13 μm with a frequency of 1 kHz and a magnetic field strength of 20 mT. The advantage of this design is that the liquid pressure above and below the cantilever arm is the same. This allows large liquid volumes to be controlled, up to 500 ml/s. The manufacture of the microactuator is also simplified due to the absence of electrical connections.

- **Membrane-type microvalve**

Figure 5.40 shows the design of a membrane-type mixing valve, the passive parts of which can be made by using the LIGA method. The valve consists of an external coil, a bimorphic magnetostrictive membrane actuator, two inlets, an outlet and the valve seat.

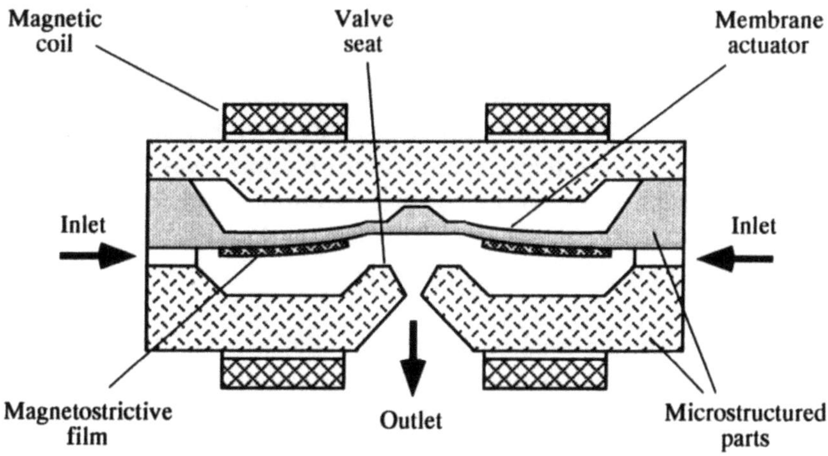

Fig. 5.40: Magnetostrictive valve. According to [Flik94]

The results achieved so far are very promising with regards to future MST applications which use thin film magnetostrictive actuators. The prerequisites that must still be met are the use of materials with a low driving voltage, excellent magnetic properties and Curie temperatures. The compatibility of magnetostrictive materials with micromanufacturing methods is also important.

5.5 Electromagnetic Microactuators

Electromagnetically driven microactuators are gaining in significance as manufacturers are improving the three-dimensional production methods for a variety of materials. With electromagnetic actuators, electric energy is transformed into mechanical energy like forces or torques. Classical examples of such actuators include electric motors and relays. Simple drives of this type can usually be directly realized by using the silicon or LIGA techniques.

5.5.1 Motion Principle and Its Properties

Electromagnetic drives which can change electric energy into a mechanical motion or force can be used for various applications due to their low cost and easy applicability. Until now they have been used in the macroworld, but attempts are being made for their miniaturization so that they can be included in MST applications. Here, miniaturized motors are the most important developments. They can produce linear and rotational motions, which are suitable for many microsystem tasks. To miniaturize electromagnetic actuators drastically the properties of magnetic materials must be improved and new manufacturing methods for microwindings must be developed.

The LIGA process can be used to make three-dimensional structures. The process allows the manufacture of magnetic materials needed for transforming magnetic energy into forces and torques (e.g. nickel). Thereby, it is possible to produce electromagnetic actuator systems in the μm range. The problem in manufacturing magnetic microactuators is the difficulty to simultaneously produce current-carrying coils and flux-conducting components by the same lithographic process. Because of this limitation, magnets and coils have to be integrated into an electromagnetic microactuator by an interconnection technology. Here, the wire bonding technique is often used.

The effectivity of an electromagnetic actuator is determined by the material properties and the volume of the used magnet. For this reason, the miniaturization is limited by the size of the magnet to be integrated. Various designs of electromagnetic motors can be used for MST. Besides the traditional rotational motor, there has been an increasing demand for linear motion drive systems operating in the micrometer range. They are needed for positioning tasks in tool machines, laser systems, scanning tunnel microscopes, robots and handling devices since they allow very exact positioning in the 100 nm range. They can be used e.g. as high-speed relays, pumps or valves.

To produce a linear motion, linear reluctance motors may be used. They have the best potential for MST applications. Reluctance drives use the magnetic resistance (reluctance) change in the magnetic field of a coil when a ferromagnetic object is brought into the field. Due to the mutual interaction of the object and the coil, a force is produced which can be used to transform electrical into mechanical energy. These drives are interesting for future applications since the properties of the used materials and the available production technologies allow the construction of many different devices.

Figure 5.41 shows the motion principle of a reluctance type stepping motor having a serrated stator and a magnetic rotor consisting of three magnetic subsystems to produce a sliding motion according to the magnetic bearing principle. The magnetic rotor takes a fixed position at which the device has the lowest magnetic resistance (reluctance). If the coils are excited with a phase-shifted current, a displacement of the rotor can be obtained due to the energy change in the magnetic field between the rotor and stator. These motors can be used for positioning or manipulation of various technical devices.

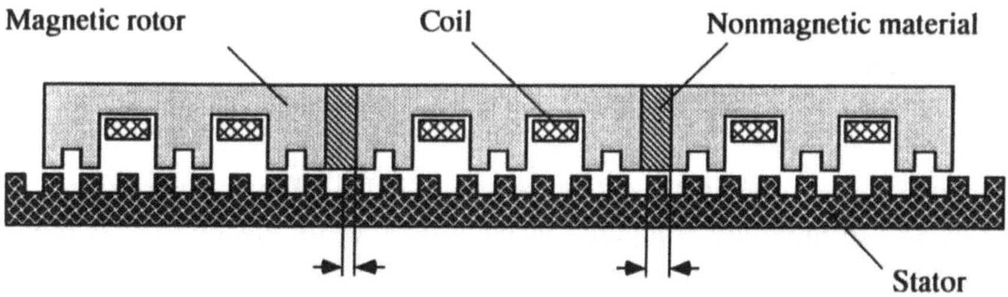

Fig. 5.41: Linear stepping motor. According to [Lehr92]

Rotary reluctance motors are also of interest to many applications, e.g. for medical purposes such as minimal invasive surgery and diagnosis; they can also be used in entertainment electronics such as miniaturized drives for portable devices. Figure 5.42 shows a scheme of a three phase rotary reluctance motor. It has a permanent magnet rotor with externally extended poles and a stator with internally extended poles; around the latter the electrical coils are wound. A magnetic field is created with these coils. For the realization of a rotary reluctance motor it is important to have a different number of teeth on the stator from that on the rotor, because otherwise all the magnetic loops have the same phase. This would make it impossible to generate a rotation. In the next section, various prototypes of miniaturized reluctance motors will be described in more detail.

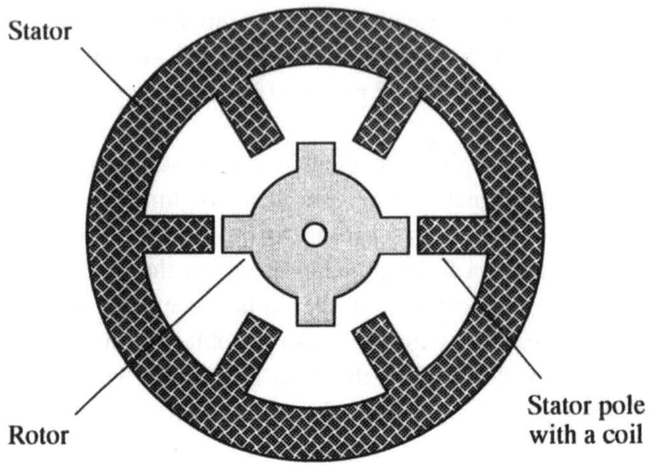

Fig. 5.42: Rotary reluctance motor

Advantages of electromagnetic actuators include the relative insensitivity to impurities and to humidity compared to electrostatic microactuators. A theoretical analysis shows that a magnetic reluctance actuator can generate usable drive energy for devices having structural dimensions less than 1 mm [Löch94]. A magnetic field cannot collapse as rapidly as an electric field and the drives only need low voltages that can be supplied by a battery. With this advantage the electromagnetic microactuator is an ideal component for many applications, especially for biomedical use in which high voltages are unacceptable. Also, the response times of such actuators are excellent. In general, electromagnetic microactuators are useful when efficiency is less important than reliability and safety.

5.5.2 Concepts and Prototypes of Electromagnetic Microactuators

Rotational electromagnetic micromotor with gears

For several years, the SLIGA technique (Section 4.2.5) has been investigated for the manufacture of electromagnetic micromotors [Guck92], [Chri92],

[Guck93]. The process allows the making of metallic structures from nickel of only 300 µm height. The precision which can be obtained makes it possible to produce various planar micromechanisms, such as electromagnetic reluctance motors with gears, Fig. 5.43 and Fig. 4.33. The structure produced from galvanically plated nickel is 100 µm high and consists of several gears and a reluctance motor. A rotational speed of 10000 rpm is possible. The gap between the motor shaft and the rotor is only 500 nm. This nickel-nickel system has very low friction, which gives the motor excellent dynamic properties. The coil windings of the stator pole are made of an aluminum alloy wire using wire bonding; direct etching of U-like bridges has proven to be problematic.

Fig. 5.43: Electromagnetic micromotor. Courtesy of the University of Wisconsin-Madison (Department of Electrical and Computer Engineering)

Another use of this design was the construction of a microelectric generator as a local energy source. A dynamometer was made which consists of a motor, a generator, a transmission, positioning sensors and a control loop. The maximum supply voltage needed was 15 V. The control signals were processed by an analog interface and a digital controller.

Micro-stepping electromagnetic motor

Another electromagnetic rotational micromotor is being investigated by designing a variable electromagnetic reluctance micromotor having six stator poles and four rotor poles [Guck93]. It is a three-phase rotating step motor with a step angle of 30°, Fig. 5.44.

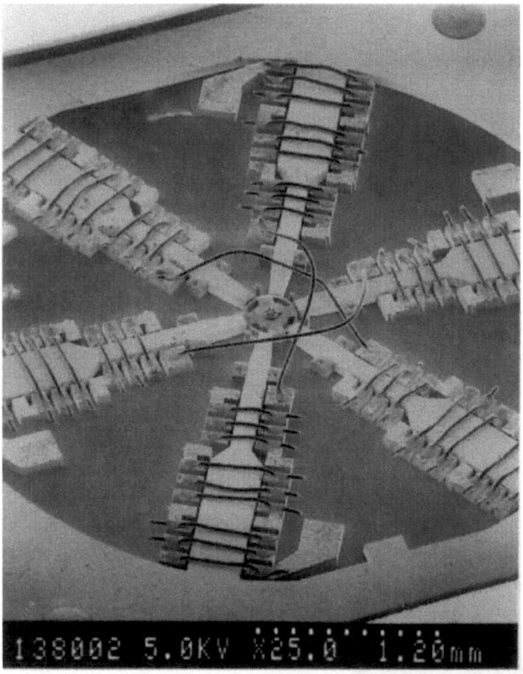

Fig. 5.44: Micro-stepping electromagnetic motor. Courtesy of the University of Wisconsin-Madison (Department of Electrical and Computer Engineering)

The shapes of the stator and rotor are defined by the X-ray lithography. With this technique a stator height of about 300 μm can be obtained – a dimension which directly influences the torque. Many different micromotor realizations are being tried out. Of particular concern is the friction between the rotor and the substrate, which is minimized by keeping the rotor relatively thin as compared with the stator. Highly permeable nickel was used as magnetic material to make the reluctance of the motor fairly independent to the size of the stator-rotor gap. The micromotor was built up on a silicon substrate, having integrated photodiodes to indicate the rotor's position and the rotational speed. The magnet coils were made by the LIGA process using galvanically deposited nickel. After the stator and rotor are mounted, the coils are made by bonding with 32 μm thick aluminum wire. The smallest motor realized had a rotor with a diameter of 285 μm and a shaft with a diameter of 72 μm. The distance between the axle and the rotor is 500 nm and the gap between the rotor and stator is 3 μm. By the application of a current of 0.6 A, the motor could reach its maximum rotational speed of 30000 rpm. A minimum current of 150 mA is necessary for a stable function of the motor. After about 50 million revolutions, no changes in the original operating behavior was noticed.

Hybrid rotational micromotor

A 3 mm long reluctance motor the components of which are partially made by the LIGA process was developed and reported in [Ehrf93]. Figure 5.45 illustrates the dimensions of this micromotor. The motor's components are shown in Figure 5.46. The stator having microrods as its main machine elements, the rotor and the spacer ring were made by the LIGA technique in order to ensure tolerances below 5 μm. The other motor parts were made by conventional precision machining techniques and are commercially available.

The motor drive uses the reluctance principle. The micromotor generates a rotating magnetic field with the help of electric current in the stator coils; this brings the low-retentive rotor into motion. The rotor is placed on a shaft; the latter has a diameter of 0.24 mm. A spacer keeps the distance between the rotor and stator columns at 20 μm, an exactly defined distance. The axle rests on two ball bearings (1.6 mm diameter). A sleeve which encompasses the stator has an outer diameter of 2 mm and functions as casing. This rather simple motor design is capable of producing a torque of about 10^{-7} Nm. The construction of this micromotor can be easily varied, e.g. the torque can be increased by using a permanent magnet in the rotor. In this case, the rotor can be produced from powdered material with the help of the lost-mold method (Section 4.2.4). Such motors, only several mm in size, are needed for many applications, e.g. microsurgery and microproduction.

Fig. 5.45: Micromotor's size compared to fennel and caraway seeds. Courtesy of the Institut für Mikrotechnik GmbH, Mainz

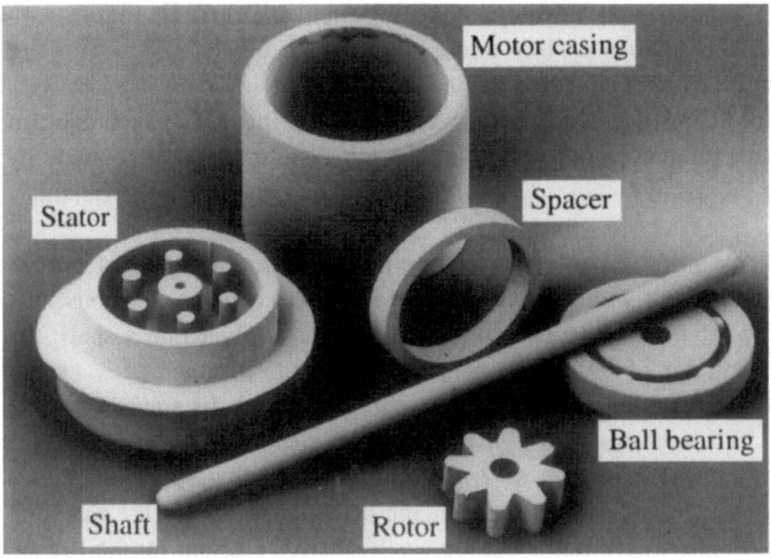

Fig. 5.46: Components of the micromotor. Courtesy of the Institut für Mikrotechnik GmbH, Mainz

Micromotor using a rare earth rotor

Some very interesting work on the development of miniaturized electromagnetic rotational motors is reported in [Itoh93]. Figure 5.47 shows a prototype of an electromagnetic d.c. motor; Fig. 5.48 shows its cross-section view.

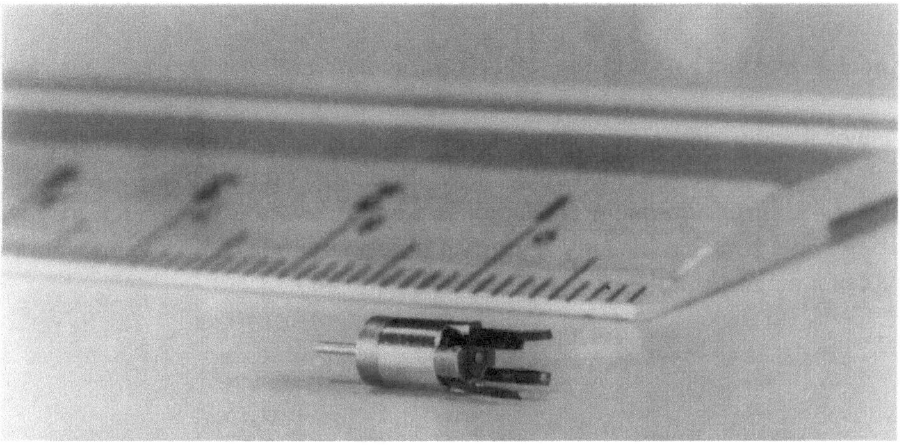

Fig. 5.47: Prototype of an electromagnetic d.c. motor. Courtesy of the Toshiba Corporation, Nagoya (Manufacturing Engineering Research Center, Small Motor Development Center)

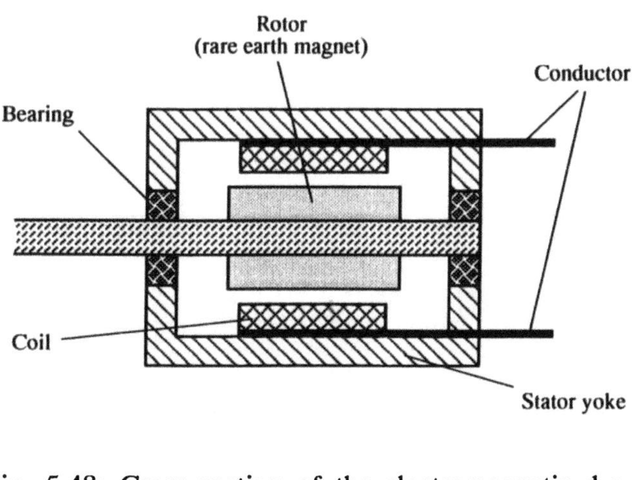

Fig. 5.48: Cross-section of the electromagnetic d.c. motor. According to [Itoh93]

The prototype has an outer radius of 3 mm and is 5 mm long. The casing, which functions as stator, contains the driving coil. The rotor, a magnet made of rare earth metal alloy (SmCo), has an inner diameter of 0.5 mm, an outer diameter of 1.5 mm and is 3 mm long; the 3-phase magnet coil is 0.2 mm thick. The motor has an operating voltage of 2 V and generates a torque of 2×10^{-5} Nm.

Special method of making electromagnetic coils

One of the difficulties in making electromagnetic micromotors is the realization of 3-dimensional wound coils. A solution to this problem is shown in Figure 5.49. In this design, a conductor is wound around a meandered magnetizing body. This significantly simplifies the integration. The conventional design of a magnetic coil is also depicted in Figure 5.49.

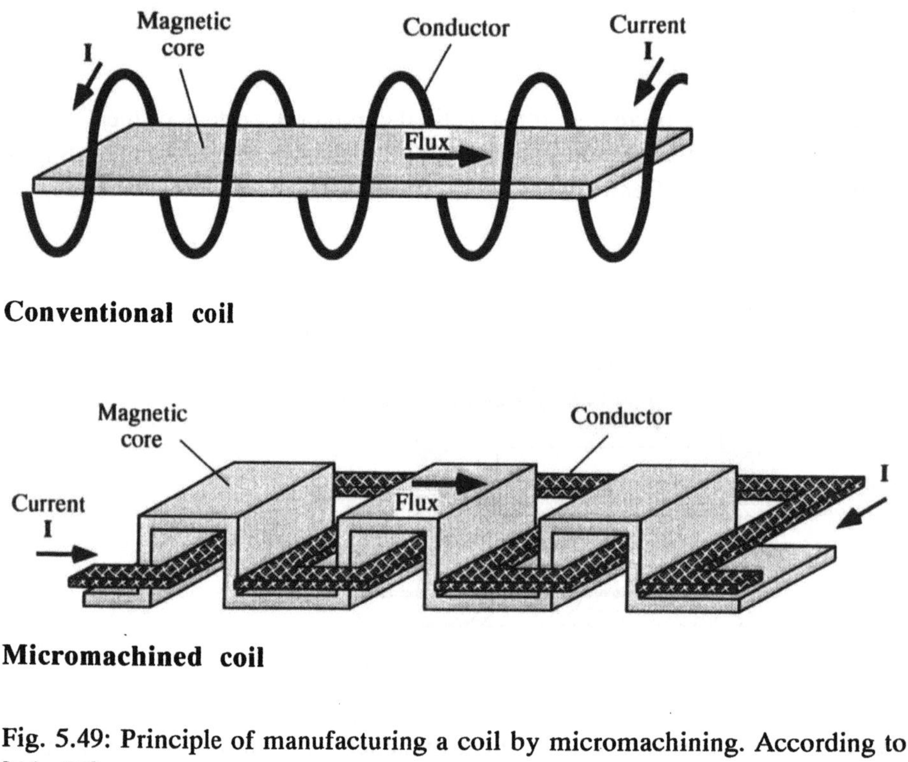

Fig. 5.49: Principle of manufacturing a coil by micromachining. According to [Ahn93]

An electromagnetic micromotor using the reluctance principle was manufactured based on this coil design [Ahn93]. This rotary motor was constructed on a silicon substrate by using several microtechniques (deposition, lithography, dry etching, microgalvanics and microassembly). The motor has a separately manufactured rotor made of an FeNi alloy; it contains 10 poles, is 40 μm thick and has a diameter of 500 μm. The stator has 12 poles and is 120 μm thick. It was also made separately and then integrated with a monolithic copper coil by successive etching/deposition steps; its outer diameter is 1.5 mm. The motor needs less than 1 V at a current of 300–500 mA and can be operated in both directions up to 500 rpms. The rotational speed can be varied by changing the switching frequency of the supply current. It is hoped that in the future the motor can be manufactured as one monolithic device, including stator and rotor. At the moment, the integrated control circuits are being developed which should help to increase the motor's performance.

Microvalve

Numerous silicon microactuators are being investigated for operating gas and liquid pumps. An electromagnetic microvalve was discussed in [Löch94]. The valve was manufactured with a new approach by using UV exposure and galvanic deposition together with new photosensitive material (AZ4000 from *Hoechst*). Figure 5.50 depicts the valve structure which consists of two silicon chips having a dimension of $6.25 \times 6.25 \times 0.5$ mm^3. The lower active chip has a 2×2 mm^2 square membrane made of a galvanically deposited FeNi alloy, it is attached to the corners of the silicon substrate. An additional strip of identical magnetic material is fastened to the middle of the membrane to increase its strength. The upper passive chip has a conical 0.2×0.2 mm^2 inlet/outlet. In its output state (current off) the valve is open and allows liquid to go through. When a magnetic field is applied, the FeNi membrane moves toward the opening in the upper chip and closes the valve. A later development integrated a magnet coil on the back side of the passive chip, which may eventually lead to the design of a complete microsystem.

Microswitch

A very promising application for cheap and energy-efficient electromagnetic actuators is the communication technology. At the moment, electromagnetic

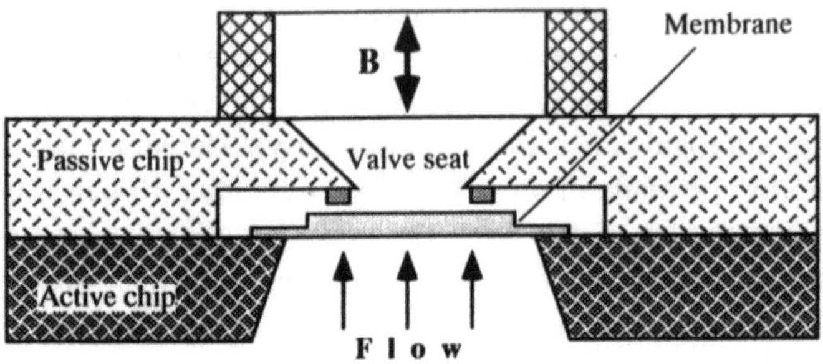

a

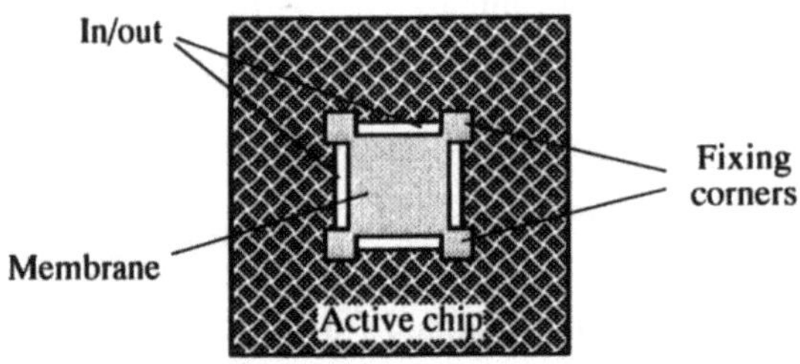

b

Fig. 5.50: Structure of an electromagnetic microvalve
a) valve cross-section; b) top view of the active chip. According to [Löch94]

microrelay principles are being researched and their application possibilities are being investigated ([Hosa93], [M²S²93]). A conventional 3-dimensional relay drive consists of an iron core around which a copper coil is wound. However, this construction is not suitable for an integrated microsystem. A miniaturized electromagnetic switch with a planar magnet coil was presented in [M²S²93], Fig. 5.51. The switch consists of a steel cup and a movable magnetic plate. In Figure 5.51 (top) the contact plate of the switch is in its neutral position. When the current is switched on, a magnetic field is built up which pulls the plate downwards, Fig. 5.51 (bottom). When the current is switched off, the plate moves back to its original position by means of a spring, not depicted here. The actuator has a diameter of 12 mm and is 2.5 mm high; its plate lift is 0.3 mm.

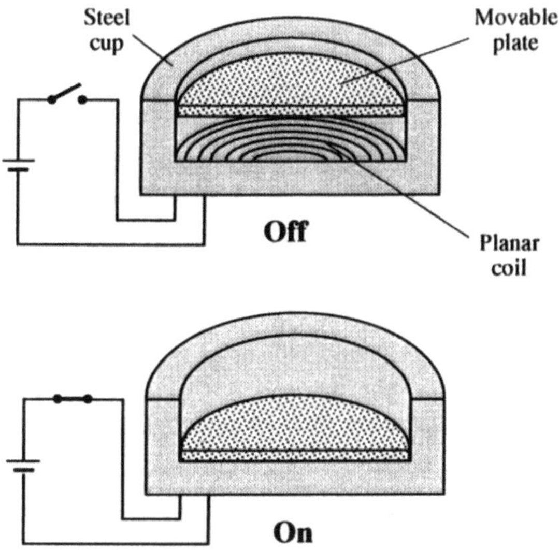

Fig. 5.51: Electromagnetic switch. According to [M2S293]

Linear micromotor

Numerous research projects are concerned with the development of electromagnetic linear actuators. Since almost all present efforts to design linear actuators are based on the silicon technology, the available structures are limited to a height of about 20 μm, which means that the forces that can be produced are very weak. There are few devices using planar coils; a linear motor with a sliding rare earth magnet is discussed in [Wagn92]. The magnet slides in a channel between two silicon chips which are attached to a glass substrate. The operating principle of this motor is depicted in Fig. 5.52. Planar coils located in the silicon chips are progressively energized to generate the linear motion of the magnet.

There are 8 pairs of planar coils, integrated in parallel to the guiding channel of the chip, Fig. 5.53. The coils opposite one another are driven sequentially with a current of the same magnitude so that a travelling perpendicular magnetic field (parallel to the magnetization of the permanent magnet) is produced. Thus, the magnet is pulled along the channel in a synchronous manner by the moving magnetic field. The magnet's dimensions are 0.7 × 1.8

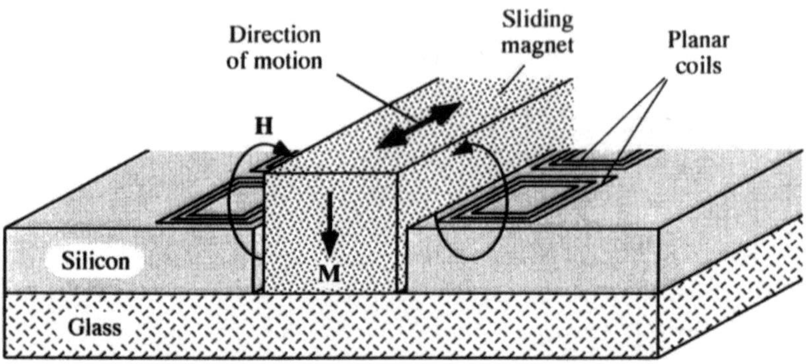

Fig. 5.52: Functioning principle of the micromotor. According to [Wagn92]

Fig. 5.53: Prototype of a linear micromotor. Courtesy of the Fraunhofer Institut für Siliziumtechnologie (ISIT), Itzehoe

× 0.9 mm^3 and the channel is 0.8 mm wide. With a maximum current of 650 mA, a motor speed of 24 cm/s can be reached; the magnetic field has a strength of 5.1 Gauss. Several other linear motors using rolling or rotating permanent magnets as rotor, were also manufactured [Wagn92].

5.6 SMA-based Microactuators

When shape memory alloys (SMAs) are deformed under a certain critical temperature and then heated up to above this critical temperature, they will "remember" their original form and assume it again. This effect can be used for generating motions or forces. Characteristic for actuators that use SMA are their low complexity, light weight, small size and large displacement; e.g. SMA components have been used for several years as active pipe connectors. However, the potential use of these alloys in MST has just recently been recognized.

5.6.1 Motion Principle and Its Properties

In the mid 50's the SMA effect was discovered in various copper alloys, in which a reversible, thermic-mechanical transformation of the atomic structure of the metal takes place at certain temperatures. When the temperature is raised or lowered, the metallurgical structure of an SMA transforms from a martensitic state (low temperatures) to the austenitic state (high temperatures), or vice versa. In Figure 5.54, the basic transformation mechanism is schematically shown. Starting from a stable and rigid austenitic state, the SMA transforms into the martensitic state as the temperature sinks under the critical temperature; thereby the shape of the SMA can be deformed by up to 8% (as for NiTi-alloys [Menz93]). In the low temperature state, the SMA keeps the desired deformed shape until it is exposed to a higher temperature. When it is warmed up above a threshold temperature, the deformed martensite is transformed back to austenite and the SMA takes on its original form (thermal shape memory). With this property, large displacements can be obtained compared to other actuator principles.

This effect can be explained by using the stress-strain relationship of an SMA wire, Fig. 5.55, left. The martensitic state below the critical temperature T_{Mf} shows an unusual behavior, which is referred to as the martensite plateau. In this temperature range the alloy increases its strength only slightly; this means that under constant stress the alloy can be deformed considerably.

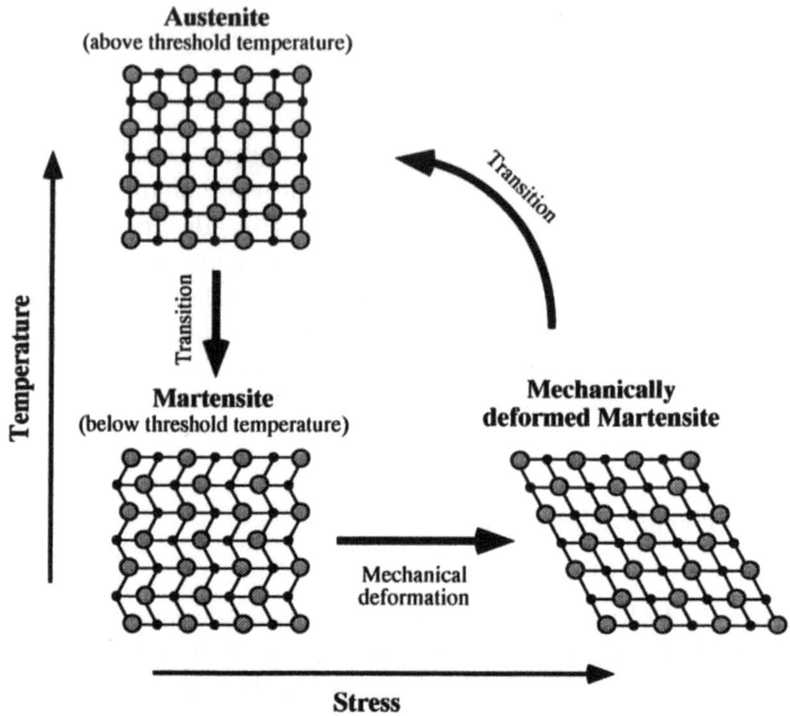

Fig. 5.54: Schematic representation of the SMA effect. According to [Tin92]

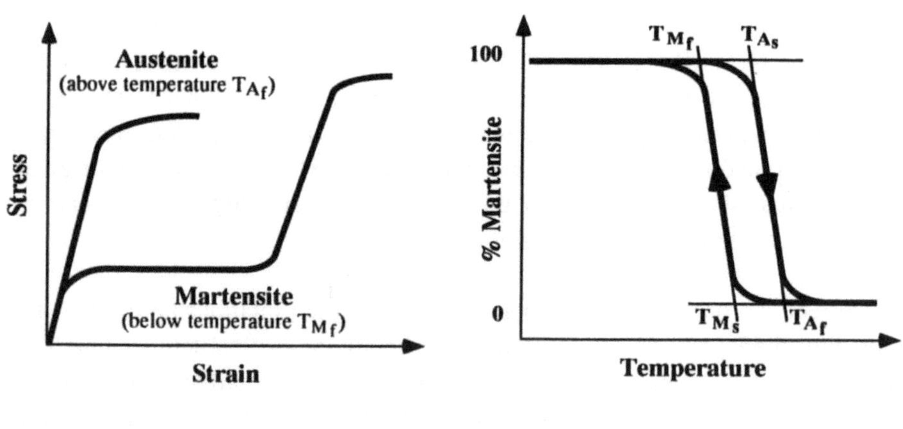

Fig. 5.55: Stress-strain diagram (left) and the hysteresis curve (right) of an SMA. According to [Stöck92]

However, once the austenitic state with the temperature $T > T_{Af}$ is reached, the deformation is restored and a hysteresis behavior typical for SMA materials can be observed, Fig. 5.55, right. Here, T_{Ms} and T_{Mf} are the martensite formation start and finish temperatures; T_{As} and T_{Af} are the austenite start and finish temperatures, respectively.

The difference in strength between the material on the martensite plateau and in the austenitic state can be considerably big; for some alloys 10 to 1. SMA actuators take advantage of this difference. In the practical use of SMAs the yield point of the metal, which is at about 8% of elongation, must be carefully observed. The loading of the metal beyond the martensite plateau can cause the material to be permanently deformed, Fig. 5.55, left.

The phenomenon described in Figures 5.54 and 5.55 is called the one-way-effect. The material is first permanently deformed under load in the martensitic state by an external force and then returns completely to its original form when heated. The so-called two-way-effect requires a certain thermal and mechanical pre-treatment of the SMA. Then, a transformation can take place by heating without an exerted force. By applying only heating-cooling cycles for deformation, the alloy "remembers" two different geometrical forms and switches between the two back and forth. The two-way-effect offers more application possibilities, especially for microactuators, but it is of less magnitude as the one-way effect (max. 5% for NiTi-Alloys [Menz93]).

The shape memory effect was observed in several alloys and even in some ceramics. The nickel-titanium alloys are of particular interest as they have excellent shape memory and mechanical properties, making them very suitable for actuators. Probably the most important parameter for SMAs is the critical temperature at which the phase transformation takes place; it is basically dependent on the material composition. NiTi alloys have their critical temperature between -100°C and +100°C, which limits the applicability of thermal SMA actuators to this temperature range. Due to the small switching temperature range (about 10–20°C), unfavorable temperature conditions can lead to an undesired transformation or can make it difficult to return the material to its original state. Part of this temperature problem can be overcome with NiTiPd alloys which have a higher transformation temperature of about 200°C. When properly applied this material may open up new opportunities for actuators based on the SMA effect.

The greatest disadvantage of thermal SMA actuators is their relatively long reaction time, which limits in many cases their practical application. The heating cycle can be controlled quite easy by adjusting the magnitude of the applied current; the cooling cycle, however, is very difficult to control. Usually, extra cooling devices must be used to significantly improve the response time of the actuator. Furthermore, the specific problems arising from the temperature hysteresis of an actuator based on the shape memory effect must be taken into consideration. For applications in which time is not a critical factor, the switching temperature can easily be reached by changing the ambient temperature of the environment of the device. In this case, the actuator works as a thermostat using the sensoric SMA property, which is characteristic of the so-called *smart* materials.

Large forces can be generated with a small SMA actuator because of a high power-weight ratio. Due to its simple construction and actuation principle, it can easily be implemented in a control system. Therefore, it is possible to integrate an SMA actuator into a complete microsystem. SMAs are of particular interest as thermal actuators with integrated sensor function and as actuators to realize complex motions in a small space. An SMA can be operated with the same voltage level as its IC control circuit. Since the mass of an SMA microactuator is usually small, relatively short reaction times can be attained, making it possible to build within certain limits actuators with dynamic response.

Microrobotics and special medical applications are interesting applications of SMAs. Typically, an SMA actuator can be used as a joint of a miniaturized robot arm, or as an end effector of a biomedical instrument, like a catheter or an endoscope. Another application could be a miniaturized joint for a hand prosthesis. However, suitable SMA actuator designs have to be found that allow good actuator control and cooling; by such means the functions of an actuator could be tailored to a problem, which would enhance the applications of such a device. One particular area of interest are underwater applications since the effect of the natural cooling medium could be taken advantage of. At present, however, it seems impossible to attain the same dynamic behavior as that of electrostatic or electromagnetic microactuators. A limiting factor of an SMA is the poor ability to microstructure its components, which is the reason why only few miniaturized SMA actuators have been realized up to now. However, new research and development results indicate that the breakthrough can be expected soon [Kohl94], [Kuri93]. In Figure 5.56 a NiTi microbeam is depicted; it is 2 mm long, 35 μm thick and 92 μm wide. This microstructure can either be manufactured by a lithographic/etching process or a laser cutting method.

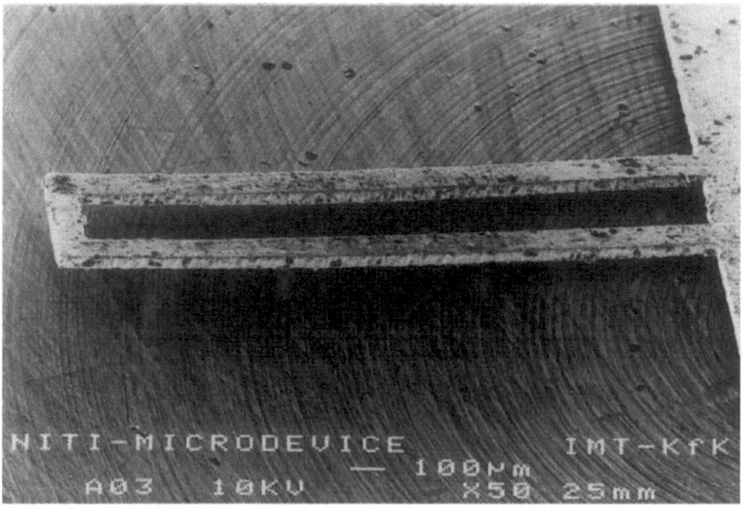

Fig. 5.56: A microbeam made of NiTi shape memory alloy. Courtesy of the Karlsruhe Research Center, IMT

There are several design principles available for the technical application of SMAs. The most common ones are expansion and compression coil springs and cantilever springs. Depending on the application, the SMA effect can be used in different ways. E. g. when no forces are acting during the reversed cycle of a deformed element, the SMA has a so-called "free" shape memory. In a multitude of applications, however, the SMA actuator is mechanically restricted from returning to its original position when heated. Because of the stresses created, high forces are produced which can be used to do work. One possible type of application is depicted in Figure 5.57.

An SMA spring can be deformed in the martensitic phase, e.g. by a weight. When heated and thereby transformed into an austenitic structure, it lifts the weight up to a certain height. By cooling the SMA spring under the "switching point", the weight deforms the "softened" martensitic spring again, thereby resetting the mechanism. A straight SMA wire can also be used instead of a spring for similar applications.

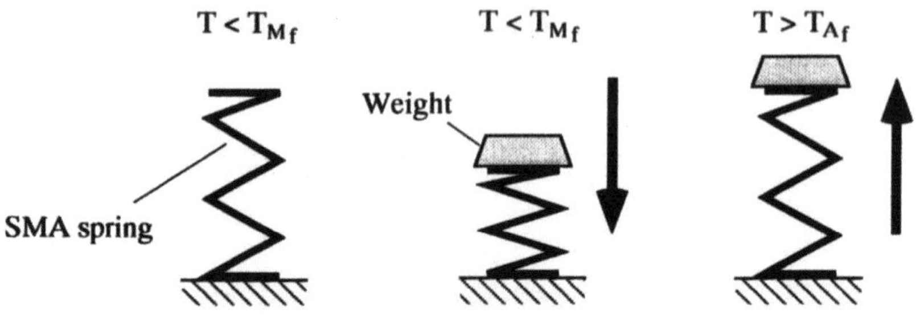

Fig. 5.57: Working principle of a shape memory alloy spring. According to [Humb94]

The experimentation with SMA actuators has brought forth many original uses. Researchers resort to nature as their model in order to adapt the special properties of an SMA to an application or to a particular solution that can be of use in the technical world. For example, a possible design of an SMA actuator for a miniaturized pipe inspection robot was suggested in [Hess92], Fig. 5.58. The worm-like SMA actuator, which can change its diameter and length as needed, can offer a good solution towards the problem of moving through a pipe. Researchers have tried to develop such inspection robots with chain or wheel drives; with little success in a small pipe, however, because of the limited space available.

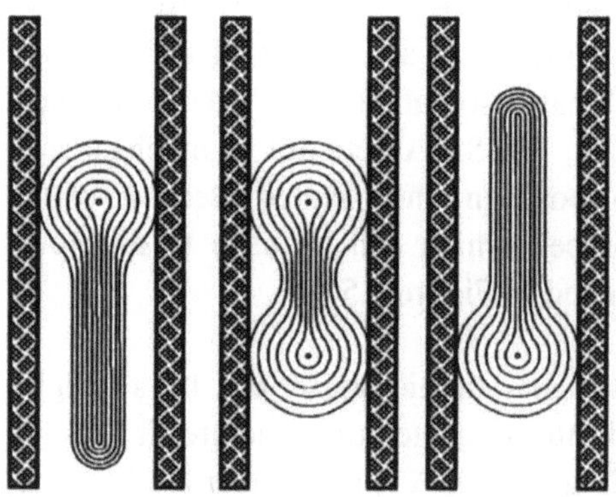

Fig. 5.58: Worm-like actuator design for a pipe inspection robot. According to [Hess92]

In summary, many solutions have to be found before widely using SMA actuators. There are still fundamental problems which have to be researched, such as high stability, high temperature performance, improved dynamic behavior, and the design of SMA-specific controls. Also affordable manufacturing methods have to be devised. Whether the SMA microactuators can be a real alternative to the proven piezoelectric or electrostatic actuators will depend on whether or not solutions for these problems can be found.

5.6.2 Concepts and Prototypes of SMA-based Microactuators

Micro endoscopes and catheters

Minimal-invasive surgery and new diagnostic techniques require the availability of a new class of micro and miniature instruments, like endoscopes and catheters, which are equipped with sensors and effectors. Here, it is important to develop instruments which can extend the physician's capabilities into the microscopic world of surgery. One goal is to develop endoscopic instruments which can be manually inserted by a surgeon and which have microtools at their tips to perform microsurgery. Since blood vessels are usually very narrow and winding, it is rather dangerous to guide a catheter around curves or into branching vessels. The present trend towards minimal invasive therapy requires that precise catheter systems with active guidance will be available to enable the surgeon to enter the various cavities of the human body or to direct them into a specific branch of a blood vessel.

In [Esashi93] the development of an active endoscope was described which uses springs made from shape memory alloy. These springs produce the force for allowing a defined trajectory which is needed to perform a complicated motion in minimal-invasive surgery, Fig. 5.59. The prototype is 215 mm long, has a diameter of 13 mm and consists of 5 active segments. Each of the segments can be moved by 1 mm diameter SMA springs which are controlled by a microprocessor.

In order to facilitate the insertion of a microcatheter, a device with two degrees of freedom was developed and described in [Fuku93]. Figure 5.60 depicts the catheter design. Three SMA wires are encased in a flexible plastic

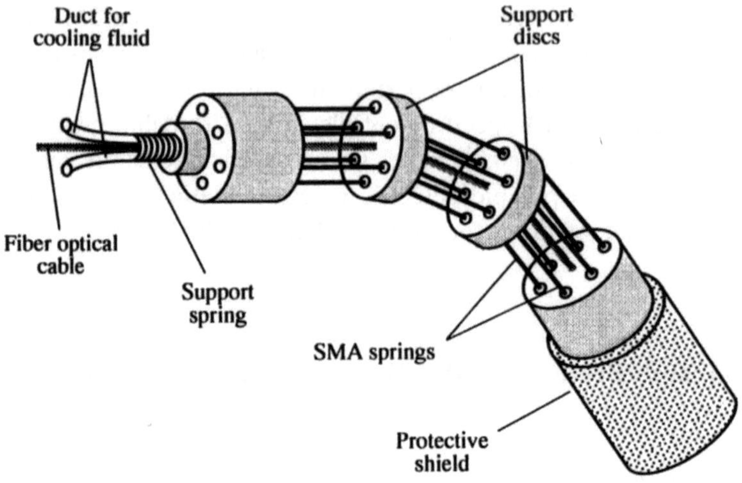

Fig. 5.59: Active endoscope using shape memory actuators. According to [Esashi93]

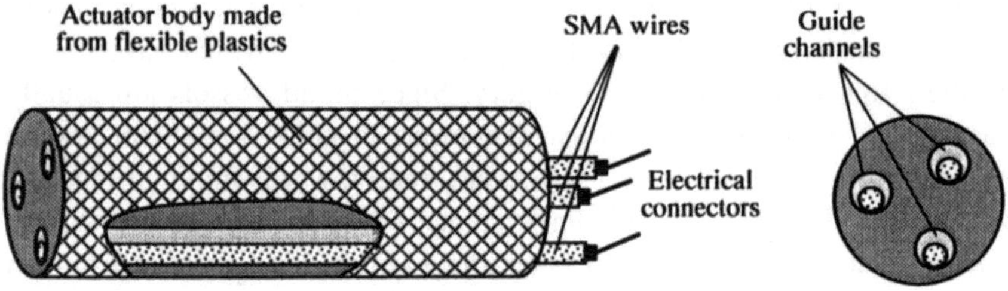

Fig. 5.60: Schematic representation of a MAC actuator. According to [Fuku93]

actuator body; each wire has a diameter of 150 µm. There is an electrical connector on each end of the SMA wires to which the electric voltage can be applied. The wires contract when an electric current is applied to them, causing a temperature increase. When the power is turned off, the wires take on their original form after cooling. The direction of motion of the endoscope and its angle of the bend can be controlled by selectively applying electric voltage to each of the three wires.

The MAC actuator was tested in air of 20°C and in a physiological solution of 36°C. Response times of 0.78 s and 0.45 s were attained, respectively. The microactuator's movement capabilities were demonstrated by *in vitro* experiments. For this, an apparatus was built consisting of a blood vessel simulator, a pump for moving the body liquid, a heater and a measuring instrument. A physiological liquid having the characteristic temperature of the human blood of 36°C was pumped through the artificial blood vessel. The goal was to investigate the efficiency of the MAC actuator at different flow speeds. The actuator with a diameter of 1.65 mm was manually pushed forward and its bending was controlled electrically in order to follow a specific branch or to enter a specified vessel. The test results were very satisfactory. Experiments are also being carried out on animals.

Micro grippers

TiNi alloys have many features which can be useful for actuators of microrobots or miniaturized robots. They can, for example, be applied in robot grippers. One of the first experiments in which a flexible NiTi microarm was mounted on a silicon wafer was described in [Kuri93]. This microactuator is shown in Figure 5.61.

Fig. 5.61: SMA microrobot arm made of NiTi. Courtesy of the Yamaguchi University (Faculty of Engineering)

To produce this beam structure on a (100) silicon substrate bulk and surface micromachining techniques were combined. This involved two successive lithographic and three wet etching steps. A 5 μm thin NiTi film was deposited using the sputter technique. By changing the actuator temperature, the microarm can bend up and down, Fig. 5.62.

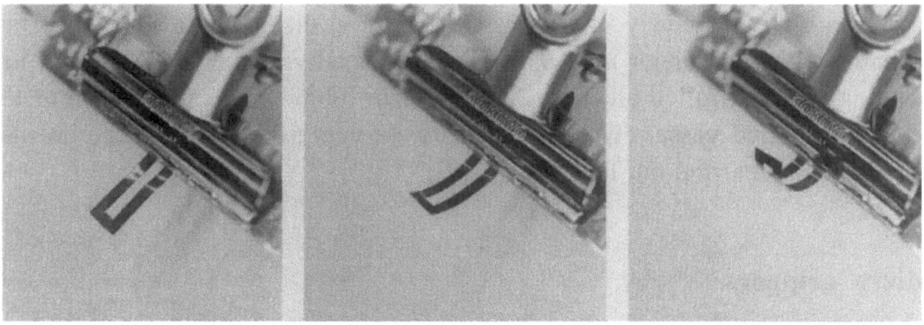

low temperature → high temperature

Fig. 5.62: The working principle of an SMA microarm. Courtesy of the Yamaguchi University (Faculty of Engineering)

The figure shows the behavior of the SMA actuator at various temperature conditions caused by the SMA two-way-effect. By heating the material above the austenite temperature T_{Af}, the SMA actuator returns to its original curved shape; when the temperature is lowered, the microarm straightens out. If the temperature falls below the martensite temperature T_{Mf}, the material bends in the opposite direction. The threshold temperatures of the depicted microarm were $T_{Af} = 57°C$ and $T_{Mf} = -9.1°C$, respectively. Two or more of these SMA structures can be integrated into one robot gripper.

5.7 Thermomechanical Actuators

Thermomechanical microactuators are based on the principles of a change of the shape or the volume of a material which takes place when heat or cold is applied to it. The best-known actuators of this type are the bimaterial actuators. They are made up of layers of different materials which have different thermal expansion coefficients. Another type of thermomechanical actuator uses the thermal expansion of gases or the liquid-gas transformation to produce an actuation. For extremely small actuators, the thermal expansion or contraction can be induced by optical energy; often such systems only need a very small amount of energy. In the following paragraphs, several thermomechanical prototypes and design principles are presented.

5.7.1 Concepts and Prototypes of Bimaterial-based Microactuators

Microactuators made of layered bimaterials function for the most part according to the cantilever principle and are used as microswitches or microvalves of temperature control systems; they are activated by the ambient temperature or an artificial heat source. They can be used in microrobot grippers or in medical tools. To produce a bimaterial actuator materials with two different thermal expansion coefficients simply have to be combined to form a bimorphic sandwich structure. The actuators are activated by a forced change of the actuator's temperature by electric heat or by optical radiation. The first method is most commonly used, but the actuator structure is complex. The necessary drive voltages are low, making it easy to integrate an actuator into another system. However, miniaturized bimaterial-based actuators have a great disadvantage. A reduction in the mass of the material also reduces the amount of electric/optical energy which is necessary to operate an actuator, whereby short response times are obtained. However, the actuator's ability to exert a force is also reduced. For this reason it is difficult to predict what place these actuators will have in MST.

Bistable cantilever actuator

In [Mato94], a bistable cantilever actuator was presented. It was made by the surface micromechanics method from a (100) silicon substrate. This microstructure consists of a U-shaped three-layered thin-film cantilever beam (polycrystalline silicon – silicon dioxide – polycrystalline silicon), a tension band (silicon nitride) and a three-layered anchor structure (polycrystalline silicon – silicon nitride – polycrystalline silicon), Fig. 5.63. The tension band is mounted in the middle of the anchor structure. The silicon nitride layer of the anchor structure and that of the tension band are made from a single layer by structuring techniques. The overall length of the actuator is 187 μm; the length of the cantilever is 82 μm, its thickness is 30 μm and the gap between the two strips of the cantilever beam is 30 μm. The tension band is 70 μm long, 6 μm wide and 500 nm thick. The anchor has a length of 105 μm.

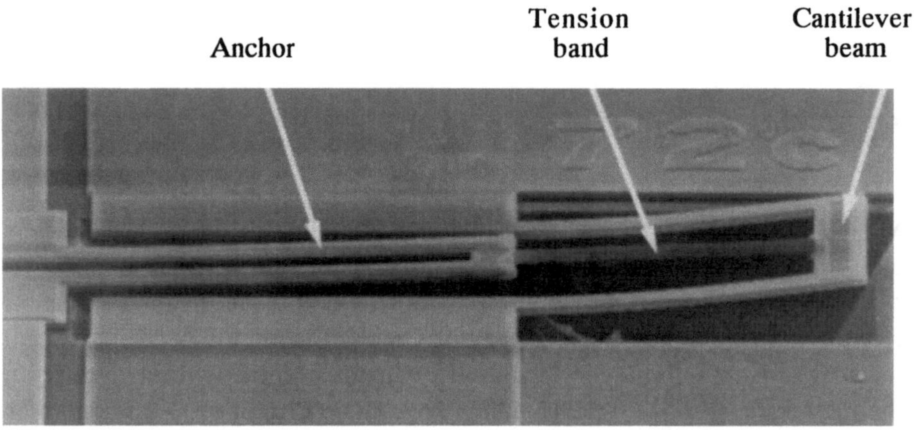

Fig. 5.63: A scanning electron microscope photo of a bistable actuator. Courtesy of the Sharp Corp. (Precision Technology Development Center), Nara

The cantilever beam can be bent up or down by the forces created by the tension band causing a "snapping" motion. It can be controlled by alternating the heating-cooling cycles of the beam and that of the anchor structure. Heating is done by applying short 7 mA pulses (voltage of 24 V) to the cantilever beam and 3 mA pulses (voltage of 7.5 V) to the anchor structure.

Figure 5.64 shows the motion sequence of the actuator. There are three currents which can be switched on and off selectively to operate the microactuator. A current I_1 can be applied to the anchor and I_2 and I_3 to the upper and lower polysilicon layers, respectively. In the initial state, no energy is applied and the cantilever beam is pulled up by the tension band. In the second step, the polysilicon layers of the anchor structure are heated by the current I_1. This results in a thermal expansion of the anchor decreasing the force exerted by the tension band and with this the deformation of the cantilever beam. The cantilever beam is further deformed by simultaneously switching

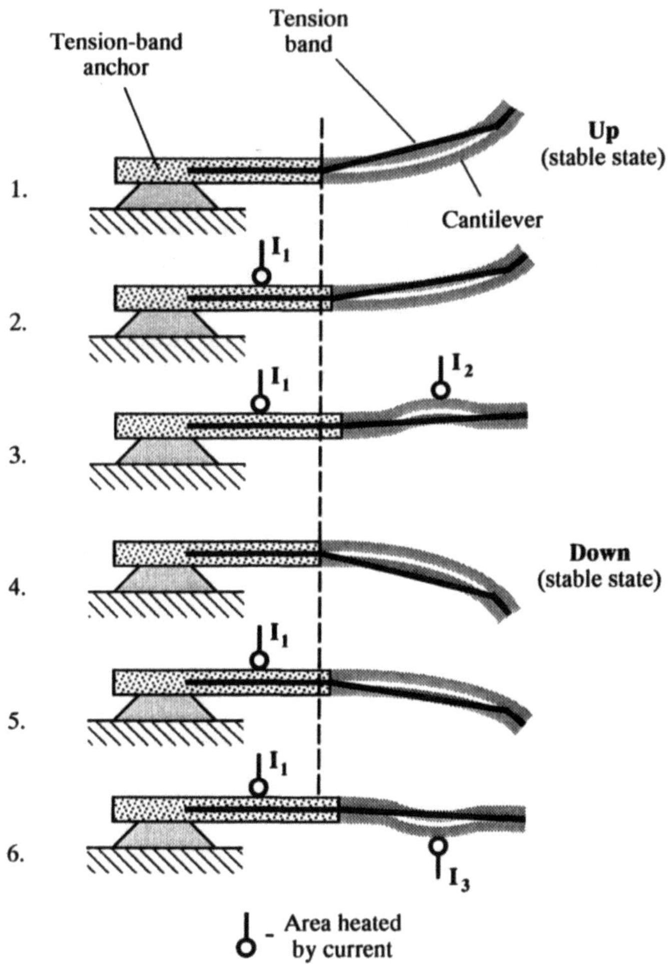

Fig. 5.64: Function principle of the bistable actuator. According to [Mato94]

on the current I_2 which flows through the upper polysilicon layer of the beam (step 3). Here, the end of the beam is bent down and pivots below the actuator base plane. In step 4, both currents are switched off; the material layers cool down and the cantilever beam is pulled into the second stable state by the tension band. The actuator reaches its initial state by the consecutive steps 5, 6 and 1. Here, the current I_3 flows through the lower polysilicon layer.

Ciliary microactuator system for locomotion

The development of a ciliary type locomotion system was described in [Fuji93]. It consists of a multitude of flat bimaterial microactuators forming an actuator array that moves smoothly according to a coordinated motion principle. This actuator array imitates the movement of ciliates to generate locomotion. The principle was adopted from biology; it can also be found in the human respiratory tract. When many of these cilia vibrate synchronously they can move objects or liquids. Since one single actuator can only carry out a simple motion, several actuators have to be combined to perform a more complicated task, such as transporting a plate as shown in Fig. 5.65.

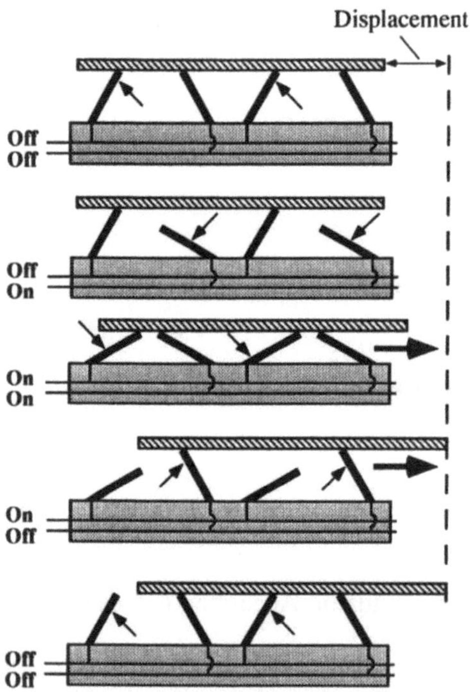

Fig. 5.65: The bimaterial microactuator principle to cause locomotion using the ciliary motion principle. According to [Fuji93]

The individual step of a cilium is very small, but when many cilia are arranged sequentially, relatively long distances can be covered. By distributing a heavy load over many weak actuators, a flexible mechanism is obtained which can easily be expanded and which is insensitive against defects of individual elements. Of particular interest is the low friction of this actuator system. Friction is one of the basic problems in the microworld; often it makes it impossible to design good transmissions and movable joints.

To operate the microactuator cilia, a micro-heating element is embedded between its two polyimide layers, which have different thermal expansion coefficients. When a voltage is applied to the heating element, its temperature increases and the microactuator bends. By alternatively heating neighbouring cilia, an object lying on the actuator system is moved. Each cilium microactuator is 500 μm long, 100 μm wide and 6 μm thick. 512 of these elements were arranged on a 1 cm^2 substrate, Fig. 5.66.

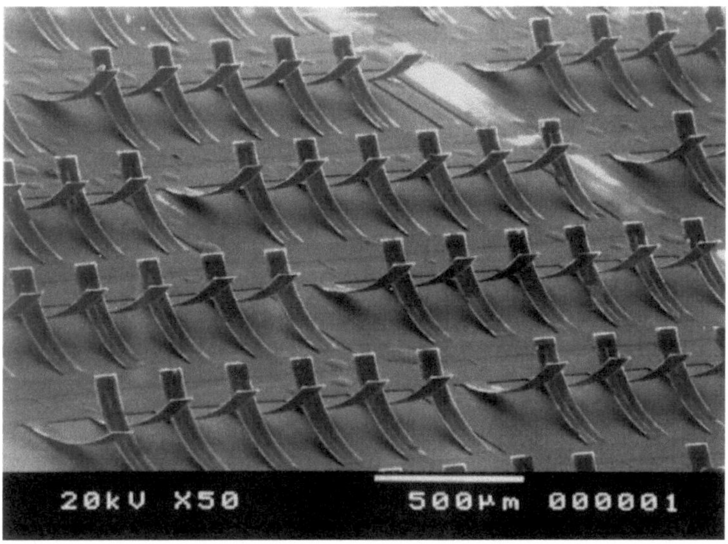

Fig. 5.66: A photograph of a ciliary microactuator system for locomotion. Courtesy of the University of Tokyo (Institute of Industrial Science)

A prototype of this microactuator was able to transport small silicon plates weighing 2.4 mg at a speed of 27 μm/s and an operating frequency of 1 Hz. The supply current was 22.5 mA, the power consumption 33 mW and the temperature 200°C. More complex transport or positioning tasks are thinkable with a corresponding control scheme.

Microvalve

Devices for handling tiny amounts of gas and liquid are very important products of MST. The thermal expansion of material can be used for microvalves, attaining easily relatively large deflections and forces. The development of a thermally actuated microvalve was discussed in [Lisec94], Fig. 5.67.

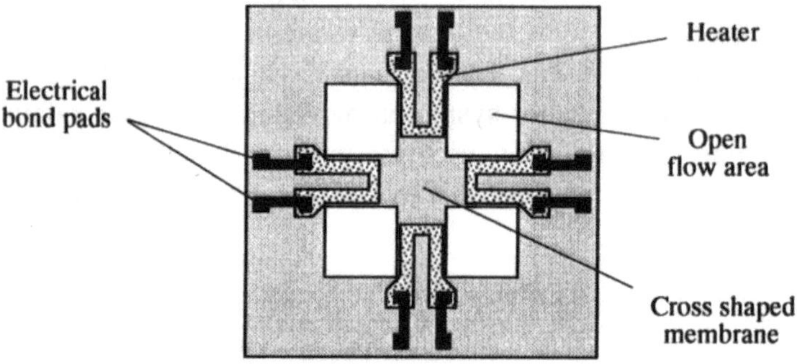

a

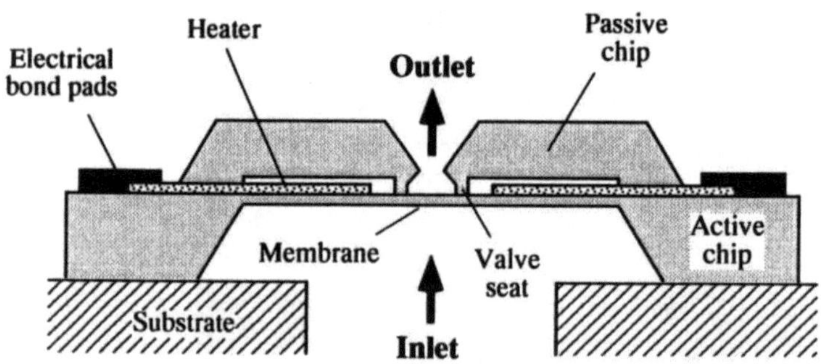

b

Fig. 5.67: Schematic drawing of a microactuated valve
a) top view of the membrane chip; b) cross-section of the valve. According to [Lisec94].

The microvalve consists of two $5 \times 5 \times 0.5$ mm^3 silicon chips, an active and a passive one. The lower, active chip has a cross-like membrane with a dimension of 2.6 mm × 2.6 mm. There are four n^+-doped, 600 μm wide clip-like polysilicon heating strips bonded to the membrane. The upper, passive chip has a valve seat with a 360 × 360 μm² outlet. When a current is applied, the polysilicon heating strip and silicon membrane are heated up, whereas the bridge supports on the substrate remain relatively cold, causing the membrane to buckle downward and to open the valve.

The chips are produced by seven lithographic steps, including wet and plasma etching and galvanic deposition. A membrane thickness of 12 μm and a valve seat height of 10 μm were attained. With a power consumption of 3 W, the microvalve can handle fluids with a pressure of 0.75 bar. The corresponding gas flow is 700 ml/min and the operating time constant is under 20 ms. The valve prototype has proven to be very stable. After 2 million work cycles with a switching frequency of 3.3 Hz (100 ms on, 200 ms off), a flow pressure of 0.5 bar and an electric power of 3 W, the valve showed no signs of wear.

5.7.2 Concepts and Prototypes of Thermopneumatic Actuators

As mentioned earlier, the thermal expansion of gas or the thermal transformation from liquid to gas or vice versa can be used to drive micropumps and valves; thereby short response times and large forces can be produced. Actuators operating with either of these two principles will be introduced below.

Light-operated microactuator

Light-operated microactuators, which are operated by optical energy via glass fibers, are of interest to many applications. They are not subject to electromagnetic disturbances and only have negligible power losses. A light-operated actuator which uses thermal expansion of a fluid was described in [Mizu93]. A cross-section of this microactuator is shown in Figure 5.68, consisting of a silicon microcell which contains a liquid and a light absorber. The dimensions of the device are shown in the figure. The microcell is hermetically sealed with a flexible square membrane; the latter is under mechanical tension and bent inward in its initial state. When laser light is guided into the cell through a glass fiber, the light absorber heats the liquid, the inner cell pressure increases and the membrane is pushed outward, Fig. 5.69.

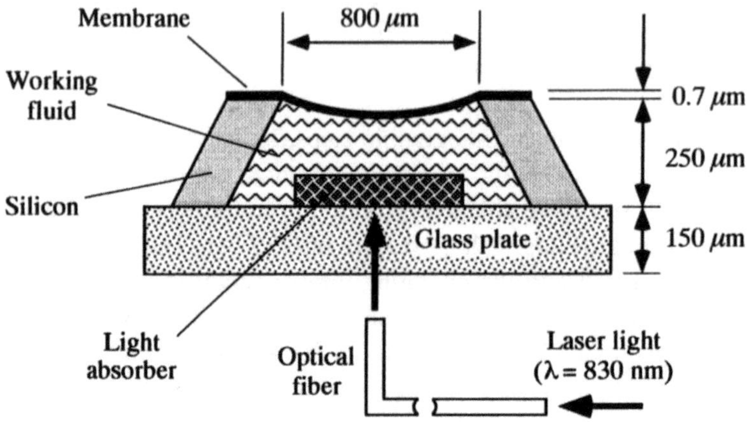

Fig. 5.68: Light-operated microactuator. According to [Mizu93]

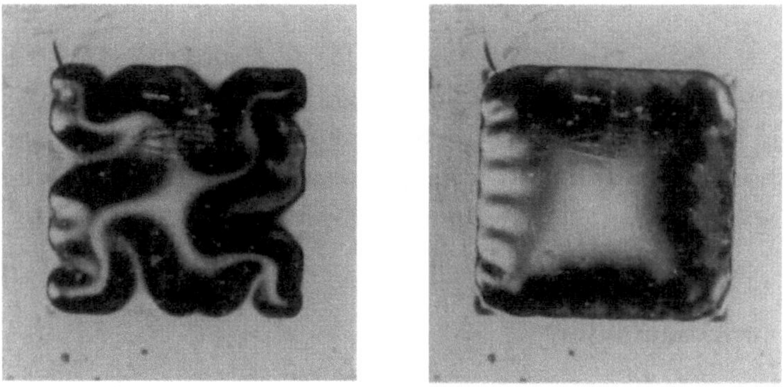

Fig. 5.69: The two states of the micromembrane. Courtesy of the Aisin Cosmos R&D Co. (Electronics Field), Tokyo

The microactuator was manufactured by the bulk micromachining technique and the sputtering technique using (100) silicon. The membrane consisting of a SiO_2 layer and a NiCrSi alloy layer will buckle out by 35 μm when a pressure of 1 kPa is generated. A prototype was subjected to 50,000 work cycles, showing no wear problem. To prove the practicability of this light-operated actuator, researchers have developed and built an original micropump which uses several of these microactuators to realize a worm-like motion, Fig. 5.70.

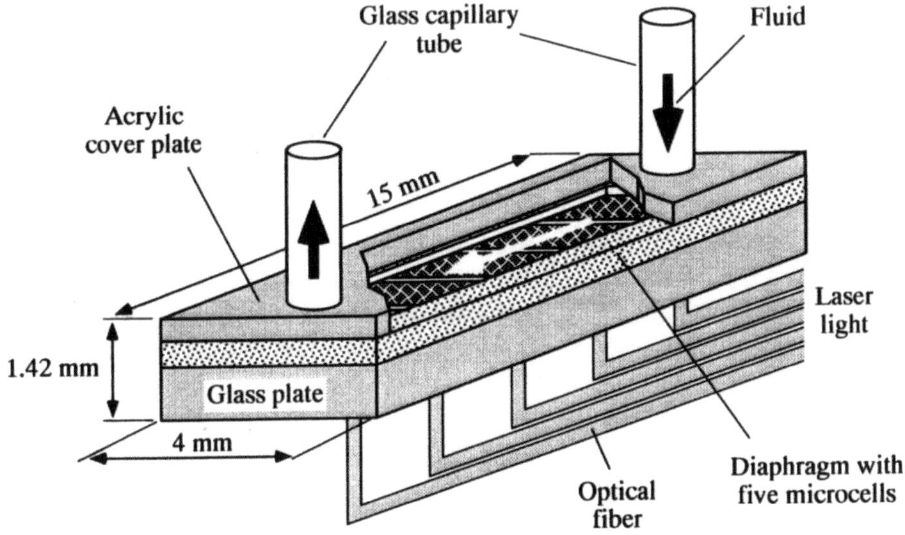

Fig. 5.70: A pump driven by 5 microactuators. According to [Mizu93]

Five microcells were serially arranged in a diaphragm structure and covered by an acrylic plate; the whole device was mounted onto a glass plate. Two glass capillaries with an inner diameter of 0.9 mm are connected to both sides of the channel to transport the liquid to and from the pump. If there is liquid in the inlet capillary, the diaphragm can be moved in a fashion of passing fire buckets by supplying laser light successively to each microcell. Thereby the liquid is pumped through the system. At a pump frequency of 3 Hz, the micropump can deliver a maximum volume of 0.58 μl/min under laboratory conditions using water. In the outlet capillary, the head pressure can reach a maximum of 10 mm of water.

Thermopneumatic micropumps

As discussed in Chapter 2, on-line analyses of chemical substances for environmental protection or medical work are very important MST applications. It is of advantage to equip these measurement systems with micropumps so that only a small amount of the test substance has to be used and that the power consumption is low for selfcontained units; this also reduces cost and analysis time. A thermomechanic micropump which can be used for both gases and liquids was described in [Büst94], Fig. 5.71.

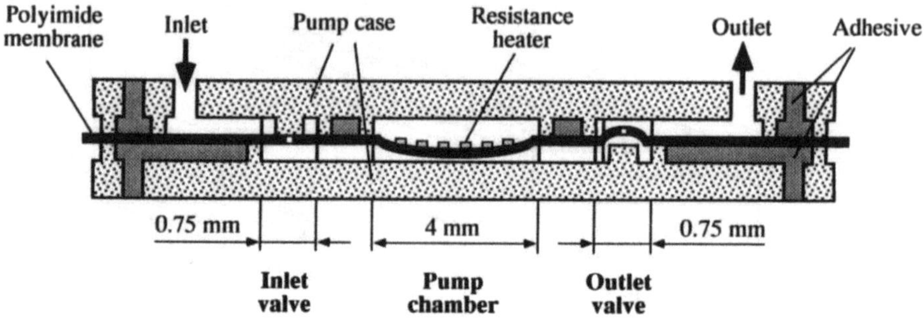

Fig. 5.71: Schematic representation of a micropump. According to [Büst94]

The pump consists of two passive valves and a polyimide membrane between two plastic substrates (made from polysulfone), forming the pump casing. The polyimide layer functions both as a valve diaphragm and as a pump membrane. The latter separates the gas/liquid in the pump chamber from the actuator chamber. The actuator chamber is filled with a fluid (or gas) which can be heated electrically and which causes the membrane to move up and down. A metal wire serves as the heating element and is fixed to the membrane on the side of the actuator chamber. The overall size of the pump is about $9 \times 10 \times 1$ mm^3, Fig. 5.72.

When electric current is applied, the membrane thermally expands and lifts, causing the pressure to increase in the pump chamber. Once the pressure has reached a critical point, the outlet valve is opened and discharges its contents. The pump was made by using the LIGA method. To make the pump casing, a plastic mold was used. The polyimide membrane with the resistance wire was fabricated on a silicon substrate and then integrated into the pump casing. Three membranes were constructed with a thickness of 2.5 μm, 1.5 μm and 1 μm, respectively. Two wires were used, one made of titanium, the other of copper with the respective thicknesses of 2 μm and 0.25 μm. They were bonded and assembled using a special high-precision adhesion technique. The prototype with the 2.5 μm thick membrane and the titanium wire was extensively tested. With a pumping frequency of 30 Hz and an operating current of 100 mA, the pump was able to pump 220 μl/min of air. The micropump is already commercially available.

Fig. 5.72: The size of a micropump compared with that of an ant. Courtesy of the Karlsruhe Research Center, IMT

For medical and other applications, a microdosing system having a dimension of a few mm^2 was developed for handling minute amounts of liquid, Fig. 5.73. Its basic components are two pumps and flow sensors. When the two pumps are combined with a detector and a mixer, a dosing system can be built to be used for chemical analysis.

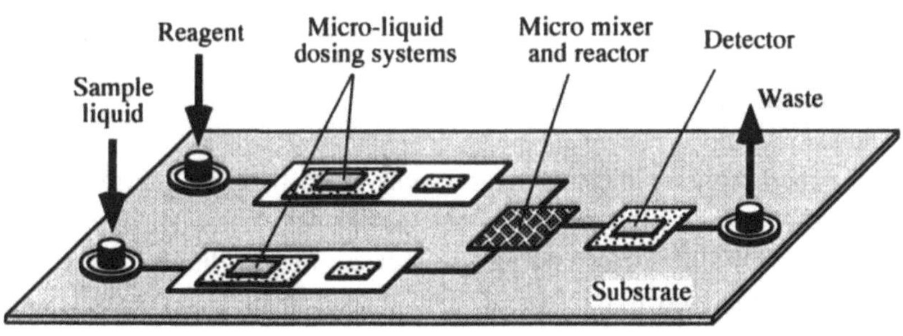

Fig. 5.73: A chemical analysis instrument. According to [Lamm93]

The heart of this system is the thermally actuated micropump, Fig. 5.74. It consists of two passive valves, a pump chamber and a pump actuator, which is provided with an air chamber and an integrated heating element. The heating element can be periodically turned on and off. When it is turned on, the air in the chamber is heated up and the membrane below expands. The volume of the pump chamber decreases, and the liquid is forced out by the resulting pressure increase. When the heating element is turned off, the membrane assumes its original position and the liquid is drawn into the pump chamber.

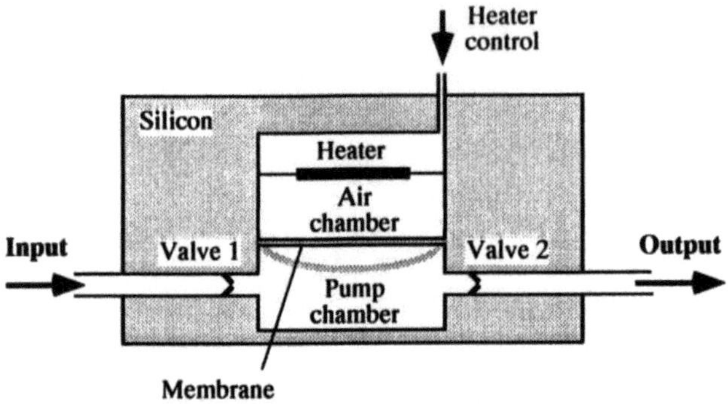

Fig. 5.74: The thermally actuated micropump. According to [Lamm93]

Both the bulk micromachining and surface micromachining techniques were combined to produce this system. The thermal relaxation time of the pump is 0.1 s, the hydraulic relaxation time 1 s. The power consumption of the pump is about 5 W; the maximum pump volume 0.9 μl. At low frequencies (about 3–5 Hz), a maximum pump volume of 50 μl/min can be attained.

Thermopneumatic microvalves

A concept of a thermopneumatic microvalve was described in [Zdeb94], Fig. 5.75. The key component of this normally open microvalve is a silicon diaphragm made of a (100) silicon disc using the wet etching technique. It is enclosed by two preprocessed pyrex plates. The direction of flow is indicated by arrows in the figure. The upper disc and the membrane form a closed chamber which is filled with liquid and is then hermetically sealed. Several resistors are embedded on the pyrex disc and function as heating elements

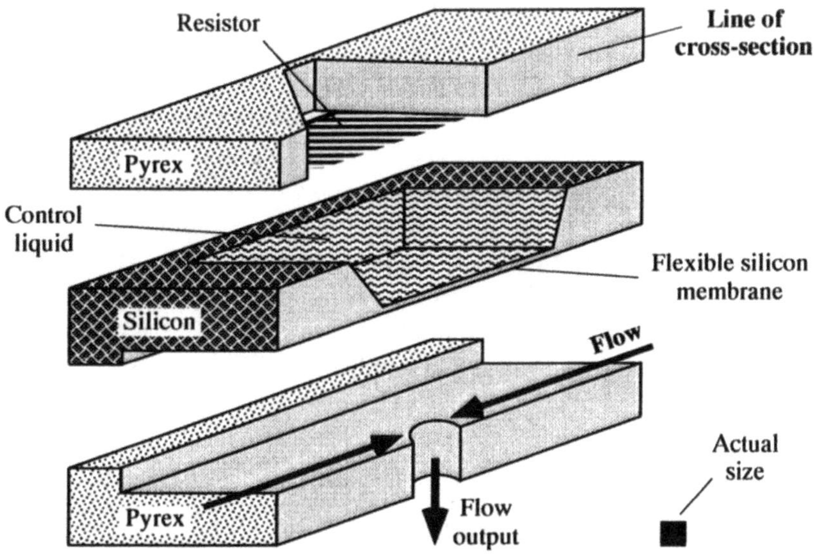

Fig. 5.75: A design of the normally open thermally actuated microvalve. According to [Zdeb94]

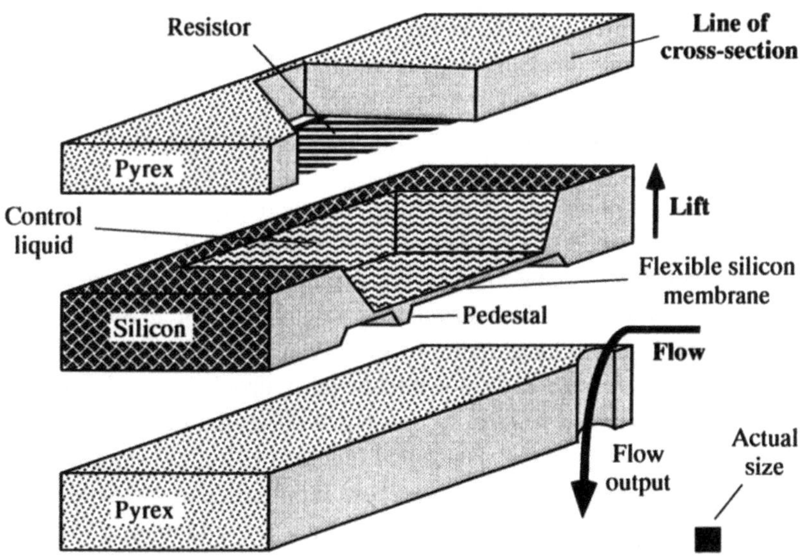

Fig. 5.76: A design of the normally closed thermally actuated microvalve. According to [Zdeb94]

when supplied with current. In the figure, the upper pyrex plate is cut open for visualization purposes. When the resistors are activated, the liquid in the closed chamber expands and the diaphragm bulges out to decrease or stop the flow in the channel by covering the outlet in the lower pyrex plate.

A design of the normally closed microvalve is shown in Figure 5.76. Here, a convex diaphragm is placed in the flow channel: when the liquid in the valve chamber is heated, the right membrane support (pedestal) lifts off the valve seat and opens the outlet. The bulk micromachining technique was used to make both prototypes. The membrane thickness is 40 µm. The actual size of both the microvalves is shown in the right hand corner of the figures.

5.8 Electrorheological Microactuators

The electrorheological effect was discovered in the forties. Electrorheological liquids change their flow properties under the influence of an electric field. This effect can be used to operate mechanical clutches, vibration absorbers, valves, etc. The first electrorheological liquids were used in 1947 for couplings [Block92]. In the meantime, the improvement of the liquid properties has led to many new applications and research activities. The applicability of these liquids for MST systems is also being intensively investigated.

The most important property of an electrorheological liquid is that it has different flow behaviors under the influence of an electric field. Within a few milliseconds an electric field can solidify the liquid to a plastic body by increasing its dynamic viscosity. This process is reversible: when the electric field is removed, the liquid returns to its original viscosity. The liquids used today are made from suspensions of solid non-metal hydrophilic particles, e.g. silicic acid anhydrides or metal oxides, in non-conductive oils like transformer oil or paraffin. In [Bayer94] newly developed electrorheological liquids consisting of polymer particles and silicone oils were reported. The solid particles have dimensions ranging from 1 to 100 µm and they are mixed in a ratio of 15% to 50% with the liquid, depending on the application [Rech93]. A scanning electron microscope photo showing a typical particle distribution in an electrorheological liquid is shown in Figure 5.77 [Wolff94].

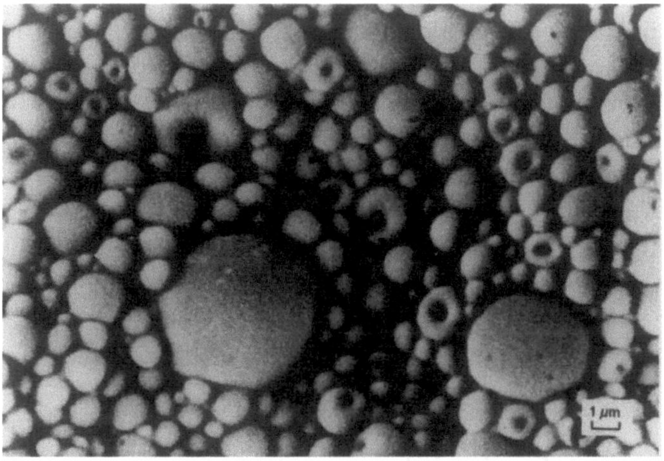

Fig. 5.77: A scanning electron microscope photo of solid particles in a electrorheological liquid. Courtesy of the Rheinisch-Westfälische Technische Hochschule Aachen (Institut für fluidtechnische Antriebe und Steuerungen) and the Bayer AG, Leverkusen

The characteristic increase of viscosity of electrorheological liquids is caused by the polarization of the solid particles when an electric field is applied. The greater the polarization, the stronger the liquid is with regard to a shearing action. The suspended particles form chains along the field lines, Fig. 5.78.

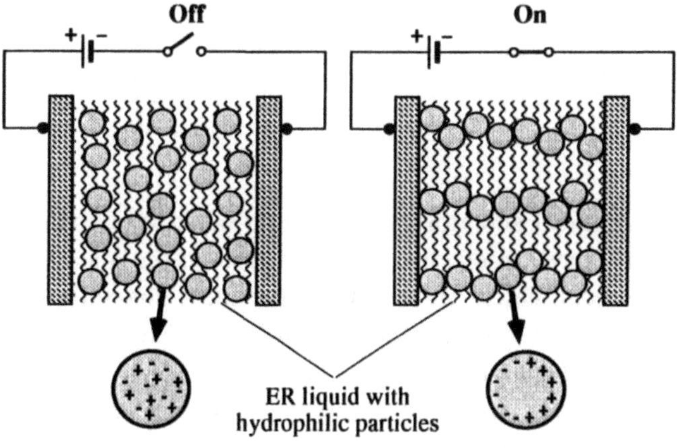

Fig. 5.78: Simplified model of the electrorheological effect

Thereby, the particle interactions create an increase in flow resistance. The rheogram in Figure 5.79 shows the transition of a Newtonian liquid to a Bingham body under the influence of an external field; here: τ_0 – the static yield stress, $\Delta\tau$ – the field induced stress enhancement. The effect of applying an electric field to the suspension is very characteristic. If the liquid is not exposed to an electric field, it behaves as an ordinary liquid, i.e. the shear rate γ increases linearly with the yield stresses τ. When an electric field is applied to the liquid, it solidifies. It returns to the liquid state when the electric field is turned off or when the static yield stress τ_0 is overcome.

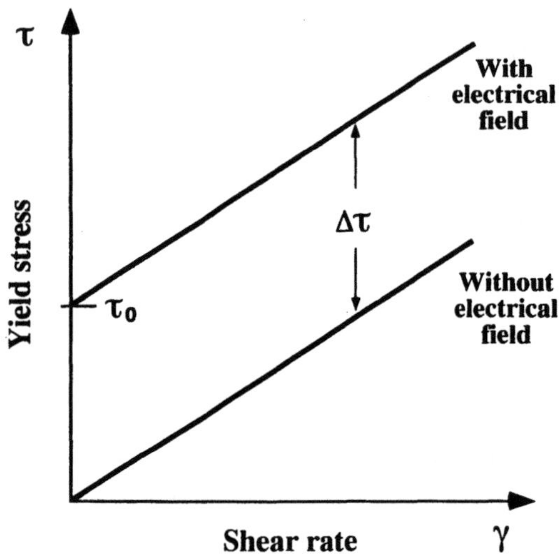

Fig. 5.79: Typical rheogram of an electrorheological liquid. According to [Block92]

When designing electrorheological actuators, it should be taken into account that the used liquids not only have an electrorheological effect, but also have a multitude of secondary effects which can influence the ideal characteristic curve in different ways. This leads to unexpected non-linear actuator behavior and must be considered in the control design. Due to this effect, it is usually impossible to make a general statement about the behavior of electrorheological liquids. There is also no general formula about the composition of such liquids due to the lack of knowledge about them. The liquid properties can be changed by the amount of solid material used. An electrorheological liquid becomes more viscous with increasing particle concentrations; this also holds true if no electric field is present. But in most applications, viscosity itself is not what is important, but the viscosity difference between the two states of the liquid. Therefore, it is often not desirable to have a large amount of solid particles in the liquid; usually the amount is kept at about 40% by volume. The switching times that can be reached are in the range of a few milliseconds with an applied voltage in the kilovolt range.

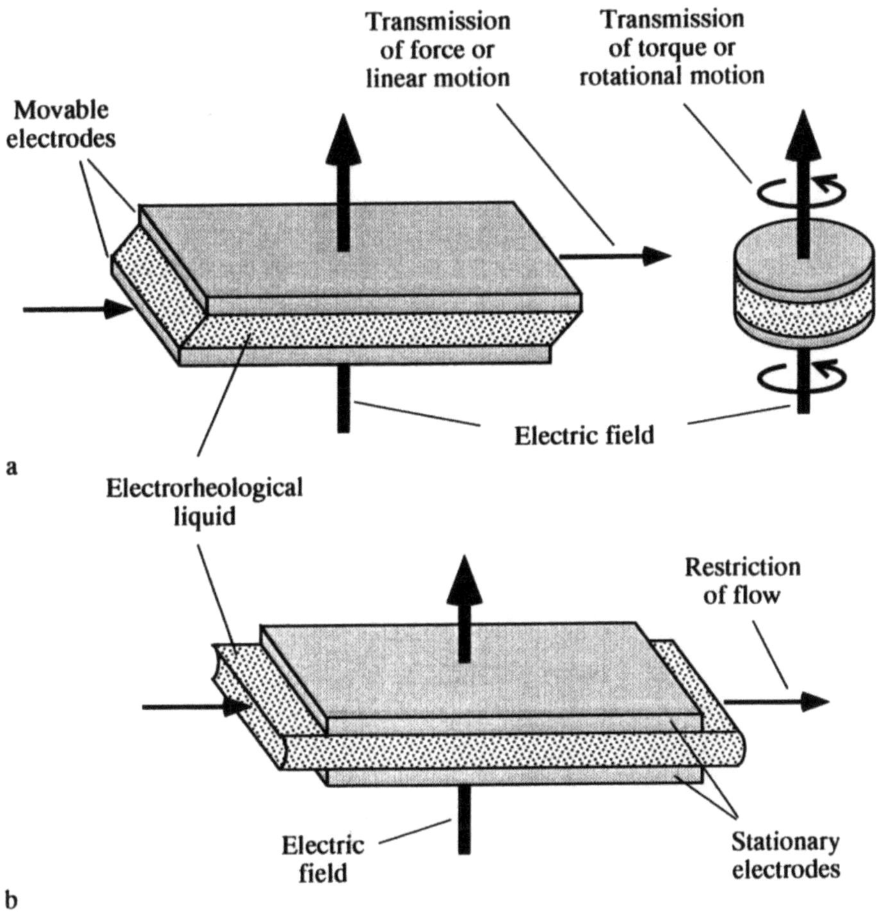

Fig. 5.80: Fundamental functions of electrorheological actuators
a) using the shear force principle to obtain transmission functions; b) using the flow restriction principle to obtain valve functions. According to [Jano92]

The great advantage of electrorheological actuators is their simple design compared with other actuators. No additional mechanical parts are necessary, such as a flow restricter in a valve or a complex gear mechanism in a clutch. Electrorheological drives operate either according to the shear force principle or flow restriction principle, Fig. 5.80. In both cases an electrorheological liquid is placed between two or more electrodes. With the shear force principle, oppositely charged electrodes are moved towards each other (force or torque transmission principle). Rotational movements may also be transmitted, controlling the output torque by an electric field. With the flow restriction principle, the applied electric field influences the flow resistance of the electrorheological liquid which moves between fixed electrodes (valve principle). In some cases a combination of both principles can be applied.

The shear force principle of electrorheological actuation was used by [Hoss92] for designing and producing a rotational clutch, Fig. 5.81. It was experimented that the device must be relatively long in order to create a large contact area between the liquid and the rotor, and to avoid the effect of inertia particular at high rotational speed. Compared to electromagnetically operated couplings, electrorheological couplings have a time constant which is lower by a factor of 10.

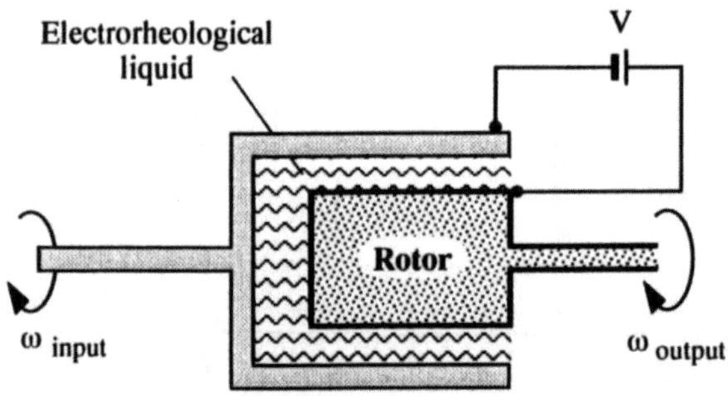

Fig. 5.81: An electrorheological coupling system. According to [Hoss92]

Since electrorheological actuators are simple and have no complicated and expensive mechanical parts, they are easy to control and relatively cheap. A well known design of an electrorheological actuator, the so-called 4-valve bridge, is shown in Figure 5.82. It is based on the flow restriction principle.

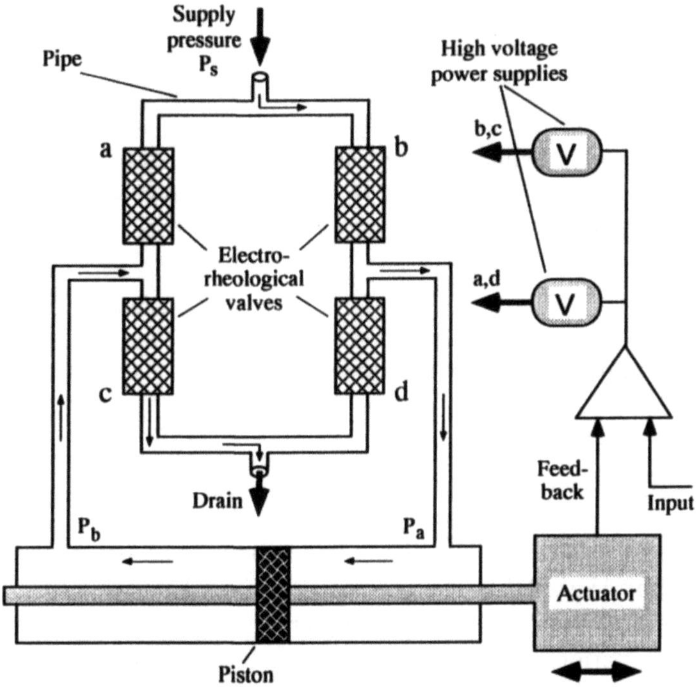

Fig. 5.82: Electrorheological actuator using the 4-valve bridge concept. According to [Broo92]

The bridge is balanced by the application of the same voltage to all four valves; the pressure drop is the same across each valve and the pressure is in equilibrium on both sides of the piston ($P_a = P_b$). If the voltage of the valves "a" and "d" is increased and decreased for the valves "b" and "c", the pressure is out of equilibrium and the liquid flows from the supply through the open valves "b" and "c". In this case, $P_a > P_b$ and the piston actuator moves to the left. The displacement depends on the active surface area and the piston pressure. Electrorheological actuators can also be realized in a 2-valve bridge arrangement, which works in a similar fashion. In both arrangements, the valves may be directly connected to the piston or coupled to it by conventional hydraulic means. The most important control parameters are the position of the piston and its speed. Since the liquid can change its state within a few milliseconds, the actuator system can be operated with frequencies of 200 Hz to 300 Hz [Broo92].

Electrorheological microactuators are not yet on the market as standard products. In order to make them more useful for microsystems, the electrorheological effect and the liquid's stability must be improved. Other important parameters to be investigated are higher ranges of operating temperature, higher shearing forces, shorter switching times and a reduction of power consumption. New types of magnetorheological liquids have been discussed in [Carl94], but there are no assessable results yet.

5.9 Hydraulic and Pneumatic Microactuators

Flexible rubber microactuators

A flexible microactuator to be used by miniaturized robots was reported in [Suzu91], [Suzu91a] and [Suzu91b]. The actuator is driven by hydraulic or pneumatic pressure, can be bent in every direction and is designed for use as robot hands or legs for various applications. The structure of this device is shown in Figure 5.83. It is made of rubber reinforced with nylon fibers and has three autonomous actuator chambers. The internal pressure in every chamber can be controlled individually by flexible hoses and valves leading to them. The device can be expanded along its longitudinal axis when the pressure is increased equally in all three chambers. If the pressure is only increased in one chamber, the device bends in the opposite direction. The pro-

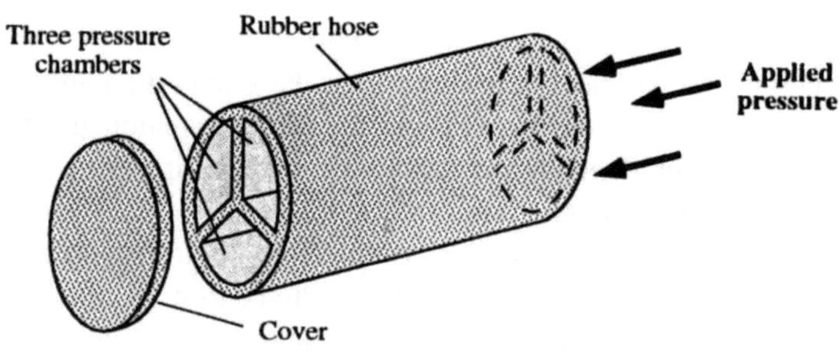

Fig. 5.83: Principle of the pneumatically-driven flexible microactuator. According to [Suzu91]

totypes developed so far have a diameter ranging from 1 to 20 mm. Several robot hands and walking machines were produced with them, Fig. 5.84.

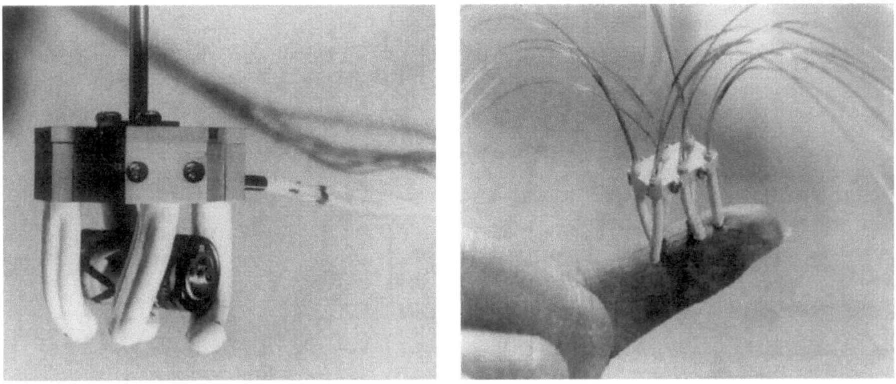

Fig. 5.84: A robot hand (left) and a walking machine (right), both based on the flexible microactuator design. Courtesy of the Toshiba Corp. (Research and Development Center), Kawasaki

The robot hand consists of 4 fingers, each having a diameter of 12 mm. The walking robot has 6 legs, each with a diameter of 2 mm and a length of 12 mm. The actuator design can easily be miniaturized due to its simple structure. Such actuators are cheap, quiet and can carry out precise manipulations with a suitable control scheme. Research is being done about other materials and specific production processes for making this design [Suzu94]. Because of the features discussed above, the principle of the actuator is of high interest for microrobot applications. Concerning autonomous robots the main difficulty is the miniaturizing of the energy supply. The many connecting tubes necessary for remotely controlling a microrobot are also a problem.

Pneumatic multiactuator microsystem

The principle of a pneumatic micropositioning system was presented in [Fuji93]. It allows flat objects to be lifted, transported and positioned by air currents coming from several microjets, Fig. 5.85.

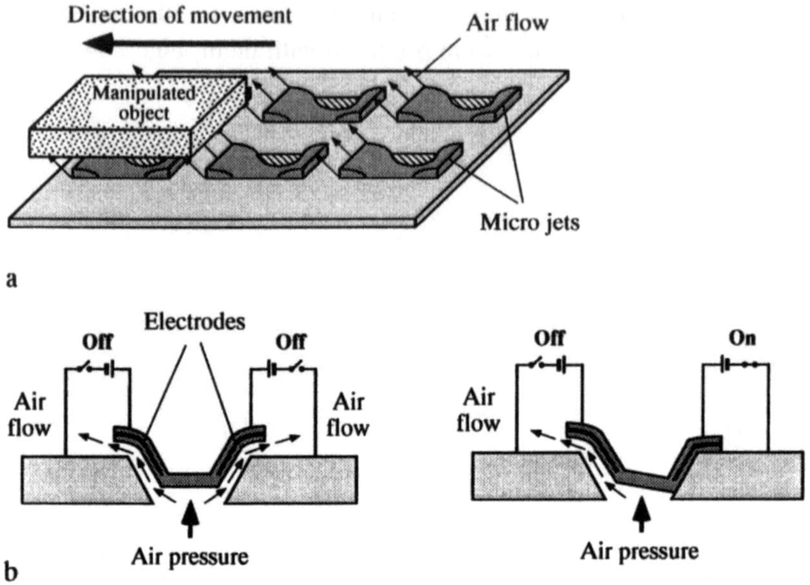

Fig. 5.85: A multiactuator microsystem for object positioning and transportation: a) detail of the system; b) functional principle. According to [Fuji93]

The device has several air jets, each of which is equipped with two air channels. The air currents can be guided in either one or both directions. Each air channel is covered by a soft polyimide layer, into which two electrodes are embedded. If a voltage is applied to one of the electrodes, the corresponding air channel closes, Fig. 5.85b. All jets can be individually controlled this way. A prototype system was able to accurately move a 1 mm^2, 300 µm thick silicon plate; the operating voltage was 90 V and the air pressure 2 kPa. The main body of the device was a silicon wafer into which a 7 × 9 matrix of 63 rectangular air channels (100 µm × 200 µm) were etched anisotropically. The size of the entire microsystem is 2 mm × 3 mm.

Hydraulic piston microactuator

An interesting hydraulic microactuator system was presented in [Ruth95], [Wall95]. The piston actuator and its integrated calibration system is depicted in Figure 5.86. The actuator chamber with its inlet for its operating fluid, e.g.

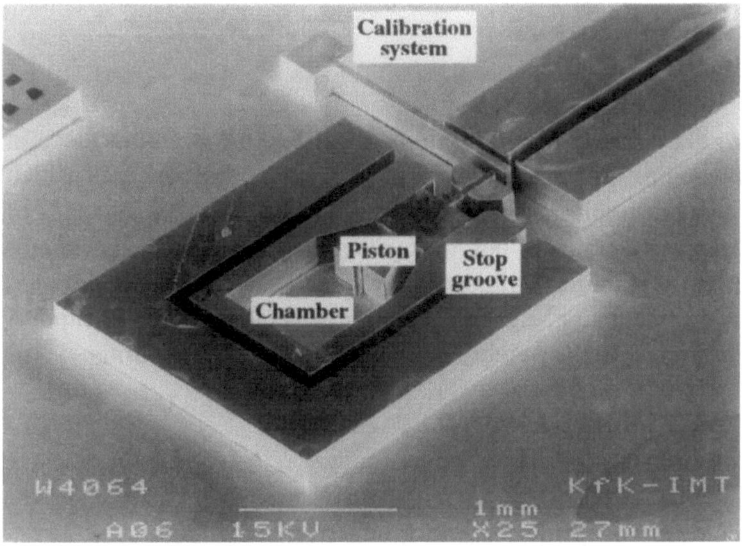

Fig. 5.86: A hydraulic piston microactuator. Courtesy of the Karlsruhe Research Center, IMT

water, was made by the LIGA process. The unit contains a force-transmitting piston which can be moved along the side walls of the chamber by a fluid. The device is covered by a glass plate (not shown in the figure). A stop groove is added to absorb excessive adhesive which may ooze out when the glass cover plate is being fixed; this is necessary to prevent the piston from sticking to the walls of the chamber. The actuator made from copper has dimensions of 2 mm × 2 mm × 0.2 mm, the gap between the piston and the chamber walls is only 1–3 μm. The friction of the piston is reduced by the lubrication effect of the driving fluid.

This microactuator will be integrated in an active heart catheter (Chapter 2.2) to drive a linear cutting tool for removing deposits in blood vessels mechanically. In this case, the device consists of two symmetric actuators (of the type described above) having an opposed piston between them. The cutting tool is fastened to the piston which is driven by an oscillating fluid jet: a fluidic micro switch (Figure 4.36) alternately directs the fluid into one of the two actuator chambers with a frequency of several 100 Hz. The attainable stroke of the piston is several 100 μm; a force of 5 mN can be reached with a fluid pressure of 0.54 bar.

5.10 Chemical Microactuators

Chemical actuators are based on different chemical processes taking place in fluid or gaseous media. E.g. many chemical reactions produce gases which can be used to create a high pressure in a chamber [Jend93a]. Substances of interest for chemical actuators are polymer gels. Various polymer gel microactuators are currently being investigated; they use the swelling of a polymer gel as the principle of actuation, e.g. for pumping of small amounts of liquids. Such actuators are robust, light-weight and can take on different forms.

Polymer micropump

The uni-directional microcapsule polymer pump shown in Figure 5.87 was introduced in [Hatt91] and [Hatt92]. It can be used as a medicine dosing system implanted in a patient. The pump cylinder has a semipermeable membrane on its inlet side, which only allows a substance to flow in the direction indicated; a one-way valve is located on the outlet side. The pump is separated by a thin film and filled with a pharmaceutical solution which is on the left side of the chamber. On the right side of the chamber a highly concentrated water-absorbing polyacrylamide gel is located.

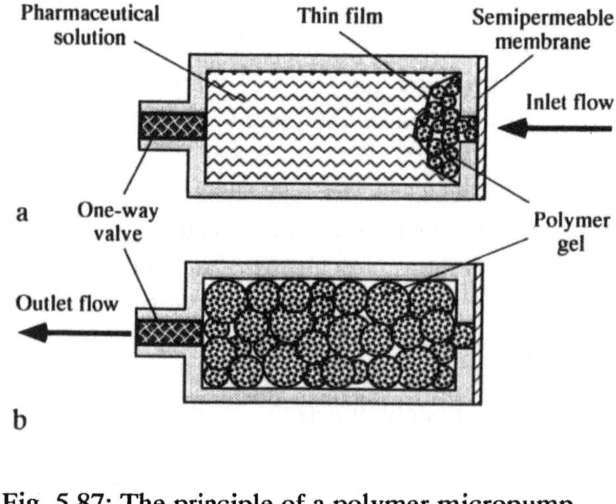

Fig. 5.87: The principle of a polymer micropump
a) initial state; b) final state. According to [Hatt92]

The work cycle of the pump is as follows: the osmotic pressure difference across the membrane takes water from the ambient solution, i.e. from the patient's blood, and drives it through the membrane into the right side of the pump chamber, causing the polymer gel to swell. The pharmaceutical in the microcapsule is pushed out due to the volume increase of the polymer gel. When the pump space is completely occupied by the gel, the medicine is fully injected and the cycle is completed. The duration of the cycle depends on the concentration difference between the polymer gel solution and the ambient solution. A pump prototype had a diameter of 5 mm, was 9.8 mm long and had a weight of 0.7 g. The semipermeable membrane was made of a 70 μm thick cellulose sheet.

Polymer microcatheter

For the microcatheter presented in Figure 5.60, alternative actuation principles are being investigated. A prototype of a microcatheter was developed with two integrated guide wires and a liquid channel [Guo95]. Through the channel physiologic solutions or contrast media can be transported. Attached to each guide wire is a polymer film actuator which enables the catheter to be bent, Fig. 5.88.

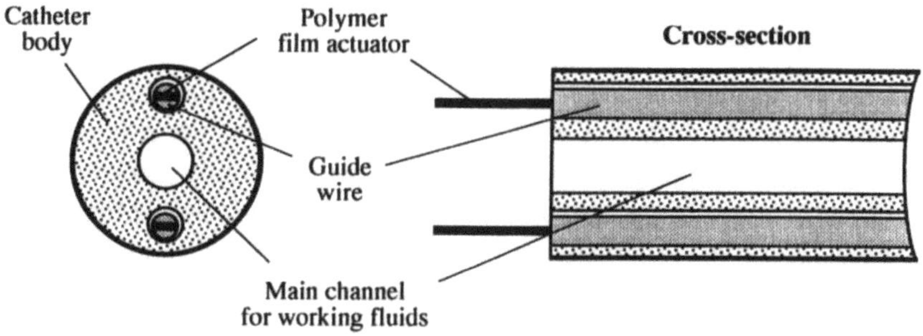

Fig. 5.88: Structure of the microcatheter. According to [Guo95]

The guide wire consists of two electric conductors covered by a wire sheath and the actual actuator, an ionic conducting polymer film. The end of this film is embedded between two flat platinum electrodes, Fig. 5.89. In order to bend the tip of the catheter, a defined voltage is applied to both electrodes of the desired actuator. This voltage causes the polymer gel to expand at the cathode side, which results in a movement of the film actuator towards the anode side. The actuator tip bends in a circular manner, making it possible to control the radius of this bend (and with this the position of the catheter) with the voltage applied to the electrodes. The special features of these actuators, compared to other actuation principles like SMA or bimorphic materials are quick response and low driving voltage, about 1.5 V. The latter is advantageous in wet surroundings, e.g. in blood, because harmful electrolytic disintegration of the surrounding fluid can be avoided.

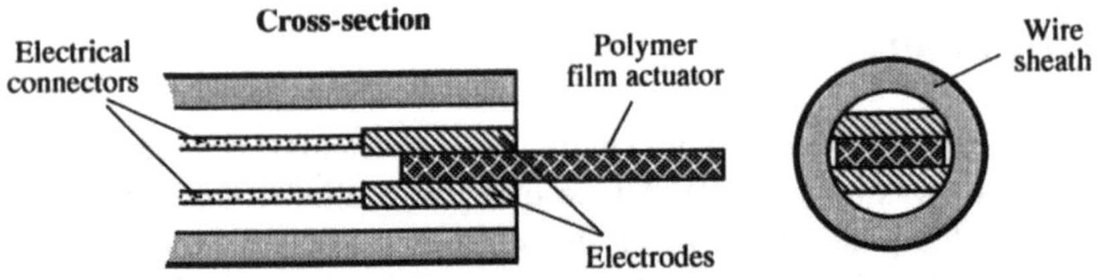

Fig. 5.89: Design of the polymer actuator. According to [Guo95]

Prototypes of the catheter had a diameter from 1 to 2 mm; the catheter tips were 20 mm long. The catheter system was tested with a blood vessel simulator by using a physiological NaCl solution. The actual movement of the catheter was consistent with that of the computer model of the microcatheter. By using a good strategy control and rotating the whole catheter manually, it could be moved into body cavities or enter selected vessel branches. With the help of this microcatheter, it is possible to perform diagnostic and surgical tasks in very thin vessels, e.g. in the human brain.

This section must have vividly shown the application potential of microactuators. Many other interesting devices had been developed but were not introduced here. There is also much work going on in the development of microsensors and system integration. The interested reader can refer to the literature to broaden his knowledge.

6 Microsensors: Principles and Examples

6.1 Introduction

Sensors are increasingly being used for many technical applications. A sensor is a vital organ of an artificial system, forming the interface between the controller and the environment. Sensors can smell, taste, see and feel by measuring mechanical, biochemical, thermal, magnetic and radiation parameters. They are usually classified according to the signals that are measured. An easy to understand sensor classification can be seen in Table 6.1.

Table 6.1: Classification of sensors. According to [Gard94]

Form of signal	Measurands
Thermal	Temperature, heat, heat flow, entropy, heat capacity etc.
Radiation	Gamma rays, X-rays, ultra-violet, visible and infrared light, micro-waves, radio waves etc.
Mechanical	Displacement, velocity, acceleration, force, pressure, mass flow, acoustic wavelength and amplitude etc.
Magnetic	Magnetic field, flux, magnetic moment, magnetisation, magnetic permeability etc.
Chemical	Humidity, pH level and ions, gas concentration, toxic and flammable materials, concentration of vapours and odours, pollutants etc.
Biological	Sugars, proteins, hormones, antigens etc.

Presently, there is a trend to make sensors smaller and smaller. Initial stages show an evolution from a single sensor element to an intelligent sensor system with extremely small dimensions by MST. The so-called smart (or integrated) sensing devices can be developed by integrating sensor components with those for signal processing. This integration also decreases the noise that is often created by the transmission of signals to an external data processing unit. Researchers are trying to combine different miniaturized sensors by using the new production methods which were introduced in Chapter 4; the trend is to integrate the whole sensor system on one chip. Thus it will be possible to measure and evaluate for a certain task all interesting parameters at one place and at one time. An important step toward the further development of microsensors is the conception and design of intelligent electronic signal processors. This will lead to advanced distributed sensor systems in which noisy sensor signals, resulting from cross-talk or insufficient selectivity, can be successfully evaluated. The signal processing system of humans is very advanced; sensor signals are received over the nervous system and transferred to the brain which reliably evaluates them by a natural "parallel computing system". Scientists are trying to copy the system by innovative information processing approaches based on neural networks and fuzzy logic; both technologies look promising for this application.

Today, numerous microsensors are emerging; they have a dynamic development and an excellent market prognosis. The yearly increase of the worldwide market volume is now about 20% [Hein93]. There are many competitors, however, and capturing an attractive niche of the market is only possible by a full commitment to MST. Microsensors must have high reliability, low weight and volume, and low mass-production cost. Present and future microsensor applications have their greatest potential in the automobile industry, environmental protection, production and process technology and military sector. The demands put on them are high accuracy, high safety for man and material and the ability to deliver reliable results in real-time.

Microsensors have become an indispensable tool for medicial applications. E.g. the continuous measurement of various physical/chemical parameters of the blood like temperature, pressure, pH value, flow volume and respiratory values like oxygen, carbon dioxide or anesthetics are especially important [Gamb91]. Miniaturized sensors to determine the radioactivity of waste material have also gained in importance. Presently, to measure simultaneously different parameters of a patient, usually several catheters must be inserted into the body; this increases the risk of infection. Therefore, integrated sensor chips are desirable, which can measure many different values at the same time. The incorporation of sensors has the advantage of measuring *in situ* and of allowing signals to be recorded continuously. Sensors must be biocompatible, have an excellent signal stability and be of small size. They should be able to function without an external power supply and must not be toxic. So far, with these stringent requirements only a few microsensors have been developed, especially for pressure and temperature measurement. This can be relatively easy accomplished *in situ*. Measuring biochemical parameters *in situ* is much more complicated since the required chemically sensitive layers of the sensor must be biocompatible and stable over a long period of time.

Another important microsensor application is environmental protection. Chemical and biosensors are primarily used here for quantitative determination of the concentration of substances. Such sensors are especially needed for measuring the solid contents of fluids and gases, the concentration of carbon monoxide, nitrogen oxides, oxygen, heavy metals etc. Reliable carbon dioxide sensors have also become very important to study the effect of CO_2 on our climate. A very detailed compilation of chemical quantities important for environmental protection can be found in [Ange92].

Sensors are frequently exposed to agressive environments like sewage or garbage. Their operation can result in high maintenance costs, short lifetimes and signal transmission errors. Such applications usually require a multitude of sensors being distributed over a large area. This is only feasible with inexpensive, highly reliable miniaturized sensors which are insensitive to occasional failures. For the use in adverse environments, measurement systems realized on one chip would preferably process information locally; making signal transmission simple or even unnecessary.

Chemical and biosensors are also important in food processing, where all different kinds of contaminants or impurities must be observed constantly. In process engineering, thorough control of technological processes depends on the availability of miniaturized sensors. Here control parameters have to be measured often at many places with different types of sensors.

The automobile industry shows a high interest in the further development of microsensors. They need controllers in a car which can take over many control functions the driver is currently doing. By using a variety of intelligent sensors and signal processors it is possible to optimize many of the driving chores. The main developments are in the areas of increased safety, reduction of fuel consumption and pollution. For this purpose, numerous physical and chemical parameters must be monitored within the automobile; for this various miniaturized sensors are essential. The spectrum of microsensors needed includes distance, acceleration, pressure, vibration, temperature and chemical sensors. Until now, most of these sensors have only been available in luxury cars. However, they are of interest to all cars and the increasing mass-production of them will lead to lower sensor cost.

In robotics, especially in microrobotics, microsensors show great promise, since they are an important component of an intelligent system. Depending on the application, microrobots are equipped with distance, acceleration, force, torque, tactile, pressure and temperature sensors. Several prototypes of such microsensors will be introduced later. The further development of microsensors to nanosensors will open up many applications or robots that are unimaginable today. Readers who want to know more about future applications of nanosensors are referred to the book [Drex91].

The availability of many materials is a basic requirement of the microsensor technology. Besides the well-established semiconductor materials there must be available special metals, plastics, ceramics and glasses. Various biomaterials, like different enzymes and antibodies, are also being used for biochemical microsensors since these protein compounds are capable of sensing chemical and physical properties. The development of various MST production methods and new types of design principles (Chapter 4) have lead to the conception of small and accurate sensors. There is also a trend to design three-dimensional sensors. Besides the silicon technology, which makes use of the excellent electrical, chemical and mechanical properties of silicon and of silicon dioxide (Section 4.1.1), the LIGA technique has been increasingly used in the manufacture of microsensors. The advantage of the LIGA technique is its capability of mass-producing microsensors, of using molding techniques, and the ability to work with a wide variety of materials.

Microoptical systems make use of measurement principles entirely different from electrical ones. Since electrical cables are no longer necessary, electromagnetic noise is eliminated. The micromanufacturing techniques can be excellently used to produce optical sensors. Many physical and chemical parameters can be transformed into optical sensor signals. They are converted into optical parameters, such as amplitude, phase shift, spectral distribution, frequency or time. Figure 6.1 shows a general structure of optical microsensors and a schematic representation of how they work.

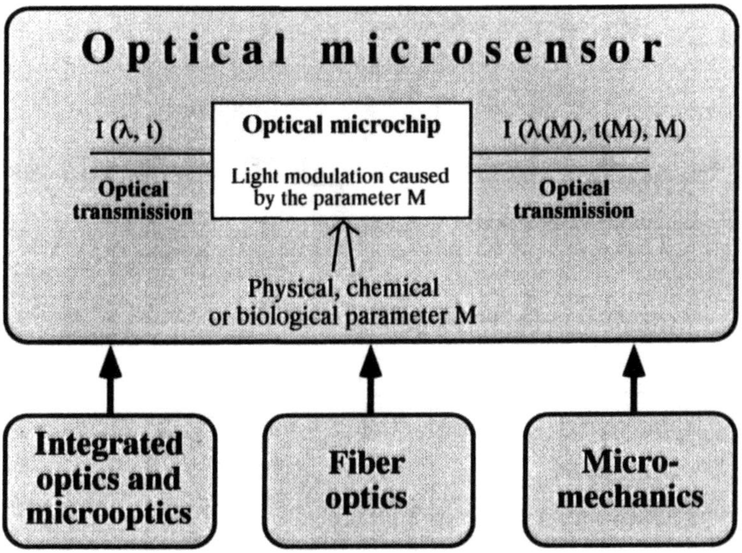

Fig. 6.1: The structure of optical microsensors. According to [Lore93]

Analoguous to integrated electronics, optical data transmission devices like splitters, couplings, grids and frequency mixers, are made by planar silicon techniques and miniaturized. Here, the term integrated optics is used. When light is transmitted through the channel of a chip, the local index of refraction of the inner waveguide relative to its outer layer is selected so that the light waves are totally reflected on the interface of the layers. The optical waveguide principle is depicted in Figure 6.2. Different kinds of waveguides or light transformers can be produced on or in a substrate. In optical sensor systems using the silicon technology, the layer structure SiO_2/SiON/SiO_2 has excellent optical behavior and can be used as a sensor element.

Light is brought into or out of an integrated optical element by fiber optics, usually made of glass (Section 3.1.4). A glass fiber consists of a light-conducting core and an outer mantle or coating. The light is guided along the fiber due to the total reflection within the glass interfaces, Fig. 6.3. Glass fibers can easily be sterilized and therefore are well-suited for medical applications.

As we will see later in this chapter, so-called extrinsic fiber sensors are quite common. Here, the fiber is only used as a transportation medium for light to

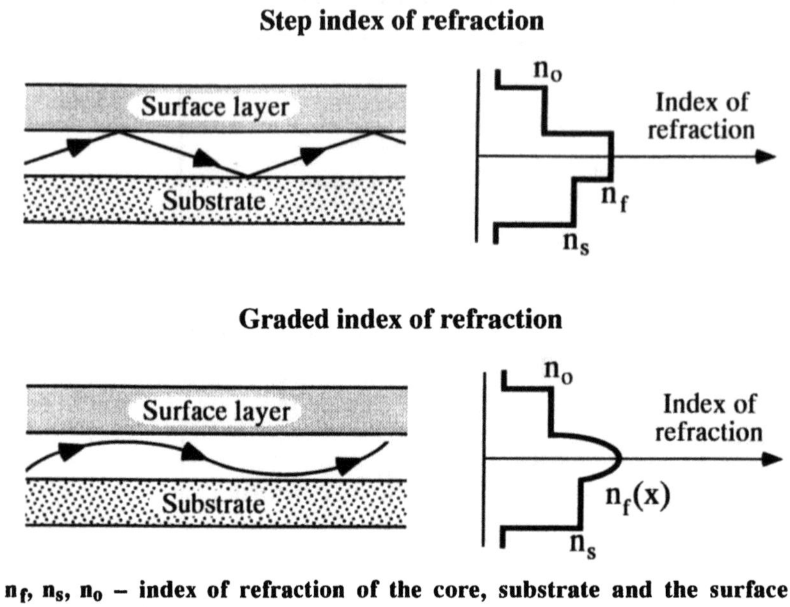

Fig. 6.2: Waveguide in an optical microchip

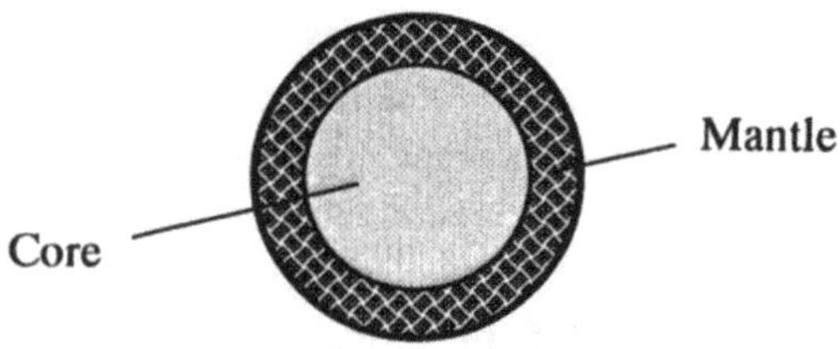

Fig. 6.3: Cross-section of a glass fiber

the actual sensor chip where the light is modulated. However, the various sensor capabilities of fibers are a very important aspect for the design of microsensors using several properties of light waves as an information carrier. Many external effects may influence the intensity, phase, polarization, time dependency, or spectral distribution of a light wave in a glass fiber. The changes of the light parameter are transformed into current or voltage signals that can be used to determine the current value of the quantity to be measured.

The multitude of modulation methods allow many parameters to be measured with intrinsic fiber sensors, such as temperature and pressure, electrical voltage, radiation doses, pH value, angular speed, etc.

Further advantages of glass fiber sensors include the possibility of detecting sensor signals in a difficult environment (i.e. when the medium is chemically aggressive or has a high temperature), and transmitting signals with practically no noise over a long distance. The sensors are highly sensitive, free of electrical potential and can be miniaturized. It is possible to make very accurate measurements by using infrared wavelengths in the range of 600–1500 nm. Later on, when we will be presenting application-specific microsensors, the concept of fiber optical sensors will appear over and over again.

In this chapter, we will be recapitulating many special effects which were explained in the chapter on microactuators. As opposed to actuators, these sensors simply reverse the basic actuation effect. In other words, an external physical effect causes a change in a certain property of the sensor, which can be used to determine the current value of the quantity to be measured.

The design of sensors, production techniques and signal processing components are constantly being improved; that is why the microsensor technology has become the furthest developed activity of MST. Problems with microsensors are identical to those of other MST components. Important are the development of interconnection technologies, the conception of suitable interfaces to the external surrounding (the transition from a micro- to a macrosystem) and the construction of intelligent information processing components, e.g. using fuzzy control or neural network methods. The microsensor technology has gained more than the microactuator technology from conventional techniques. Most measurement principles were taken from already existing macro- or minisensors. Therefore, we will not go into great detail on this subject, but will put more stress on sensor principles relevant for microrobotics.

6.2 Force and Pressure Microsensors

Due to their simple construction and wide applicability, mechanical sensors play the most important part in MST. Pressure microsensors were the first ones developed and used by industry. Miniaturized pressure sensors must be inexpensive and have a high resolution, accuracy, linearity and stability. Presently, silicon-based pressure sensors are most often used; they can easily be integrated with their signal processing electronics on one chip. Their advantages include low production costs, high sensitivity and low hysteresis.

Pressure is most often measured via a thin membrane which deflects when pressure is applied. Either the deflection of the membrane or its change in resonance frequency is measured, both of these values are proportional to the pressure applied. These mechanical changes are transformed into electric signals. Membranes can be manufactured by bulk micromachining of a (100) silicon substrate, whereby the membrane is produced with one of the etch stop techniques. Pressure sensors usually employ capacitive or piezoresistive measuring principles.

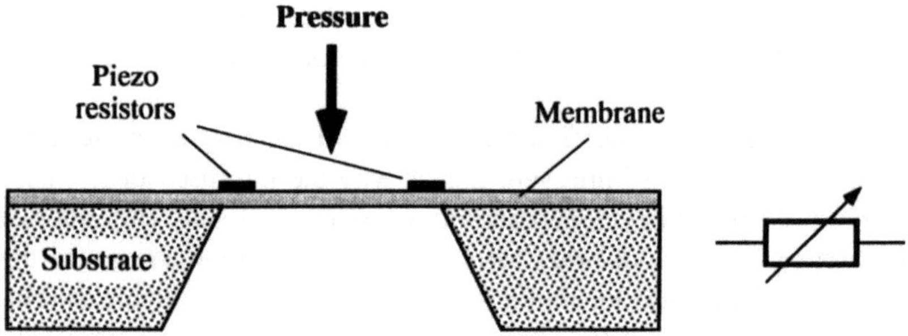

Fig. 6.4: Principle of a piezoresistive pressure sensor. According to [Heub91]

Figure 6.4 shows the design of a piezoresistive pressure sensor. The piezoresistors are integrated in the membrane, they change their resistance proportionally to the applied pressure. The resistance change indicates how far the membrane is deflected and is measured with a Wheatstone bridge. The deflection value is proportional to the pressure.

Capacitive sensors make use of the change of the capacitance between two metal plates. The membrane deflects when pressure is applied, which causes the distance between the two electrodes to be changed. Through this the capacitance increases or decreases. From the amount of membrane deflection the capacitance change is measured and the pressure value can be calculated. A silicon-based capacitive pressure sensor with integrated CMOS components including sensor, transformer, amplifier and temperature compensator was described in [Mehl92].

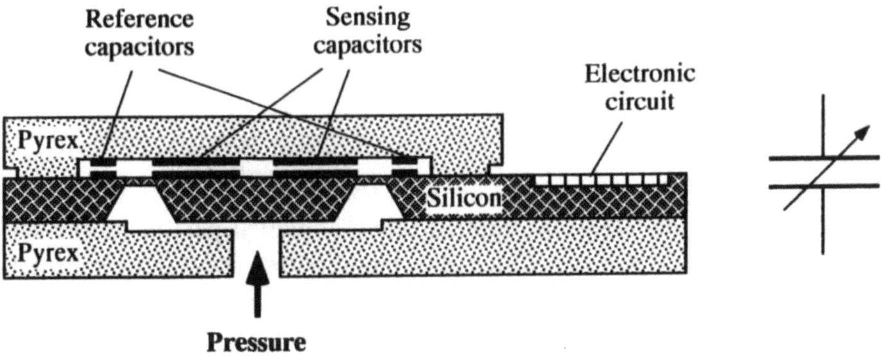

Fig. 6.5: Capacitive pressure sensor made of silicon. According to [Mehl92]

To produce this device, a square silicon substrate was etched anisotropically with a KOH solution. The resulting chip containing the membrane was then sandwitched between two pyrex chips through anodic bonding, Fig. 6.5. The sensor capacitors were fastened to the membrane and the upper pyrex chip, and the reference capacitors were located away from the pressure sensitive area. The sensor chip has a dimension of 8.4 mm × 6.2 mm.

Compared to piezoresistive signal transformers, capacitive pressure sensors have no hysteresis, a better long-term stability and a higher sensitivity. The latter depends mainly on the electrode gap that can be produced. However, the advantages of capacitive pressure sensors go along with higher production costs. As compared to the piezoresistive sensor, three wafers must be independently structured and then precisely combined to the capacitive sensor, which is expensive.

In both of the sensing principles introduced above, the sensor signal is generated by a deflecting membrane or a displaced mass. It is also possible to get a signal from a change of resonance frequency of the membrane caused by the pressure. By applying cyclically thermal energy to the device, the membrane starts to resonate. The main advantage of this measurement principle is that the transmission of the measured value in form of a frequency is practically noiseless and the signals can be digitally processed. In the next sections we will introduce some interesting prototypes and construction concepts of miniaturized pressure sensors. The order in which they are brought does not suggest a rating.

Capacitive pressure sensor

A capacitive sensor is shown in Figure 6.6. The electrodes are made up of a planar comb structure. Here, the applied force is exerted parallel to the sensor surface. In force sensors which use membranes, the force is usually applied perpendicular to the sensor surface. Here, nonlinearity and cross-sensitivity may cause problems. In the device described here, the sensor element mainly consists of two parts: first, a movable elastic structure which transforms a force into a displacement, and second, a transformation unit consisting of the electrodes which transform the displacement into a measurable

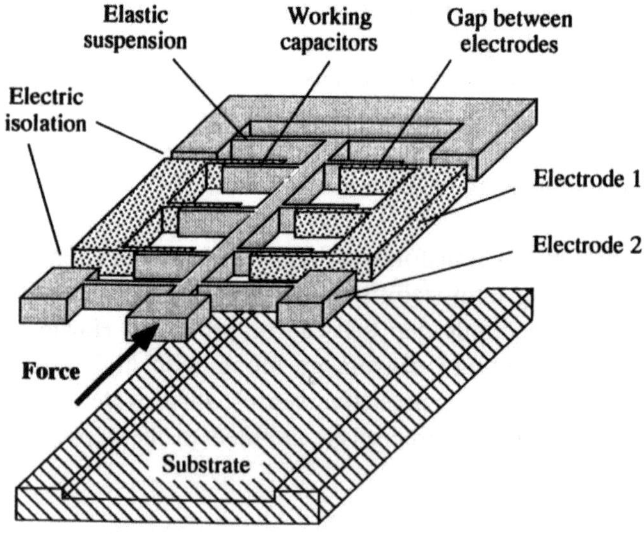

Fig. 6.6: Capacitive force sensor made of silicon. According to [Desp93]

change of capacitance. The displacement is restored by an elastic suspension beam. The capacitors consist of two electrically insulated thin electrodes with a very narrow gap between them (approximately 10 µm). They are placed on both sides of the sensor chip, making the capacitance on one side increase and decrease on the other side. By the separate measurement of the capacitance changes on both sides a high linearity and sensitivity is obtained. The sensor unit is made by anisotropically etching (110) silicon and then fastening it to a pyrex substrate through anodic bonding.

The prototype of the capacitive microsensor had a nominal capacitance of 1 pF. Measurements in this range can easily be handeled by commercially available microelectronic measuring devices. It was possible to measure very small forces with a resolution of 20 nm (0.01–10 N). The same structure can be used as a positioning unit for nanorobots.

Resonance sensor for measuring pressure

In Figure 6.7, a pressure resonance sensor is shown. The device consists of a silicon substrate, a diaphragm and three transducers equally spaced on the annular diaphragm. Each transducer consists of two resonators which oppose each other. If a pressure is applied to the diaphragm, the deformation causes the resonant frequencies of the resonators 1 and 2 to increase or decrease,

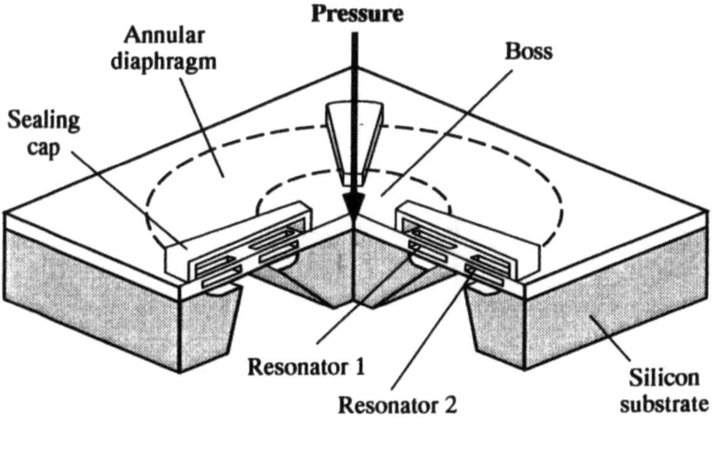

Fig. 6.7: Pressure sensor using the resonance principle. According to [Tilm93]

respectively. The frequency difference between the two resonators serves as the output signal of the sensor. By averaging the measurements of the three resonator transducers, it is possible to compensate for errors. The prototype of the pressure sensor was made from silicon using the surface micromachining technique. The following dimensions and performance data were reported: a diaphragm diameter of 1.2 mm and a thickness of 3 μm; a resonator length of 100 μm and a thickness of 0.5 μm; a maximum diaphragm lift of 0.7 μm; a pressure range of up to 1000 Pa; an accuracy of 0.01 Pa and a slight non-linearity of 0.1%.

Mach-Zehnder interferometer

Many physical quantities can be measured by optical sensors, making use of the change of light which is sent through fiber optical cables, Fig. 6.2. A so-called Mach-Zehnder interferometer is proposed as a pressure sensor, Fig. 6.8. Laser light is brought into the device by a fiber optical cable. The light is split and channeled via two waveguides to a photodiode. One of the light branches crosses a microstructured membrane which can be exposed to external pressure. The light beam in the other branch remains unchanged and serves as a reference signal. The two light beams are brought back together again to the photodiode. When the sensor membrane is actuated by pressure,

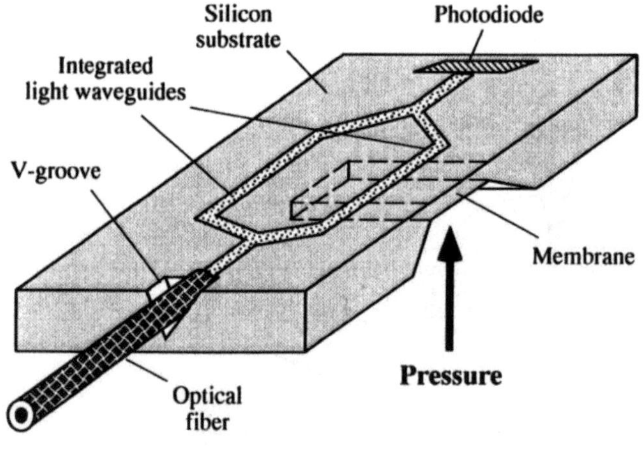

Fig. 6.8: Mach-Zehnder interferometer. According to [Fisch91]

the waveguide deforms and changes the properties of the light beam. The modulated light beam has a different propagation speed than the reference light beam, resulting in a phase shift which is registered by the integrated photodiodes. The measurement range of the device can be influenced by a change of the membrane thickness. The interferometer is made from silicon. When building such a device, particular attention must be paid to the optical properties of the waveguide.

The production of this pressure sensor was reported in [Fisch91] and [Hill94]. Here, the CVD and Plasma-CVD methods were used (Section 3.1.1.1). The device consisted of three layers placed on a silicon substrate, the light guiding layer had a high index of refraction to allow the passage of light, Fig. 6.9.

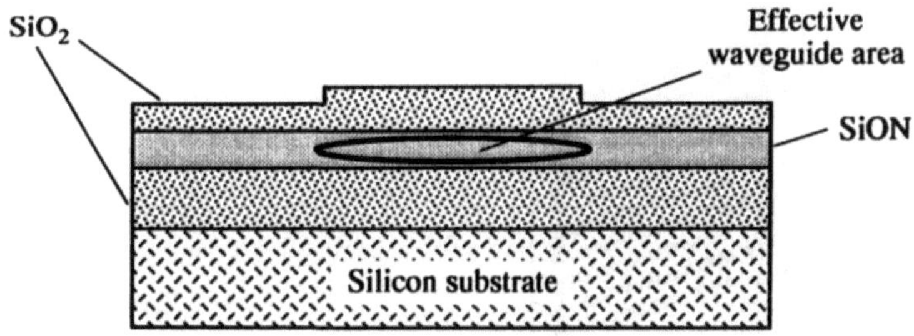

Fig 6.9: Pressure sensor waveguide. According to [Hill94]

First, a 2 to 3 μm thick silicon dioxide layer with an index of refraction of 1.46 was applied to a (100) silicon wafer to keep the laser beam from being absorbed by the silicon substrate. Second, a 0.5 μm thick light guiding layer made of SiON was deposited; it had an index of refraction of 1.52. Third, a 0.6 μm thick silicon dioxide layer was added and structured using a dry etching process. By this design, the light is guided in the middle of the SiON layer to avoid parasitic modulation from changes in the surroundings. The entire sensor system consisting of the SiON waveguides, sensitive silicon membranes, photodiodes and the CMOS amplifier were integrated on one silicon substrate [Hill94]. This system was found to be more sensitive than conventional pressure sensors. A sensor prototype with four membranes produced an output signal of 14 μV/mbar. The entire chip size was 0.3 mm × 5 mm and the size of the individual membranes was 200 μm × 200 μm.

A force sensing resistor

A new measurement principle was realized by using a so-called force sensing resistor [Witte92]. The device is fundametally different from capacitive, piezoresistive and resonant sensors, since here the resistance is inversely proportional to the pressure. The sensor consists of a polymer foil to which planar electrodes are fastened, on top of this a semiconductor polymer film is placed, Fig. 6.10. If a voltage is applied to the electrodes and there is no force, the resistance is at least 1 MOhm. When a force is applied, the resistance decreases due to current that flows across the shunting polymer foil.

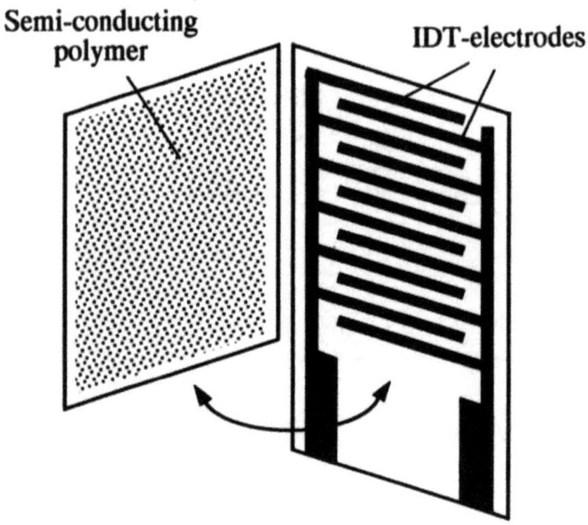

Fig. 6.10: A force sensing resistor. According to [Witte92]

The dynamic range of the sensor can be influenced by producing a finer electrode structure. This, however, is accompanied by increasing production costs due to a lower yield rate. The sensitivity can be increased by varying the foil thickness. The device can be operated at temperatures of up to 400°C and is very durable, e.g. over 10 million repeated measurements were made with a 5% deviation. The measurement range is between 10 g and 10 kg. A major disadvantage is the hysteresis, which appears during pressure changes. Despite of this fact, the device can be usefully employed for many dynamic measuring applications. It is inexpensive, compact, robust and resistant to external influences.

Capacitive tactile sensor

Tactile microsensors play an important part in microrobotics since they can be used for recognizing objects, for detecting their location and orientation and for measuring them. The sensors can be mass-produced by silicon technologies in combination with classical materials, such as plastic or rubber foils. There is an ongoing effort to develop an artificial human-like sense of touch which needs highly integrated sensors of good flexibility and stability. In [Yamad91] such a sensor is described. For its "skin", polyimide is used, which is highly flexible and has excellent electrical, mechanical and chemical properties. The structure of such a sensor is shown in Figure 6.11.

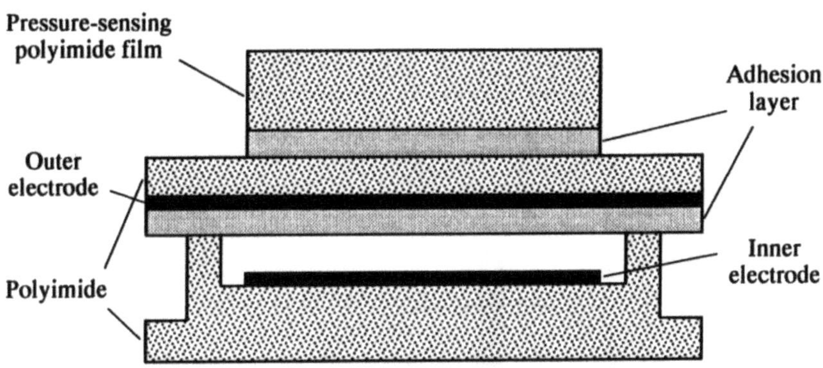

Fig. 6.11: Tactile microsensor. According to [Yamad91]

The sensor consists of a polyimide base to which a 4 mm^2 inner electrode is deposited. Over this there is an outer electrode seperated from the inner electrode by a 25 μm thick air gap. The actual pressure-sensitive polyimide foil is located on the top. It is attached via a polyimide layer to an outer electrode. When a force is applied, the capacitance changes allowing the force and location of the force to be determined. It is planned to integrate the entire signal processing circuit on the substrate.

6.3 Position and Speed Microsensors

Position and speed microsensors are essential for many applications, especially for use in automobiles, robots and medical instruments. Position and speed control is also a major concern in microrobotics in order to e.g. determine the exact position of an endeffector at any point of time. At the moment, there are various principles being used for these measurements; the contact-free optical and magnetic methods are the most significant ones for MST.

Magnetic sensor to measure angular displacement

In robotics it is necessary to exactly control the movements of robot arms and legs or other components having rotating joints. Source [Ueno91] describes a microsensor which can measure very exactly an angular displacement by using the Hall effect. The concept of the entire system is shown in Figure 6.12. The sensor consists of a rotor which has a row of teeth on its bottom. The rotor faces a stator which contains several Hall sensors and electronic circuits. A permanent magnet is located under the Hall sensors, producing a magnetic field. When the rotor moves, the teeth passing by the

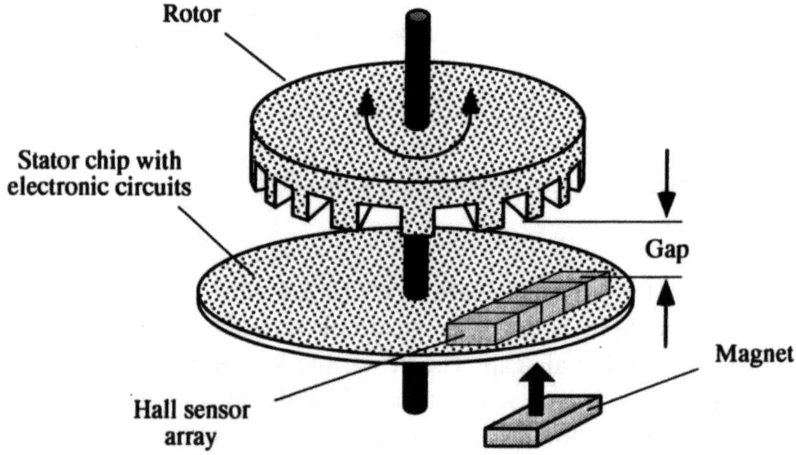

Fig. 6.12: Angle-measuring microsensor. According to [Ueno91]

Hall sensor change the magnetic field. This change is picked up by the Hall sensors and they produce voltage signals.

The measuring process is shown in the form of a block diagram in Figure 6.13. The sensor field covers exactly one notch and one tooth of the rotor. A multiplexer constantly scans all *n* Hall elements which creates a step function representing the distribution of the magnetic field. This signal is then passed through a filter and smoothed to form a sine curve. The rotor of the sensor is connected to the rotational joint of which the angular displacement is to be determined. A rotation causes a phase shift in the curve, which is determined by a comparator. This procedure is relatively independent from the distance of the rotor to the sensor matrix, which makes the system insensitive to vibrations. The system is also very stable.

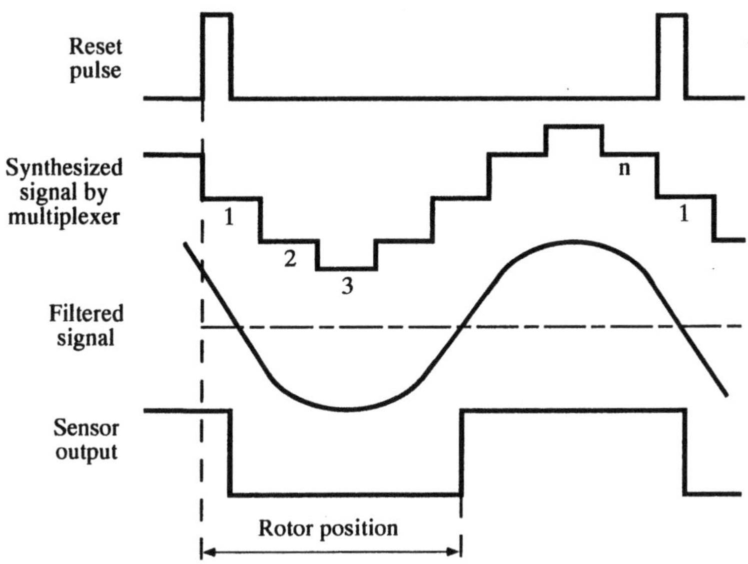

Fig. 6.13: Block diagram of the measuring procedure. According to [Ueno91]

The developed prototype of the sensor matrix was produced on a GaAs substrate having a 1 μm thick silicon dioxide layer. The prototype is about 4 mm long and can measure the rotational angle with an accuracy of 0.028 degrees at temperatures between -10°C and +80 °C.

Inclination sensor

There is a trend to develop small inspection robots for various applications (Section 1.4.2). E.g. they are missioned to go into complex machinery or inaccessible pipes to locate and repair defects. Many of these devices need inclination sensors to calculate the exact tilting of the robot with respect to a reference plane. A very novel inclination sensor and its components are shown in Figure 6.14.

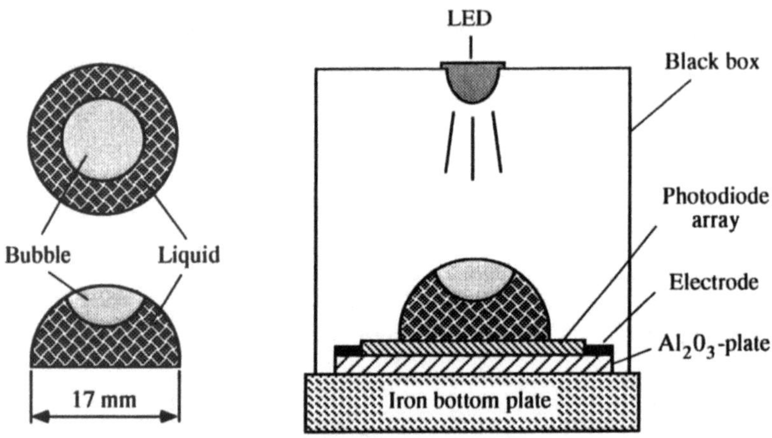

Fig. 6.14: Inclination sensor. According to [Kato91]

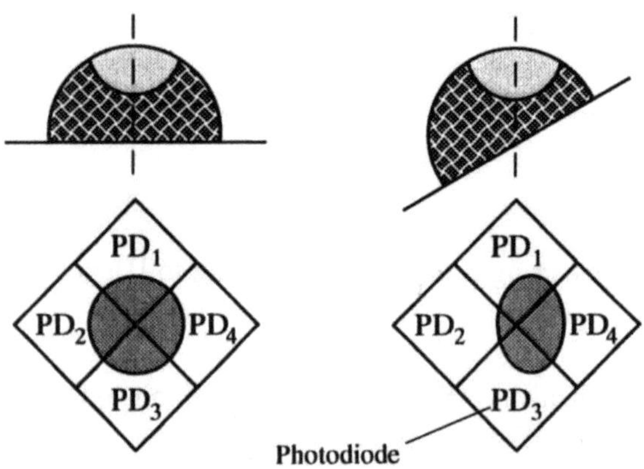

Fig. 6.15: Measurement principle of the sensor. According to [Kato91]

The sensor system consists of an LED and a semi-spherical glass cup, which is mounted on a photodiode matrix array. The glass cup is filled with liquid and has an enclosed air bubble. The LED sends light through the cup which casts shadows on the four photodiodes underneath. If the sensor is tilted in one direction, the shadow moves across the diode matrix, Fig. 6.15. The output currents of the photodiodes are transformed into voltages and are amplified, and from this information the inclination angle and tilt direction can be determined.

The sensor system was tested in two applications. The first application was the determination of the position of a small moving vehicle in a pipe system. The measuring principle made it necessary that the speed be kept as constant as possible, since sudden acceleration or deceleration would falsify the results by an unpredictable movement of the air bubble. In the second application a topological study was to be made of a small area and a topographical map was drawn. The sensor was mounted on a small vehicle which drove across the area. The measured values were compared to the known ones. It was found that the sensor could measure the inclination very accurately up to 10 degrees. If the inclination was steeper, the error increased due to the elliptical shape of the shadow and the non-linearity encountered.

Ultrasound distance sensors

Ultrasound distance sensors are well suited as position sensors for microrobots, since they do not depend on the optical properties of the object being detected and they are robust and can obtain reproducible results. Ultrasound distance sensors use the pulse-echo principle. Here a pulse sequence is emitted with the help of an ultrasound transducer which is usually made from piezoceramics. The signals reflected by objects as echos are received by a sensor and evaluated using the propagation time of the sound signal. Since the transducer needs some time for recovery after transmission, a "blind spot" appears when the detector is too close to an object, which means that an object might not be detected. The results obtained with a new concept of an ultrasound microtransformer were reported in [M²S²93]. In this device, two identical independent ultrasound membranes were integrated next to each other on a silicon substrate; one served as a transmitter and the other one as a receiver. The schematic design of a single sensor membrane and the measurement principle are shown in Figure 6.16.

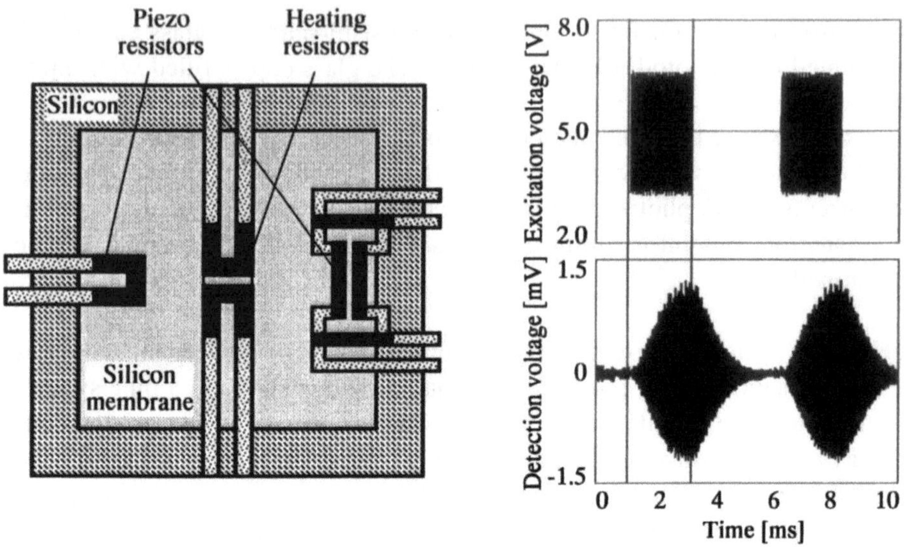

a b

Fig. 6.16: Ultrasound distance microsensor
a) design of the sensor; b) measurement principle of the sensor. According to [M2S293]

The transmitter membrane is brought to resonance electrothermally with integrated heating resistors. The acoustic pressure response is then detected by piezoresistors, integrated in the form of a Wheatstone bridge in the receiver membrane. The sensitivity of this prototype was about 3 μV/mPa at a bridge voltage of 5V.

Capacitive rotational speed sensor

In many technical systems like navigation and landing gear controllers, compact and inexpensive angular speed sensors are required. Conventional sensors using piezoelectric resonators or optical glass fibers are very sensitive, but are usually expensive. The following described silicon sensor was produced using a batch fabrication method [Hash94]. The operating principle of the resonating sensor is presented in Figure 6.17.

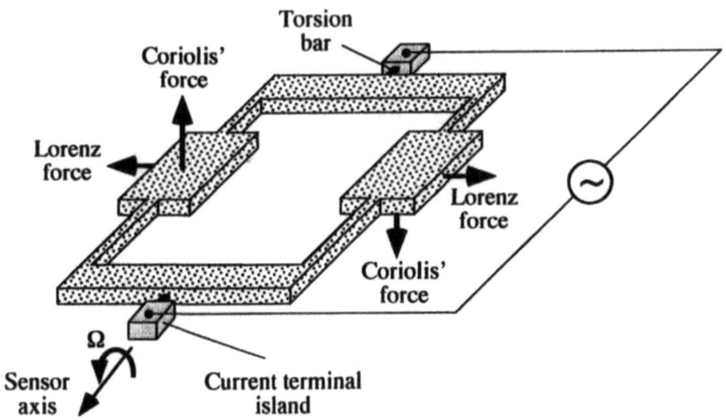

Fig. 6.17: Operating principle of a rotational speed microsensor. According to [Hash94]

A 200 μm thick tuning fork arrangement made of (110) silicon is used as the resonator. It is positioned by two torsion bars which also serve as electrical terminals. When the resonator is introduced into a magnetic field and alternating current is applied it starts to oscillate due to Lorenz forces. If the sensor rotates at Ω degrees about its longitudinal axis, the Coriolis forces induce a rotational movement in the opposite direction about this axis, this movement is proportional to the swing angle Ω. The amplitude of the swing is detected by the capacitance change between the fork prongs (movable electrodes) and the fixed detection electrodes, not shown in Figure 6.17. The latter are integrated into a glass casing consisting of two pyrex glass layers, each 250 μm thick. A sensor prototype with a base area of 2 cm × 2 cm was built, it had a sensitivity of 0.5 mVsec/deg at an exciting frequency of 470 Hz.

Fiber optical swing angle sensor

For monitoring swing angles the device shown in Figure 6.18 was developed [Ecke93]. In this sensor a very steady light is lead into a glass fiber via a polarization filter. A torque applied to the mechanical measuring section, causes a torsion in the glass fiber and a change of the direction of polarization of the light. A corresponding change of the light intensity is detected by the second polarization filter and is then evaluated by a photodiode.

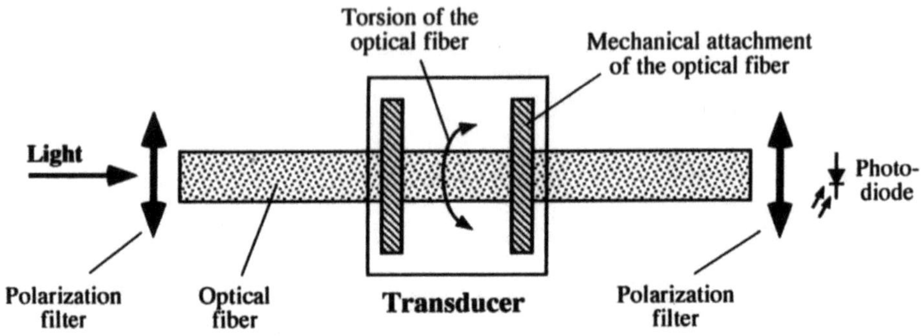

Fig. 6.18: Optical swing angle sensor. According to [Ecke93]

With a prototype of the sensor very promising results were obtained. A future goal is to improve the sensor accuracy to 0.1 degree for an angular measuring range of 100 degrees at a surrounding temperature of about 20 K. The device will then be suitable for space applications.

6.4 Acceleration Microsensors

Cantilever principle

Miniaturized acceleration sensors will mostly find their place in the automotive industry. They are also of interest to the air and space industries and for many other applications. Acceleration microsensors will help to improve the comfort, safety and driving quality of automobiles. However, in order for them to become a product of general interest, their production costs must be drastically lowered. As with pressure (Fig. 6.4 and Fig. 6.5), acceleration is usually detected by piezoresistive or capacitive methods. Mostly an elastic cantilever is used to which a mass is attached. When the sensor is accelerated the mass displaces the cantilever and the displacement is picked up by a sensor. Such a sensor is shown in Figure 6.19. It uses the capacitive measuring method to record deflection. From the deflection the acceleration can be calculated.

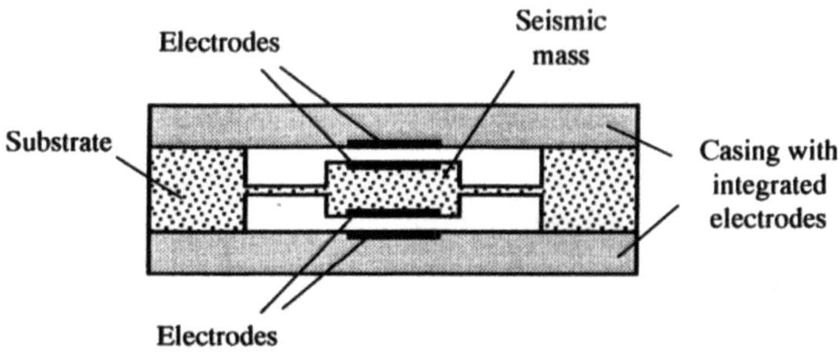

Fig. 6.19: Capacitive measurement of accelerations

Piezoresistive principle

To effectively measure acceleration with this principle, piezoresistors are placed at points of the cantilever where the largest deformation takes place. The stability and accuracy of the sensor improves with increasing number of piezoelements. If a mass moves due to acceleration, it deforms the piezoresistors, thereby changing their resistance, Fig. 6.20. The acceleration is determined from the resistance change.

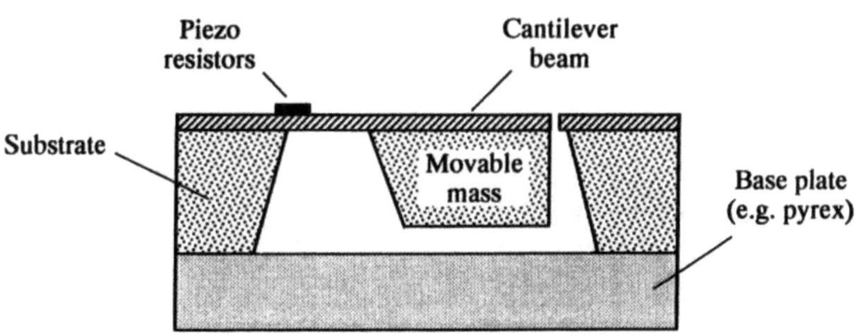

Fig. 6.20: Piezoresistive acceleration sensor

By increasing the movable mass the sensitivity of the sensor will be improved. The mass's center of gravity should be as close to the end of the cantilever as possible. Piezoresistive acceleration microsensors are usually produced using the silicon technology described in Chapter 4. This allows the microelectronic processing unit to be integrated onto the sensor chip, making the system compact and robust.

Fully integrated capacitive accelerometer

A capacitive acceleration sensor ready for mass production was already presented in 1991 [Good91]. The sensor chip with a diameter of about 9 mm was made from polysilicon by surface micromachining. The microelectronic circuits for signal preamplification, temperature compensation and system selftest purposes were integrated into the sensor, Fig. 6.21. A differential capacitor serves as the sensitive part of the device. It consists of independently fixed plates and a movable comblike microstructure; the latter changes its position in response to a change in relative motion, Fig. 6.21b.

a b

Fig. 6.21: Acceleration sensor ADXL50
a) the integrated sensor chip; b) surface micromachined differential capacitor made from polysilicon. Courtesy of Analog Devices, Inc., Wilmington, MA

This sensor was one of the first examples of a successful transfer of a MST device from research to industry, as it was the first fully integrated accelerometer to be produced in high volume. Accelerations up to ± 50 g can be measured with a sensitivity of 19 mV/g by this device. It is currently being used in the airbag systems on several car models. The measurement range of a newer sensor is ± 5 g; it can detect minute changes in acceleration with a sensitivity of 0.005 g [Ajlu95].

Capacitive cantilever microsensor

An acceleration sensor produced by the surface micromachining technique was described in [Fricke93]. A sketch of this sensor is shown in Figure 6.22. The sensor consists of one or more cantilevers acting as one electrode; they are suspended freely over an opposite electrode and a contact strip. There is only a small gap between the cantilever and the electrode to maximize the electrostatic forces and to keep the mechanical stresses as small as possible.

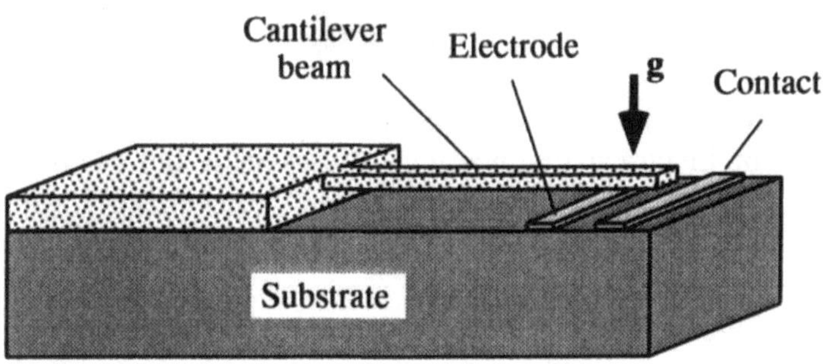

Fig. 6.22: Design of the accelerometer. According to [Fricke93]

As opposed to conventional capacitive sensors, a so-called threshold voltage is applied to offset the forces caused by the acceleration, it will give an indication of the current acceleration. With this device, a sawtooth voltage is applied in defined steps across the cantilever and the electrode, which gradually increases the electrostatic force acting on the cantilever. When the critical voltage is reached, the system becomes unstable and the cantilever bends towards the contacts and finally touches them. The voltage falls to zero, and the sawtooth voltage is applied again. The actual value of the threshold voltage to be applied depends on the magnitude of the acceleration.

The level of the threshold voltage is used to compute the acceleration; it is indirectly determined by measuring the time between the starting point of applying the sawtooth voltage and the time of contact.

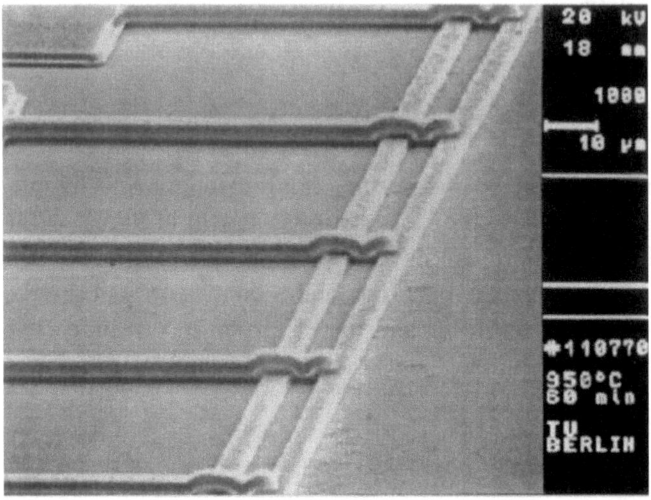

Fig. 6.23: Microstructure of the sensor. Courtesy of the Technical University of Berlin (Department of Electrical Engineering, Microsensor and Microactuator Technology)

The cantilever, the opposite electrode and the contact strip are made of polysilicon by the dry etching process, Fig. 6.23. The microstructure height is 2.2 µm (polysilicon layer) and the gap attained between the cantilever and the contact strip is 1.5 µm. For a cantilever length of 120–500 µm, the microsensor sensitivity is in a range of 0.6–100 mV/g.

Piezoresistive microsensor with oil damping

In publication [Muro92] the design of a prototype of an integrated, piezoresistive acceleration sensor with oil damping was described. The sensor chip was fabricated on a (100) silicon substrate and contained a seismic mass suspended on a thin cantilever and also an integrated signal processing circuit, Fig. 6.24. The circuit consisted of a piezoresistor bridge and amplifiers. Oil is used to dampen the resonance of the suspended mass. As opposed to an air damping mechanism, this design did not require an extra fabrication step. However, the damping is influenced by the viscosity of the oil, the dimensions of the sensor and the temperature of the device.

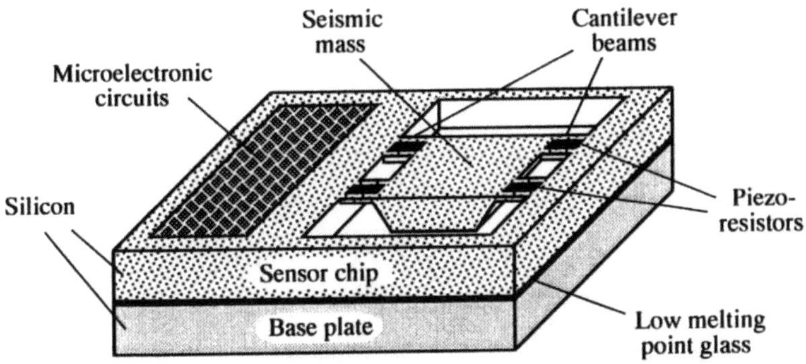

Fig. 6.24: Design of the microsensor. According to [Muro92]

Various silicon techniques were used to make the sensor. The micromechanical parts of the sensor which are the movable mass and the cantilevers were made by wet etching of the silicon substrate. The cantilever is 480 μm long, 200 μm wide and 12 μm thick; the seismic mass weighs 2 mg. The measurement range of the device is between 20 and 50 g, which makes it quite useful for many industrial and automotive applications.

Multi-sensor acceleration measurement system

The development of a multi sensor system using the LIGA process was reported in [Stro93] and [Kröm95]. The device is based on the capacitive measurement principle; it has a sensor array consisting of several inexpensive sensors integrated on one chip. With this design, the reliability and accuracy of the measurement system could be improved. The first result of this project, a two-dimensional acceleration sensing device was already presented in Chapter 2, Fig. 2.13. The central components of this system are micromechanical capacitance sensors made from nickel by the SLIGA technique. The sensor is a differential capacitor; it consists of two fixed electrodes between which a movable mass is attached to a flexible cantilever.

A newer development of a fully integrated three axis acceleration measurement system using this principal design was presented in [Kröm95]. The system can measure a three dimensional acceleration vector. In addition to the LIGA microsensors which are responsible for the measurements in the x- and y-direction, a silicon acceleration sensor for the z-direction was integrated, Fig. 6.25.

Fig. 6.25: A three axis acceleration measurement system. Courtesy of the Karlsruhe Research Center, IMT

The signal conditioning is done online with the help of integrated circuitry. To compensate the temperature sensitivity of the device, a temperature sensor was integrated on the chip. The device with all its microelectronic components has overall dimensions of only 2 mm × 25 mm × 35 mm. The measurement range of a prototype was ±3 g and the sensitivity 2.5 V/g. Here, the combination of the LIGA technique with a silicon-based technology offered a good design solution.

6.5 Chemical Sensors

Chemical sensors detect the presence or concentration of a chemical substance in a solution. They may be used for qualitative and quantitative measurements. In medical diagnostics, nutritional science, environmental protection and the automobile industry, many different chemical quantities are to be measured. The importance of chemical sensors for various applications is illustrated in Figure 6.26.

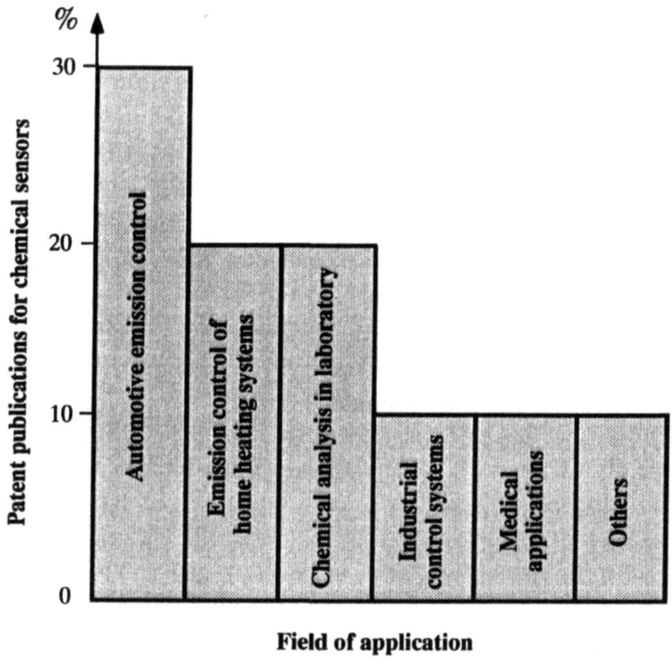

Fig. 6.26: Fields of application of chemical microsensors. According to [Gron93]

About 60% of all chemical sensors are gas sensors. The rest is used to detect substances and concentrations in liquids. An important application potential of chemical sensors is in environmental protection, medical applications and process engineering. Many industrial countries will soon be adopting very strict environmental standards and laws that will rapidly increase the demand for gas and liquid sensors. Present research is concentrated on the integration of these sensors in measurement systems. They can substantially increase the functionality of measurement systems when used together with intelligent signal processing components to make very complex environmental analyses possible.

Many conventional measurement methods, like mass spectrometry, atomic spectroscopy and gas and liquid chromatography, achieve very good results in detecting the presence and amount of a variety of substances. The measurement methods are often very complicated and expensive, which limits their widespread use. In most cases, the analyses must be carried out under ideal laboratory conditions. E.g. a complete water analysis for environmental protection requires the continuous measurement and testing of about 150 parameters, which is not feasible with conventional instruments and methods.

These and many other problems of conventional measurement techniques have led to intensified research to conceive reliable and fast microsensors and microsensor systems. The goal is to find very small and inexpensive sensors that can be produced easily, that are accurate and robust, use only small amounts of reagents and have short response times. By integrating several types of sensors, many conventional measurement methods can be improved and new applications for chemical analysis can be conceived. Intelligent processing of signals will play an important role here. With the help of a multisensor system, a comlex chemical compound may be tested easily and n substances with their n concentrations may be determined and evaluated, e.g. by using cluster analyses in conjunction with artificial neural networks.

Chemical microsystems which have many identical sensors are also being developed. This allows the reliable interpretation of results, and, if the sensor matrices are appropriately placed and distributed, parallel measurements. Contrary to this, conventional chemical analysis methods are often limited to the global detection of substances and concentrations. A local analysis of a substance with respect to its distribution over a certain domain, or hypothesizing a 2-dimensional reaction cannot be done. A possibility of an array of microsensors was illustrated by the work reported in [Camm94]. Here, 1024 oxygen sensors were fabricated on one silicon substrate and were connected to a control circuit. The resulting sensor matrix, consisting of 32 rows and 32 columns, was 3 cm x 3 cm and was used for medical measurement, e.g. to control a transplantation. A diagram of local oxygen measurement with this sensor is shown in Figure 6.27.

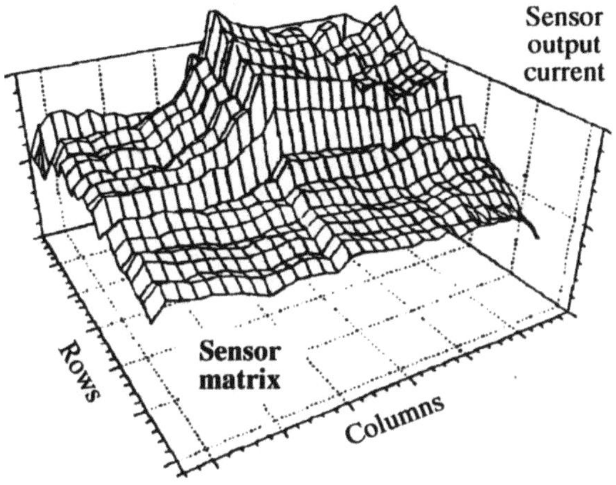

Fig. 6.27: Oxygen profile using sensor matrix. According to [Camm94]

Researchers are trying to develop chemical analysis systems for local measurement in a real-time environment. This makes it possible to immediately intervene in a monitored process if necessary. In environmental protection, a system like this can be used to determine online when a pollutant has reached or exceeded a critical value. It can also be used to monitor industrial processes. Real-time measurement can allow process optimation, which will increase the product quality. In medicine, the continuous monitoring of several blood parameters during surgery can mean a matter of life or death for the patient.

There is a broad spectrum of applicable sensor principles for chemical microsensors. The potentiometer principle in connection with field effect transistors (FET), acoustic sensors using the change of mass principle and optical sensors are most often applied. Many gas and liquid sensors are based on these principles and have similar structures. It is usually very important for chemical sensors to have a low cross-sensitivity, i.e. the measured values are not influenced significantly by other substances in the solution being analyzed.

For measuring chemical substances, a sensitive layer or a specific area of the sensor is used to contact the chemical substance. During measurements, a chemical reaction occurs on this sensitive layer/area and a transducer, of which the physical, optical, acoustical or dielectric properties are changed, transforms the recorded phenomenon into an electric signal. This signal is then amplified and evaluated by a microelectronic component. The general structure of a chemical sensor system is shown in Figure 6.28.

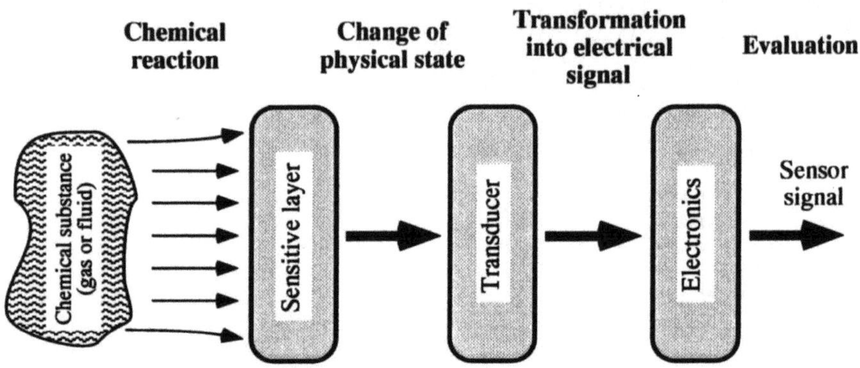

Fig. 6.28: Structure of a chemical sensor system

Several sensor principles will be discussed in the following sections.

Interdigital transducer sensor principle

Interdigital transducers using the capacitive measurement principle are often employed as chemical sensors, Fig. 6.29. The capacitance value can be adjusted either by changing the distance between the electrodes (Figure 5.4a), or by changing the dielectric properties of the sensitive layer deposited on the transducer. Interdigital transducers are well-suited to produce various chemical sensors, since the electric resistance of the sensitive layer (e.g. SnO_2) changes when it interacts with certain substances, e.g. when gases are adsorbed or liquids permeate. Interdigital transducer structures are made by the microfabrication techniques discussed earlier, Fig. 4.13. E.g. these sensors are good for measuring humidity, concentration of sulfur dioxide or ethanol.

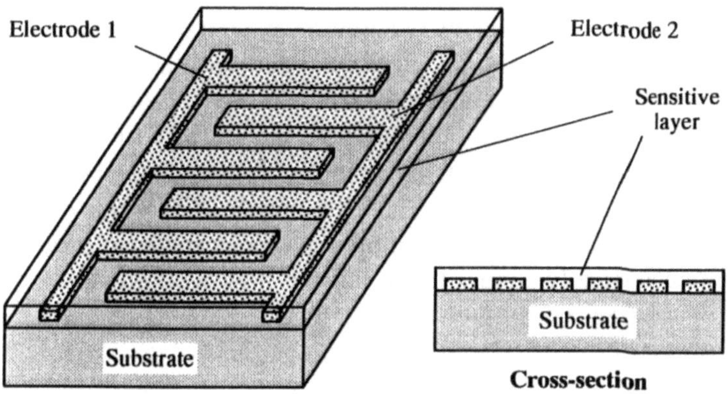

Fig. 6.29: Chemical sensor based on an interdigital transducer

Pellistor sensor principle

The concentration of many gases can be determined with the help of the so-called pellistors. These sensors are a special form of conductivity sensors which measure the increase of temperature due to a chemical reaction. A pellistor consists of a ceramic pellet plated with a catalytically active metal, e.g. Pt or Pd, and an integrated heating element made of platinum wire; the heating wire also serves as a sensitive element of the device, Fig. 6.30.

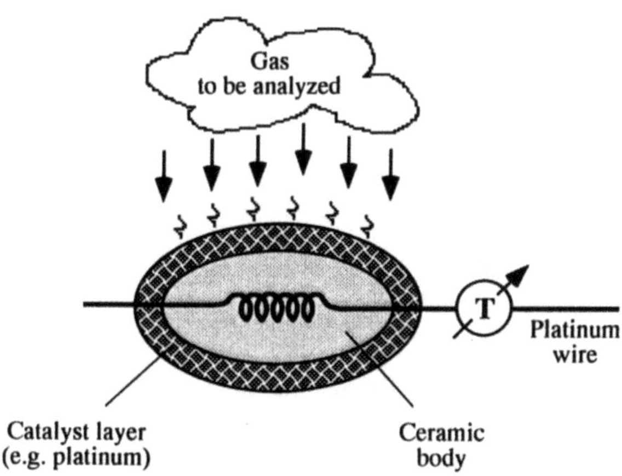

Fig. 6.30: The pellistor principle used for measuring the concentration of gas

The measurement principle of a pellistor is based on the fact that when a gas is burnt, specific activation energies are released. This causes a temperature increase specific to the gas on the catalytical surface of the sensor. The heating of the ceramics e.g. to about 400–500°C causes the gas to burn which in turn adds heat to the platinum wire. This temperature increase is proportional to the concentration of the burning gas; it can be measured through the resistance change in the platinum wire. Attempts have been made to produce the pellistor micromechanically from silicon, but the high working temperatures of up to 700°C caused serious problems [Gall91].

Optical sensor principle

As already mentioned earlier, microoptics play an important role in many microsensors. Chemical sensors can also make use of various optical principles. These sensors have several advantages, they are inexpensive, easy to sterilize, can accomodate small samples and are highly sensitive. In general, planar integrated microoptical chips in the form of an interferometer or a coupling grid are used. An optical interferometer was described in Section 6.2 (Fig. 6.8). Figure 6.31 shows the function of a coupling grid detector.

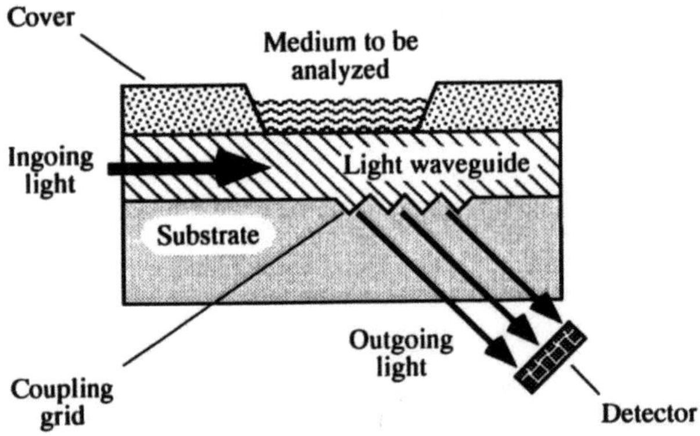

Fig. 6.31: Function of a coupling grid structure. According to [Krull93]

In this device light enters the waveguide from left and is reflected onto a photodiode sensor. The substance to be analyzed has direct contact with the waveguide, which changes its index of refraction. The amount of light striking the sensor is proportional to the concentration of the substance.

Field effect transistor sensor principle

Ion-sensitive field effect transistors are used to measure the concentration of ions of various elements such as hydrogen, sodium, potassium or calcium. The structure of an ion sensitive FET and its measurement principle can be seen in Figure 6.32.

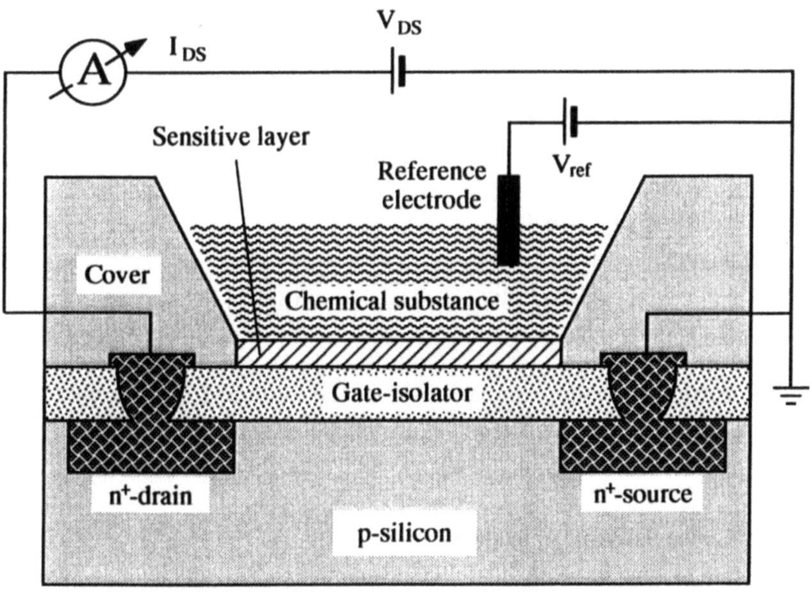

Fig. 6.32: Chemical sensor based on an ion sensitive field effect transistor

Initially, when no chemical substance is in contact with the ion-sensitive layer deposited on the gate area of the transistor, the gate potential V_{GS} is equal to V_{ref}. When the substance to be measured contacts the ion-sensitive layer, the gate potential V_{GS} changes; this voltage change is caused by the ions in the chemical substance. Thereby, the current I_{DS} between the source and drain changes as well. The gate potential V_{GS} is then corrected by adjusting the voltage V_{ref} until the original transistor current I_{DS} flows again; the voltage V_{DS} is held constant. The value ΔV_{ref} is proportional to the ion concentration of the analyzed substance to be measured, Fig. 6.33. The area of the sensitive layer can be as small as a few μm², allowing very small amounts of substance to be measured.

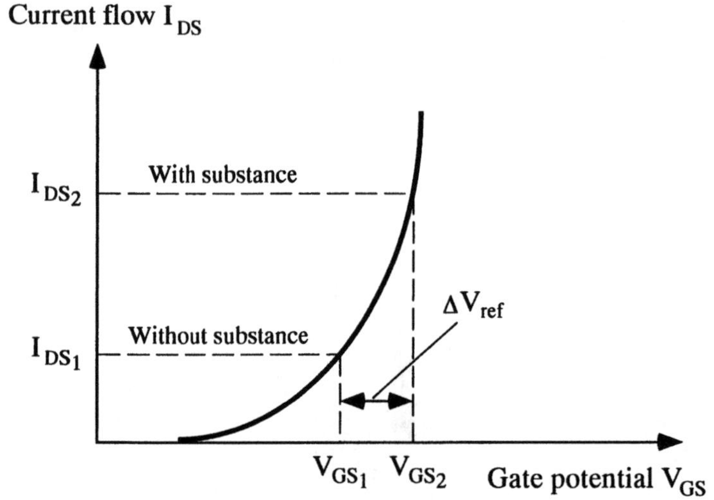

Fig. 6.33: The current flow I_{DS} as a function of the voltage V_{GS} in a chemical sensor based on an ion sensitive FET

Resonance quartz sensor principle

The acoustic-gravimetric measurement principle is applied for chemical sensors. It can be realized either with a resonating sensor or an acoustical waveguide sensor. A change of mass of the sensor, which is caused by a chemical reaction between the substance to be analyzed and the sensitive layer, is indirectly measured and the concentration is determined from this. The working principle of a resonating sensor is based on measuring the behavior of a standing acoustic wave in a mechanical structure which can resonate like a membrane or cantilever. The mass may also be a solid. A typical resonating sensor contains a piezoelectric quartz which has two metal electrodes attached to it and which is covered by a chemically sensitive material, Fig 6.34.

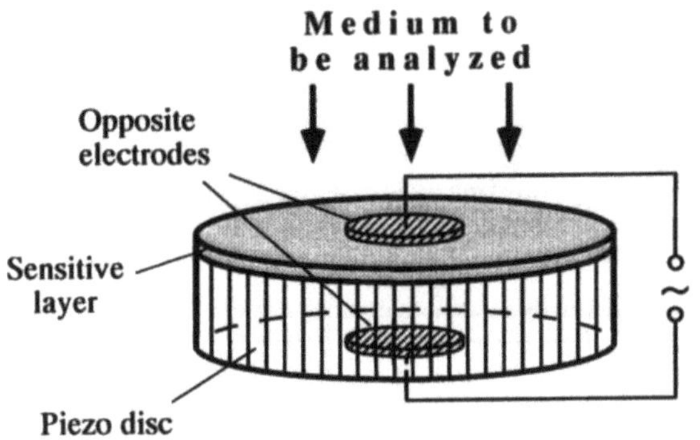

Fig. 6.34: Resonating quartz sensor as a chemical sensor

If an alternating current is applied to the electrodes at the resonance frequency, a longitudinal wave is produced in the quartz. A chemical in contact with the sensitive layer causes a change of the resonator mass, thereby the resonant frequency changes. This change indicates a certain concentration.

Waveguide sensor principle

The behavior of surface acoustic waves, like Raleigh or Lamb waves, can be used in so-called waveguide sensors to detect chemical substances. The Lamb wave detectors are the most interesting ones for MST since they have a very high sensitivity and a wide spectrum of possible applications. Here, a very thin piezoelectric film is used as a wave-guiding layer. A prototype of such a sensor is shown in Figure 6.35.

During operation of the sensor, an alternating voltage is applied to the interdigital transducer which is electromechanically transformed into an acoustic wave; it propagates along the strip towards the other transducer. If the sensitive layer interacts with the analyzed substance, the wave transmission changes due to damping. This change can be detected by the receiving transducer, whereby the acoustic signal is transformed back into an electric one. The sensor principle was tested for measuring the viscosity of liquids, the coagulation rate of blood and the concentration of substances. It could be operated with relatively small frequencies of 0.5–10 MHz. The sensitivity was very high; even a mass of 1 ng/cm^2 could be detected.

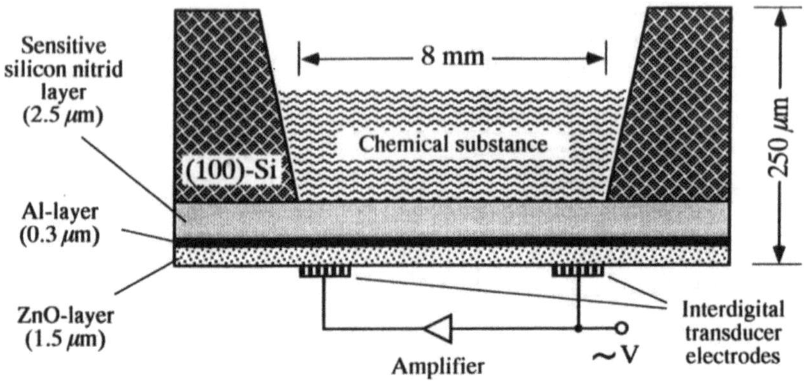

Fig. 6.35: Surface acoustic wave sensor. According to [Gies92]

The design of the device is as follows. The base element is a (100) silicon substrate onto which a silicon nitride insulation layer and an aluminum layer are applied. A 1.5 μm thin piezoelectric zinc oxide layer is sputtered onto the aluminum layer. This layer system is then wet-etched and forms a 3 mm × 8 mm membrane strip. Interdigital transducer electrodes made of aluminum are sputtered on the zinc oxide layer.

We will be discussing below various chemical microsensors which had been reported in the literature; for further reading on sensor principles we refer to [Haup91] and [Gard94].

Ion sensitive FET sensor

The continuous measurement of the gases in blood (like pO_2 or pCO_2) and of the pH value are very important in surgery. Here, invasive measurements are quite risky for the patient because of the danger of infection. A device for external use was described in [Arqu93]; it consists of a sensor and a blood sampling and processing part. Figure 6.36 shows the sensor which uses ion-sensitive field effect transistors. The sensor was realized on a 10 mm × 10 mm silicon chip and was tested in aqueous solutions and in blood for transfusion, respectively. In both cases, the sensor showed satisfactory results and had the same behavior. The sensor's lifetime was about two weeks. By using this system, a doctor can make an online diagnosis of his patient.

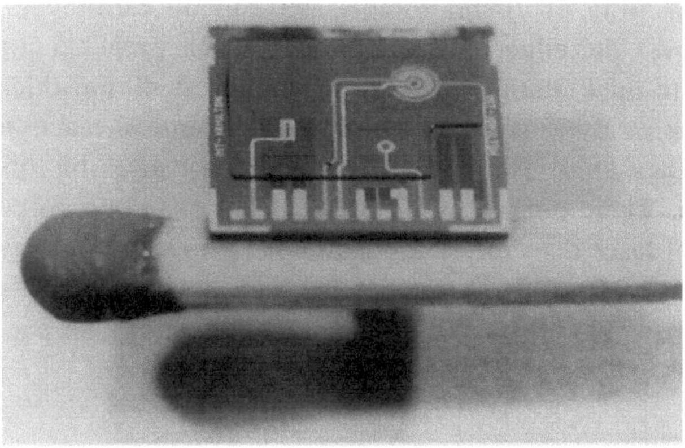

Fig. 6.36: Blood gas/pH sensor. Courtesy of the University of Neuchâtel (Institute of Microtechnology)

Bimetal sensor

Figure 6.37 shows an original measurement principle that makes use of a method where a chemical reaction is transformed into mechanical motions by using the bimetal effect.

Fig. 6.37: Prototype of the chemical bimetal sensor. Courtesy of the IBM Research Division (Zurich Research Laboratory) and of the University of Neuchâtel (Institute of Microtechnology)

The sensor structure is an array of 10 microfabricated silicon cantilevers, each extending 400 μm over the edge of a silicon wafer. On each 1.5 μm thick silicon lever a 0.4 μm thick aluminum layer is applied. A 40 nm thick platinum film served as the catalytic layer which starts the chemical reaction with the substance to be measured. The heat of this reaction causes the bimetal cantilevers to bend. Their motion is measured by a scanning force microscope with an optical laser detector. The resolution of the motion was 0.01 nm and allowed a temperature measurement as exact as 10^{-5} °C. This simple and relatively inexpensive sensor will find many applications; in particular when better methods and materials have been found to make it.

Sensors based on a zigzag interdigital transducer

CO_2 is released in all burning processes, and there is high demand for sensors used in continuous emission monitoring; the sensors must have a short response time and long-life stability. The concentrations of CO_2 in a fluid can be measured with AMO/PTMS (3-amino-propyl-trimethoxysilane/propyltrimethoxysilane). The dielectric properties of this material change with the CO_2 contents in the surrounding of the sensor [Heur92]. This change can be measured using a planar interdigital transducer, which makes an indirect CO_2 measurement possible. The sensor was constructed using a zigzag electrode design on top of which the sensitive material was deposited. The zigzag de-

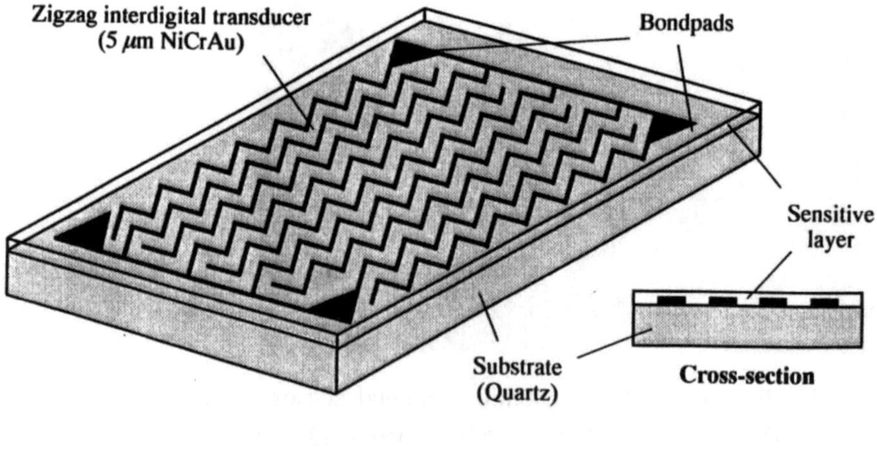

Fig. 6.38: CO_2 microsensor. According to [Heur92]

sign has better capacitive properties than the conventional interdigital finger structure, Fig. 6.38. The NiCrAu capacitor was deposited on a glass substrate. In experiments, the zigzag width was varied between 2.5 μm to 15 μm and the thickness of the sensitive AMO/PTMS layer was varied between 0.9 μm and 1.2 μm. The most effective electrode area was 21 mm^2.

Another interdigital transducer sensor for detecting CO, CO_2, NO_2 and water concentrations was introduced in source [Stei94]. The sensor was covered with a sensitive layer of SnO_2 which is the most often used oxide semiconductor material for detecting gases; it can be produced at low cost by the thin film techniques (Chapter 3) and has a high reliability. Its sensitivity and selectivity can easily be enhanced by a catalyst. A cross-section of the sensor with a catalyst made of Ca, Pt or Pt/Ca is depicted in Fig. 6.39.

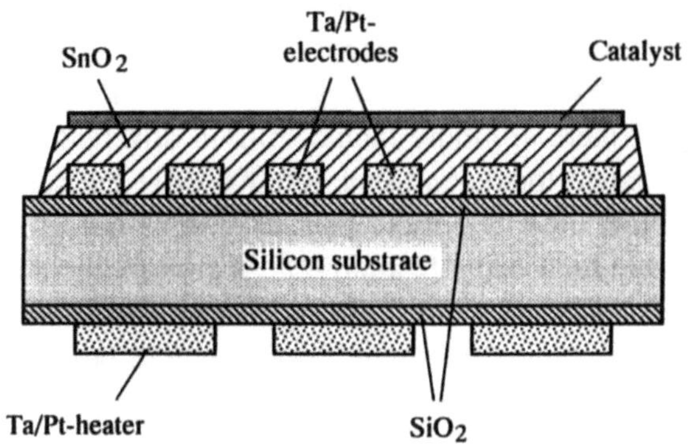

Fig. 6.39: Interdigital transducer gas sensor. According to [Stei94]

Both sides of the silicon substrate were processed by the photolithography, wet etching and sputter techniques. On the top there is an interdigital transducer made of Ta/Pt and of a 50 nm thick n-doped SnO_2 layer; an additional catalyst layer of a few nm thickness was sputtered over it. The Ta/Pt heating elements on the other side of the substrate heat up the entire sensor chip to activate the metal oxide gas reaction which causes a change of conductivity in the sensitive layer. From the change it is possible to indirectly determine the gas concentrations. A 1 μm thick SiO_2 was applied to the substrate to electrically insulate it from the functional parts of the sensor. The device is 8 mm × 8 mm and has an active area of 5 mm × 5 mm.

6.6 Biosensors

The introduced measurement principles for chemical sensors are also used for biosensors. However, in a biosensor the biologically sensitive elements such as enzymes, receptors and antibodies are integrated with the sensor. The interaction between the protein molecules of the bioelement and the molecules of the substance causes a modulation of a physical or chemical parameter. This modulation is converted into an electrical signal by a suitable transducer. The signal represents the concentration to be measured. In many molecular interactions, gases are either released or consumed, e.g. the change of oxygen can be registered by a chemical O_2 sensor. Very selective and sensitive measurements are possible with biological receptors.

There are many applications for miniaturized biosensors. In biological and nutritional research, these sensors are extremely important to analyze trace elements, especially when toxic substances like heavy metals or allergens have to be found. Considering that there are more than 5 million different inorganic and organic compounds known today, and that 100,000 different substances can be identified, it is getting clear that there is an enormous need for such small, inexpensive and reliable biosensors [Camm92]. Also in medicine, where a variety of substances are to be monitored during surgery or during an *in-situ* investigation, an increasing number of small biosensors will be used to record vital patient data for a correct and quick diagnosis.

Biosensors are divided into two groups. They are metabolism sensors and immuno-sensors. A metabolism sensor uses biosensitive enzymes as biocatalysts to detect molecules in a substance and to catalyze a chemical reaction. The analyzed substance is chemically transformed and the course of a reaction can be detected and evaluated by a chemical sensor indicating the concentration of the substance in a solution. This mechanism is illustrated in Figure 6.40 with an example of an enzyme based measurement of phosphate in waste water treatment.

The enzyme nucleoside phosphorylase (NP) is used to determine the phosphate content. This enzyme detects the phosphate and triggers a chemical reaction when inosine is added. One product of this reaction is hypoxanthine (HX). This substance then takes part in another chemical reaction and is

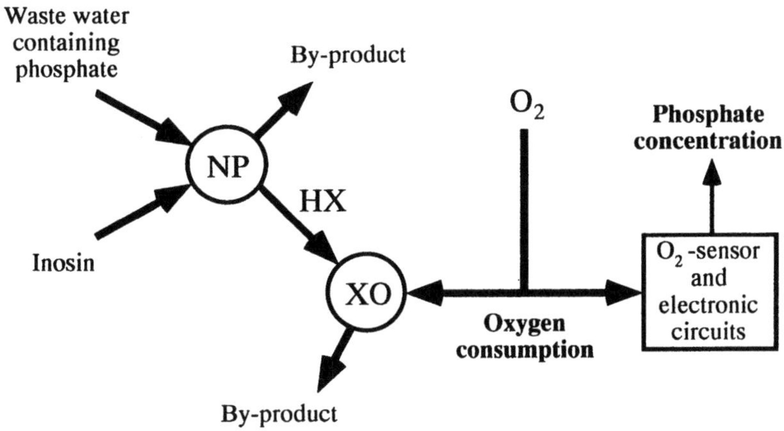

Fig. 6.40: Phosphate measurement with a metabolism sensor. According to [Nent92]

transformed into xanthine oxidase (XO) after consuming oxygen. The amount of oxygen consumed can be registered with a chemical O_2 sensor and the phosphate concentration can be determined from this.

To detect chemically inactive molecules in a substance, immuno-sensors are used; their biosensitive elements are antibodies. The detection method for an antigen molecule is known as the lock and key principle. When it interacts with the analyzed substance, immobilized antibody molecules ("lock") on the sensor surface bond with an antigen molecule "key" in the substance. No other molecule can bond with these antibodies. The bonding process can either be directly registered over a transducer or indirectly through antigen markers, e.g. using molecules of another substance; depending on the type of sensor. From this measurement the concentration can then be determined. The sensor in Figure 6.41 detects the concentration of the antigens directly with an interferometric method. The light intensity changes are here due to the bonding process.

An attempt is being made to integrate biosensors into microsystems to take advantage of the many functions they offer. The integration of biosensors with micropumps and microvalves would make it possible to manufacture very small measuring systems that only need small samples and can measure

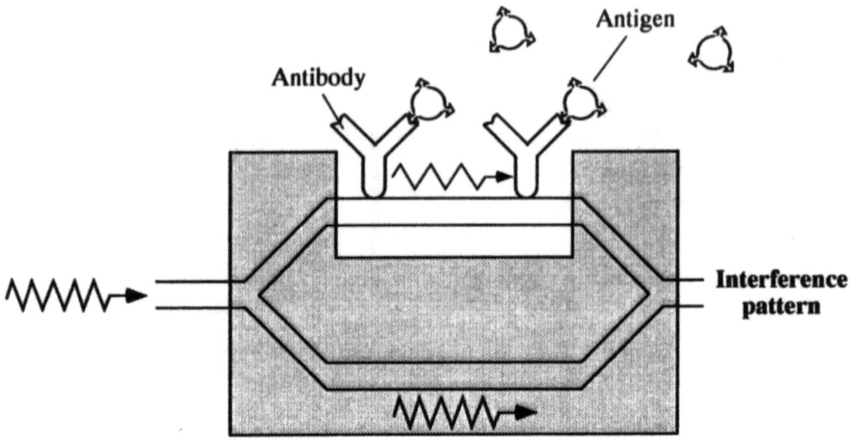

Fig. 6.41: Immuno-sensing using an optical transducer. According to [Göpel94]

quickly. One difficulty encountered, however, is that a system integration of biosensors may produce only short-lived sensors because the proteins are not very stable for a very long time. Another problem to date is the immobilization of the proteins.

Now, some practical developments in the area of miniaturized biosensors will be introduced.

Immuno-sensor with an interdigital transducer

A microanalytic measuring system for immuno-sensing is shown in Figure 6.42. The microsensor consists of five components: a cell body, a flow channel with a grating structure for holding polymeric beads with immobilized immunoproteins, inlet and outlet tubes and a transducer. The transducer uses four pairs of planar interdigital transducers made of Ti/Pt, which are integrated on one chip and simultaneously evaluate the measurement signals.

The flow channel was made of silicon by anisotropic etching. The etched grating was constructed in order to hold the pearl-like polymer beads together, which are in contact with the electrodes and which have immobilized immu-

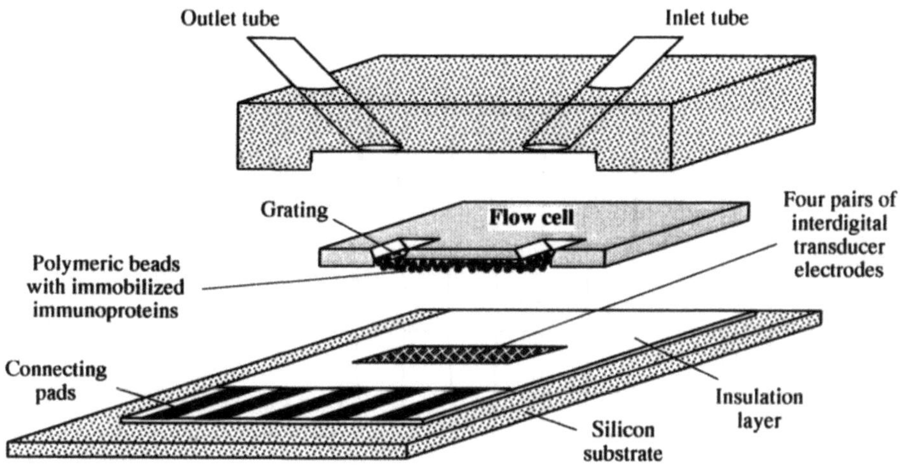

Fig. 6.42: Schematic design of a biosensor. According to [Hint94]

noproteins (antibodies) on them. The liquid to be investigated flows through the inlet and gets into contact with the beads; an immune reaction takes place between the polymer layer of the immobilized antibodies on the beads and the antigen molecules to be detected in the liquid. The antigen molecules are bound to redox (reduction/oxidation) molecules and are thereby marked. The redox reaction is reversible and substances are constantly converted, forming or consuming oxygen. When the antibodies find the "fitting" antigens and bind themselves to them, the redox substance causes several redox reactions in the polymer layer, which results in a 30-fold anodic or cathodic current increase across the interdigital transducer electrodes.

Figure 6.43 shows the output curve of the interdigital electrodes when a substance to be detected is injected into the flow channel of the microsensor, (here paraaminophenyl galactopyranoside). The antibodies on the polymer beads were β-aminogalactosidase. The sensor was able to detect a minimum concentration of about 10 nmol/l. Antibodies for other organic compounds can also be immobilized in the polymer layer. By averaging the output data of all electrode pairs or by simultaneously evaluating both output currents, the sensitivity of the sensor can be increased. This will also contribute to the improvement of the signal-to-noise ratio.

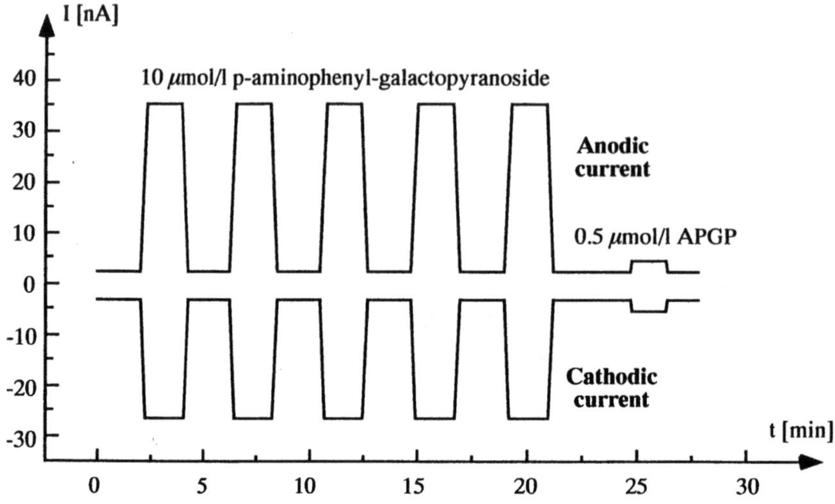

Fig. 6.43: Sensor response to the injection of APGP. According to [Hint94]

Isotropic etching and the lift-off technique (Section 4.1.2.4) were used to make the Ti/Pt electrodes. Each electrode consists of 70 fingers, each of them is 900 µm long and between 1 µm and 3.2 µm wide. The gap between neighboring fingers is between 186 and 800 nm.

Single channel metabolism sensors

Several thermal metabolism sensors can be installed in one sensor system to simultaneously detect various substances using different enzymes. For this purpose, many transducers must be integrated in the sensor system. Here microproduction techniques offer ideal solutions. Figure 6.44 shows such a microsystem; here: T_x – thermistor, E_x – enzyme reaction range.

The sensor consists of a quartz transformer chip (21 × 9 × 0.57 mm^3), a layered silicon structure (17.5 × 0.8 × 0.32 mm^3), which also serves as a spacer and flow channel, and a plexiglass cover to seal the microchannel; the latter contains inlet and outlet tubes made of steel. Several thermistors are arranged at a spacing of 3.5 mm along the flow channel; they are made by doping and etching the quartz chip at specified locations. A silicon oxide layer

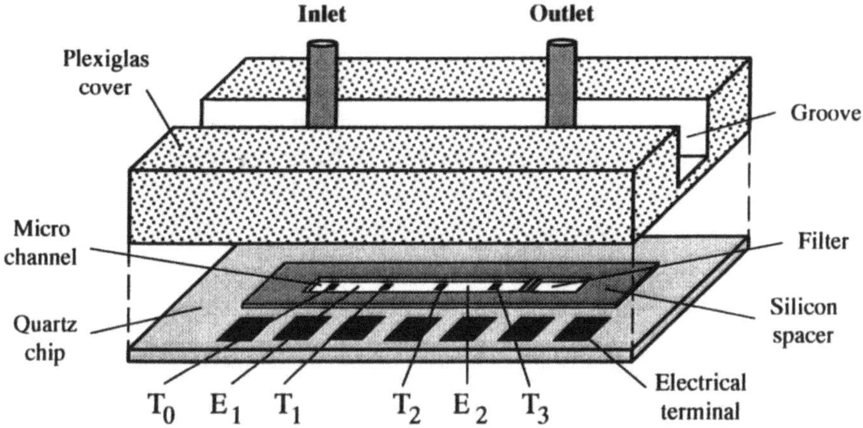

Fig. 6.44: Principle design of a biosensor (the sensor is shown in an open position). According to [Xie94]

is deposited onto the thermistors to electrically insulate them from the flow medium. The thermistors T_0 and T_1 are responsible for measuring the enzyme located in the reaction range E_1 and the thermistors T_2 and T_3 for measuring the enzyme in the reaction range E_2. In order to detect two different substances at the same time, the reaction ranges E_1 and E_2 are supplied with different enzymes (urease/penicillinase and catalase, respectively). Thereby, the reaction ranges E_1 and E_2 are separated from each other in order to avoid that E_2 is thermally influenced.

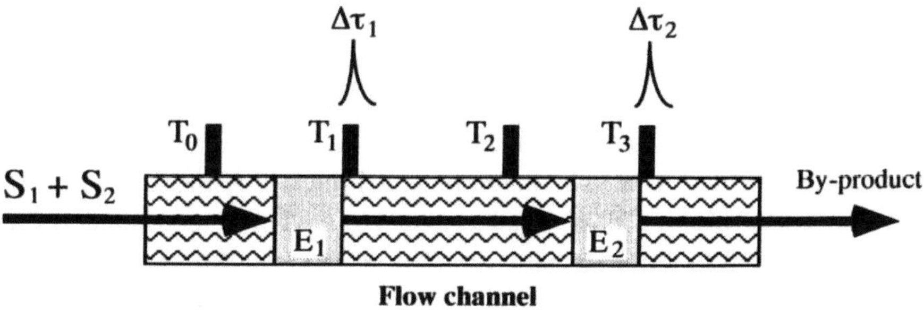

Fig. 6.45: Functional principle of the biosensor. According to [Xie94]

The functional principle of the sensor is depicted in Figure 6.45; here: S_x – substance to be measured, $\Delta\tau_x$ – temperature difference across a range E_x.

When the chemical reaction takes place between the enzymes in the reaction ranges E_1 or E_2 with the substances marked as S_1 or S_2, respectively, a relatively great amount of heat is released, which is registered by the thermistors. T_1 and T_3 serve as measurement thermistors and T_0 and T_2 as reference thermistors, respectively.

The sensor was successfully tested with a liquid containing the two chemical substances urea and penicillin-V [Xie94]. The tests were carried out at a room temperature of 22°C, with a flow rate of 30 µl/min and a sample volume of 20 µl. The concentration of both substances was 5–60 mmol/l. In the reaction range E_1 where the amount of urea was determined, a definite temperature increase $\Delta\tau_1$ of about 0.1°C could be measured. The presence of penicillin-V was detected in the reaction range E_2, the local temperature increase $\Delta\tau_2$ was about 0.1°C. Both enzymes were very selective, which means that a cross-sensitivity effect was negligible. This measurement principle can also be used in more complex metabolism microsensor systems to detect many biological substances. Such measurement systems can fundamentally be useful for clinical diagnoses, health monitoring and biological process contol.

These systems are especially useful for medical applications since many parameters often have to be measured simultaneously *in vivo*, such as glucose, lactate, creatine, urea, or blood gases. Sources [Urban95], [Kohl94b] describe a series of biosensors which were developed and individually optimized. In this context, thin-layer biosensors are being developed to measure the glucose and lactate concentrations and they are being integrated in a microflow system for use in *ex-vivo* monitoring of diabetics. The device was similar to the one described in the previous example. The sensor chip was made by photolithography and thin layer techniques. The inner volume of the flow channel was less than 1 µl. The measurement principle is based on the electrochemical oxidation of Pt-electrodes with H_2O_2. The H_2O_2 concentration which is produced by the enzymatic reaction depends on the concentration of the analyzed substance. The response time of the sensors was 25 s. The output values of the glucose and lactate sensors are linear up to 40 mmol/l and 20 mmol/l, respectively. The biosensors had a life time of about 2 weeks.

6.7 Temperature Sensors

Temperature sensors play an important role in different types of monitoring systems, especially in process, medicine and environmental protection technologies. They are also indispensable for controlling heatings or air conditioning systems and many household appliances. A temperature value may also be a parameter of indirect measurements of other parameters in gas or flow sensors or it may be used for error compensation of temperature-dependent sensors and actuators. There is a wide range of conventional temperature sensors available, like the thermoelement, thermoresistor, thermodiode, etc. A detailed description of different types of temperature sensors is given in [Gard94]. The next examples describe the development of miniaturized temperature sensors.

Fiber optical thermometer

Several glass fiber thermometers were described in [Ecke93]. Figure 6.46 shows a simplified temperature sensor. The sensor contains a light source, a glass fiber, serving both as waveguide and temperature sensor, and a photodiode. The multi-modal glass fiber is made of materials that have different temperature coefficients in the core and the mantle (quartz-silicon system). The light is introduced by an LED into the glass fiber and is propagated through the sensitive fiber area. When the temperature varies in the sensor

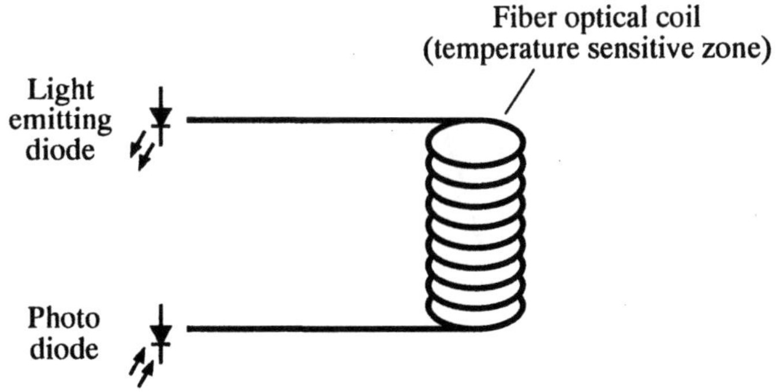

Fig. 6.46: Schema of a fiber optical thermometer. According to [Ecke93]

surrounding, the local index of refraction in the fiber changes, which results in an optical light attenuation. This, in turn, leads to a change in the intensity of light leaving the thermometer; the change of light is measured by the photodiode. This measurement can be used to calculate the temperature.

A prototype of this sensor has been used for a long time. It can measure temperatures of up to 90°C with an accuracy of 0.1°C. The thermometer is insensitive to electromagnetic noise, it costs about seven dollars.

Liquid crystal thermometer

Interesting materials for microsensors are liquid crystals since they are sensitive to light, sound, mechanical pressure, heat, electric and magnetic fields and chemical changes. The molecules of these organic substances are ordered, similar to a crystal, but are also mobile. The orientation of the molecules causes a directional dependence on their physical properties. The so-called thermotropic liquid crystals can be used for measuring temperature since their optical properties are determined by the molecular alignment, which is temperature-dependent.

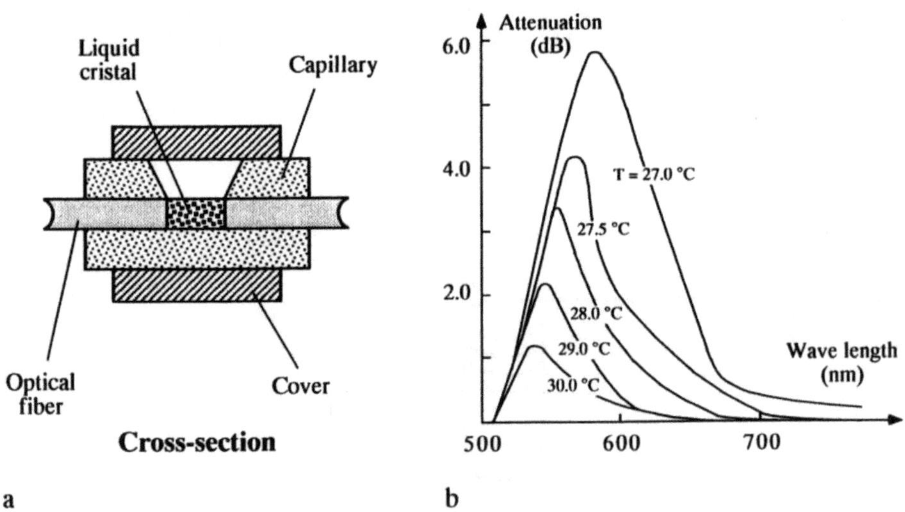

Fig. 6.47: Temperature sensor using thermotropic liquid crystals
a) schematic design; b) measurement principle. According to [Lore93]

In [Lore93], an optical temperature sensor using thermotropic liquid crystals was introduced, Fig. 6.47. The main elements of this temperature sensor are two glass fibers located within a capillary, Fig. 6.47a. The ends of the fibers are 1 mm apart and have a liquid crystal between them.

A light beam which is led into the input fiber gets attenuated depending on the molecular orientation of the liquid crystal. Thus only part of the light appears at the output fiber. A temperature change in the thermometer changes the molecule orientation in the liquid crystal and thereby effects the intensity of the light leaving the output fiber. This attenuation changes can be measured by a photodiode and this value is used to determine the temperature. The operation characteristics of the sensor are shown in Figure 6.47b.

6.8 Flow Sensors

There is a need for miniaturized sensors to measure very small liquid and gas flows, since in many applications, like in medical instruments and automobiles, microfluidic components are becoming an indispensable part of a system. Most of these sensors operate on the principle of thermal energy loss, which occurs when the heating element is located in a flowing substance (thermal dilution). Also, transit time measurements of a trace element injected into the flow can be used to determine a velocity. Another measurement principle uses the forces or torques exerted on an object which is placed in the fluid flow.

Two-mode flow sensor

In [Bran91], a flow sensor was shown which can be operated in two modes. A sketch of this sensor is depicted in Figure 6.48. In one mode, the sensor uses the elapsed time of the locally heated flow medium. A 5 Hz signal was applied to the heater and the time was measured until the temperature rise is recorded downstream by the sensor. In the second mode (thermal dilution), the heater was supplied with constant energy and the temperature difference between the upstream and downstream sensors was measured. The highest sensitivity registered was in the range of 0.05–0.2 ml/min. One disadvantage of this method would be a possible change in the liquid property due to the heat impulses.

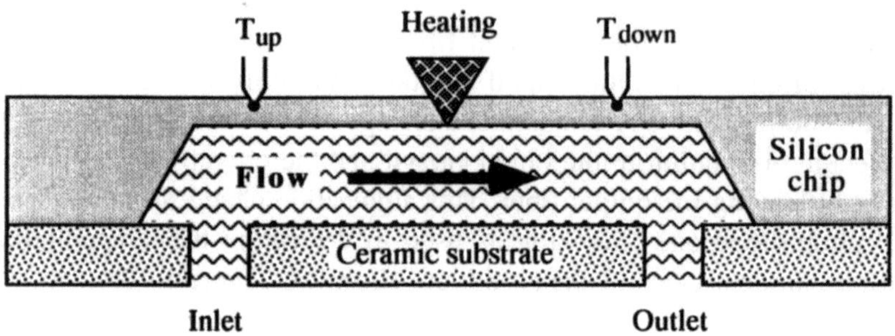

Fig. 6.48: Microflow sensor. According to [Bran91]

A similarly designed flow sensor was described in [Wiel93]. The device was made to measure both liquid and gas flows. It has two temperature sensors, one in front of the heating element and one behind it. Here, the thermal energy loss between the two temperature sensors was measured. All components of the flow sensor are integrated into one silicon chip. The flow channel is 7 mm long, 1.3 mm wide and 350 μm high; it was made by the bulk micromachining technique. Both the inlet and outlet opening have a diameter of 0.7 mm. The silicon membrane is 40 μm thick.

Force-measuring flow sensor

When an object is placed in a liquid or gas flow a force is exerted on it; from this force the flow can be calculated. This principle is commonly used in flow measuring devices. A micromechanical piezoresistive flow sensor was developed and reported in [Gass93]. The device is shown in Figure 6.49. The actual sensor consists of a cantilever with integrated piezoresistors. If a liquid flows through the sensor, a force is exerted on the cantilever. The cantilever moves in the direction of the flow and causes the piezoelements to deform. The deformation is indicated by a change of the output voltage. This change is used to determine the flow velocity.

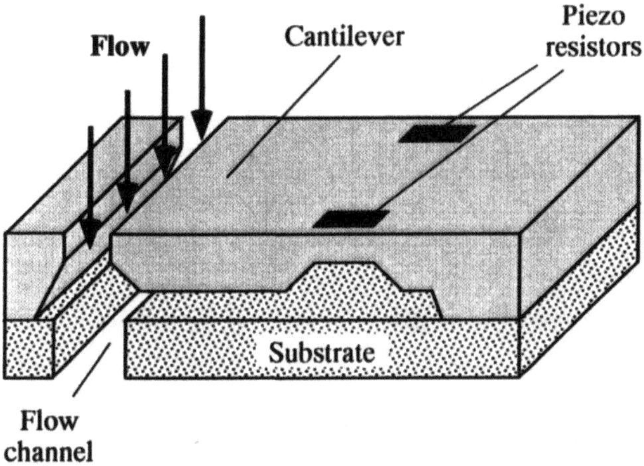

Fig. 6.49: Schematic diagram of the microflow sensor. According to [Gass93]

The cantilever was 3 mm long, 1 mm wide and only 30 μm thick. It was made using a combination of the surface and bulk micromachining silicon techniques. It operates in a range of 5 ml/min to 500 ml/min. A sensor prototype had a sensitivity of 4.3 μV/V per μl/min for water [Gass93].

Thermal flow sensor

A sensor which uses a thermal measurement principle for recording gas flows was developed and shown in [Domi93]. A schematic diagram of the sensor is shown in Figure 6.50.

The sensor has a circular silicon disc into which a heating resistor is implanted. The center of the chip is in contact with the gas and serves as a heating element. There is a ring-like silicon dioxide layer around the disc, which guarantees good thermal insulation between the hot and cold sensor parts. The chip cover was made of a 3 μm thick polyimide film. The flow speed is determined from the thermal interaction between the hot silicon and the gas. Two diodes are implanted into the silicon chip using a CMOS process (not shown in the diagram). They are the thermometers to measure the temperature difference of the gas at the heating element and at a fixed point downstream.

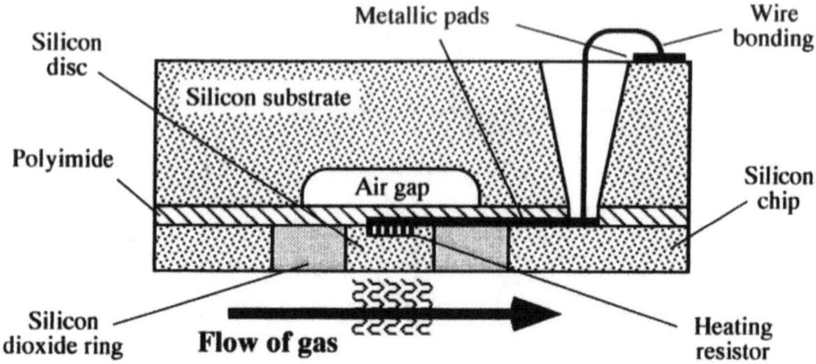

Fig. 6.50: Cross-section of a gas flow sensor. According to [Domi93]

Several prototypes with an overall dimension of 5 mm × 5 mm were produced. The heating membranes had a diameter between 75 μm and 500 μm; the membrane and SiO_2 layer thicknesses were 15 μm and 30 μm, respectively. The measured results agreed with the theoretically calculated values up to a flow speed of 2.5 m/s. The sensor sensitivity was 10 cm/s.

The development of new chemical, physical and biological sensors is a continuous process. Only a small sample of existing sensors could be covered here. A detailed description of current research in this area is given in [Gard94]. Numerous sensor principles had been developed in a variety of forms. The manufacturing technology of semiconductors and thin-films will probably play the largest role in the future to make the sensors and to integrate them with the control circuits on one chip.

7 MST and Information Processing

Already in Chapter 3 various system techniques of importance for MST were discussed. Since this research field is extremely large, we will focus on the discussion of the design and information processing encountered in microsystems. The present state of the technology for realizing interfaces between the microsystem and its surroundings as well as the many existing interconnection technologies and casing techniques will not be covered here. The interested reader may refer to [Kohl94a] and [Menz93] where these topics are described in detail.

The problems of MST concerning among others development, production and quality control are in their early research state, but are becoming more and more important. Today, the clean room activity (manufacturing) assumes about 20% of the cost of developing a microsystem compared with the cost of producing the software needed for the system design [Trau93]. As a matter of fact, microsystems cannot be conceived without computer supported design and simulation tools. Intensive government sponsoring by the BMBF and the EU (Chapter 1) focussing on system technology also reflects the growing importance of this research area.

In a microsystem sensors, actuators and information technological components are combined on little space. Due to the high degree of integration, undesired interactions between the different materials and components within the microsystem may occur. Most of the problems are the result of local and global mismatches, which makes it nearly impossible to independently develop system components. Different temperature distributions or mechanical stresses can cause strain in border regions which can negatively influence the lifetime of the materials. These problems are getting more and more significant and may no longer be neglected, since microsystems are becoming increasingly complex. The most important goal when designing a microsystem is therefore optimizing the overall system functionality, taking into account the technological and economic constraints.

MST research today focuses on component development, and this implies a bottom-up system design. In order to go from system components to complete microsystems, the present R&D strategy has to be corrected toward the solution of system-technological problems. Comprehensive, computer-supported design and quality control test methods must be available, as well as intelligent components for data processing and self-learning control algorithms. These system tools should finally support the top-down design of microsystems. However, it is still unclear whether that will be possible in the near future. Many theoretical and practical problems must be solved first, such as the analytical description of the effects that appear in a microsystem, the development of different types of data bases and the standardization of system components and interfaces.

7.1 MST Design and Simulation

With modern MST design tools, it is possible to quickly and effectively develop a microsystem, thereby reducing expenses and tedious tests. Ideally, the developer should have an interactive CAD tool available to aid him during all design phases (e.g. geometry, technology, perfomance values, etc.) and which can show him graphically each design step along with its important characteristic data. A simulation tool can help to understand all processes occuring during the manufacture of a microstructure. This could also result in a better understanding of the behavior of microactuators and microsensors (component simulation) respectively of the integration of these components in one microsystem (system simulation).

Contrary to microelectronics, where computer-supported design methods have been used for a long time (e.g. for the analysis of the electromagnetic compatibility, thermal load, and performance requirements), MST has many basic problems which are mainly due to the 3-dimensional structure of the product. The LIGA technique shows some of these problems. Here, depending on the complexity of the microstructure, up to 200 process steps must be carried out, usually in a sequential order [Trau93]. As a result of this, the processing of information is excessive for the design of the entire system. With increasing use of MST, computer-supported design and test tools are absolutely necessary to reduce the cost and time necessary to design a microsystem. Until now, the time period from the conception of a product to its completion has been much too long, in particular if handled by a small or medium-sized company.

As previously mentioned, the design of a microsystem is characterized by the necessity to take a number of parameters into account, namely the physical (especially mechanical, electrical and thermal), biological, and chemical quantities. There are also mutual influences and their effects, different technological parameters during the system manufacture, and the interaction of system components to be considered. An interdisciplinary approach is necessary to completely describe the theory behind the effects; for example, the process, chemistry or information technology are some of the research fields which are addressed.

There are various ways to build a system in MST and there are many factors to be considered. E.g. a design concept is usually different with each application (LIGA process, silicon technology or other manufacturing techniques), the abstract ability may be different, the design strategy must be consistent *(top down* or *bottom up)*, standard interfaces and extension possibilities must be provided etc. The information-related steps of designing a microsystem are shown in Figure 7.1.

The process starts with the requirement analysis, where the system capabilities and optimization criteria are determined and ordered. Then, the boundary conditions set by the technical processes and the system are analyzed. Depending on the application, a whole series of technical (performance parameters, production tolerances, material properties etc.), economical, ergonomical, environmental or security parameters are determined. A summary of the specific demands put on a microsystem can be found in [Kohl94a]. Considering the planned goal of mass production of microsystems, the production criteria, which take into account the automatic assembly of individual system components, are gaining importance. A microsystem is designed by considering the application and the constraints from a process. The first description contains all characteristic properties on an abstract level. The design provides the basic data for simulating the component and system behavior as well as adapting the manufacturing steps to meet the demand for the product. The design process can consist of several recursive cycles. After a microsystem has been completed, the product undergoes a function test which allows a variance comparison with the set goals and a possible reoptimization.

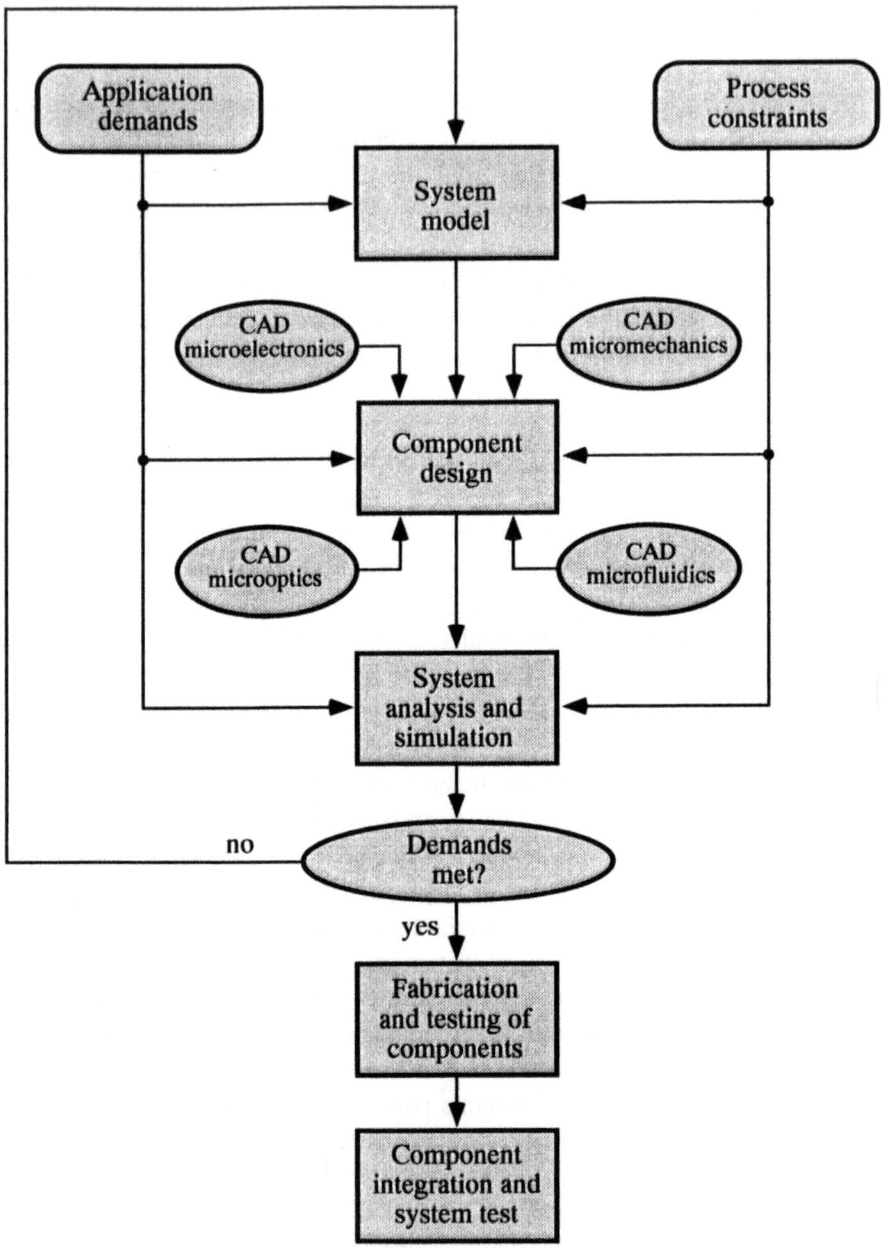

Fig. 7.1: Microsystem development steps

The pursuance of a systematic design strategy in the conceptual phase and thereby understanding the various interrelationships within the microsystem, is presently not completely possible, although there are many MST researchers working in this area. E.g. a CAD/CAE framework called CAEMEMS (*Computer-Aided Engineering of Micro-Electro-Mechanical Systems*) is being developed [CISC91]. Another design system called CAFE (*Computer-Aided Fabrication Environment*) had been implemented step-by-step during the past several years [MIT92]. It was conceived to support the design, development, planning and manufacture of an IC wafer or a microstructure.

The complexity of an interdisciplinary design system prevents its short-term realization. Due to the many degrees of freedom, a micro product can often not be controlled as an entire system, or it is only possible with great difficulty. Therefore, it is often necessary to divide a design up into relatively simple components which can be modelled (e.g. individual sensors or actuators), and integrated into submodels [Recke95]. The latter should be done as thorough as possible in order to fit the components into a complete system. A component model can serve as standardized building block for different types of microsystems. There must be continuous design support for microcomponents and manufacturing processes. Often it is attempted to adapt already existing modelling methods from other technical areas (microelectronics, mechanics) to MST.

One design example of a micromechanical structure, which involves a combination of lithographical and etching procedures, is presented in Figure 7.2. The design starts with the structuring of the layout, which may be supported by a knowledge base, but which also requires a certain degree of intuition of the designer. An appropriate mask layout must then be found for the selected structure taking into account the process constraints. Next, a 3D process simulation is carried out which is based on the characteristic process data and mask parameters. The result is the microstructure's geometry. Then, the microstructure is simulated by the finite element method (FEM). If the FEM simulation does not render acceptable parameters, the mask layout and design concept are revised. This cycle is repeated until the simulation results agree with the designer's expectations.

The described design procedure will be illustrated with an example of a real R&D project [Popp91], which had as a goal the development of a piezoresis-

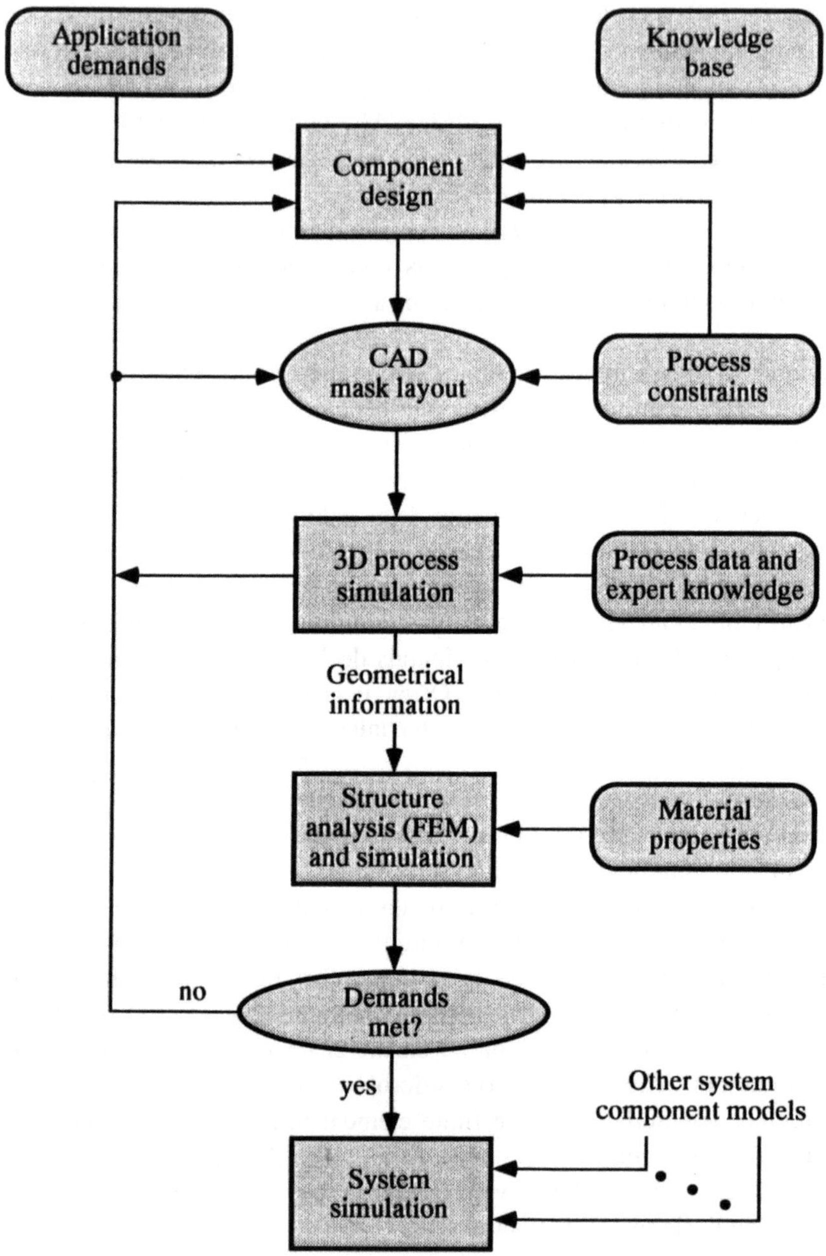

Fig. 7.2: Development of a micromechanical structure. According to [Pavl94]

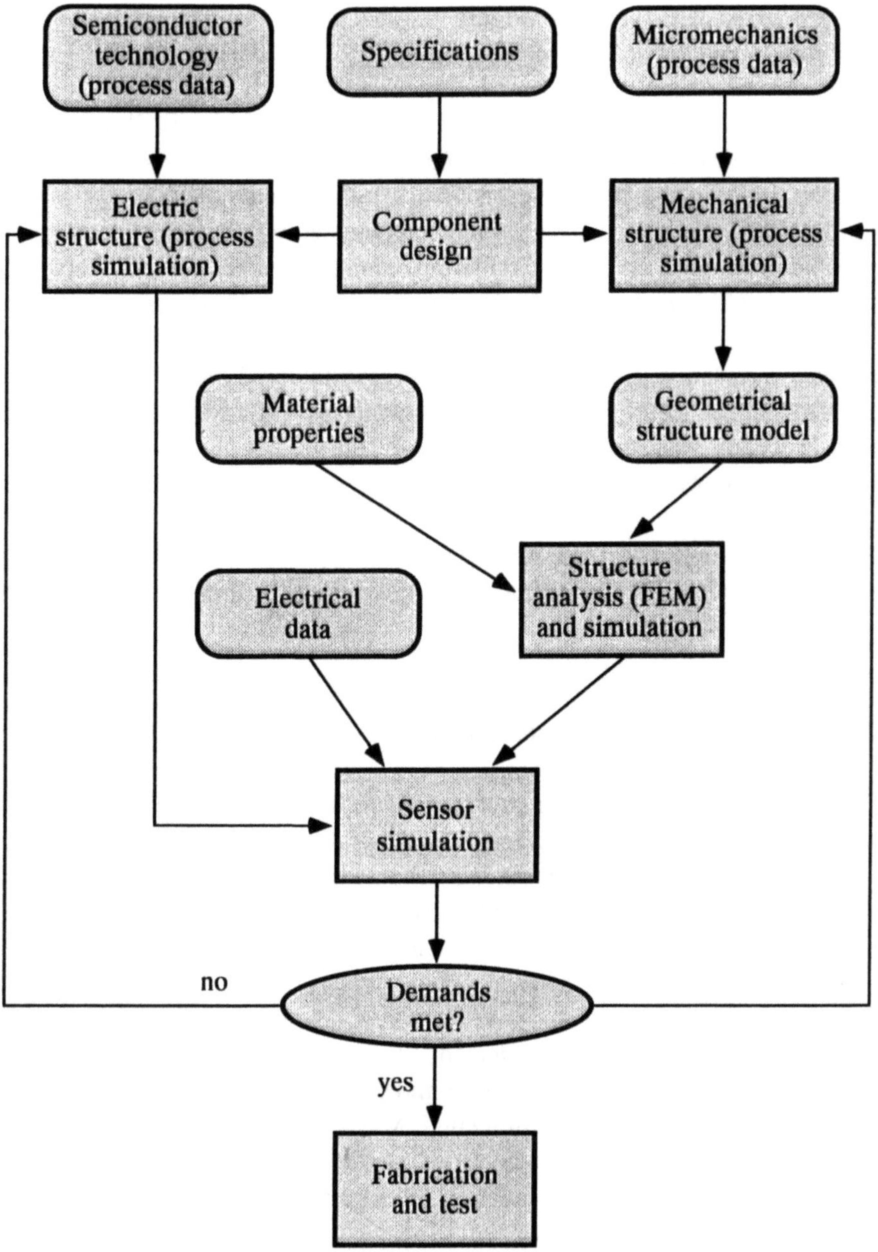

Fig. 7.3: Design procedure of a piezoresistive pressure sensor. According to [Popp91]

tive pressure sensor (see also Figure 6.4). A sketch of the computer-supported design procedure is shown in Figure 7.3. The expected sensor properties (process- and application-specific features) are first recorded in the specification from which the sensor design is derived. Afterwards, the electrical and/or mechanical structure is determined taking the constraints of the semiconductor process and the micromechanics into account. As the next step a geometrical model of the sensor structure is created which is then used as an input information of the FEM simulation, along with the material data. The result of the FEM analysis is the knowledge of the stress distribution in the entire membrane under various temperature and pressure conditions. During the sensor simulation, the membrane stress is calculated and the corresponding sensor signal resulting from the piezoresistive effect is computed. This design procedure has proven itself in the design of various pressure sensors from 2 kPa to 40 Mpa.

In the next section, the most important microsystem design steps will be discussed, including the modelling and simulation of processes and of microsystems and their components.

- **Process modelling and simulation**

A 3D simulation of structuring processes (e.g. lithography, etching and depositing processes, etc.) is necessary to generate reliable input data of the product and to better understand and use the available 3D technologies. Up til now, only wet etching and lithography can be simulated. Dry etching and depositing techniques suffer from insufficient process models and plasma-based processes are difficult to describe.

Wet etching processes are dependent on the composition and concentration of the etching solution, the temperature and the crystal orientation of the silicon; they can almost exactly be simulated. Many simulation models for anisotropic etching are being developed, such as ASEP [Buser91] and SIMODE [Früh93]. The ASEP tool can simulate the etching process of (100) silicon using KOH; whereby a user-defined mask and information concerning the etch depth is applied. Such a tool is integrated into the CAEMEMS framework mentioned earlier [Buser92]. SIMODE can simulate orientation-dependent silicon etching in any direction. SOLID is a lithography simulator used in microelectronics; it can simulate the 3D manufacturing processes of surface micromachining technology.

To model and simulate a manufacturing environment, it is necessary to have a sufficiently abstract description of the process as well as task-specific knowledge, Fig. 7.4. Silicon properties have already been well researched for microelectronics activities, but most materials used in micromechanics must still be experimentally analyzed in their various geometric forms in order to determine the corresponding manufacturing parameters.

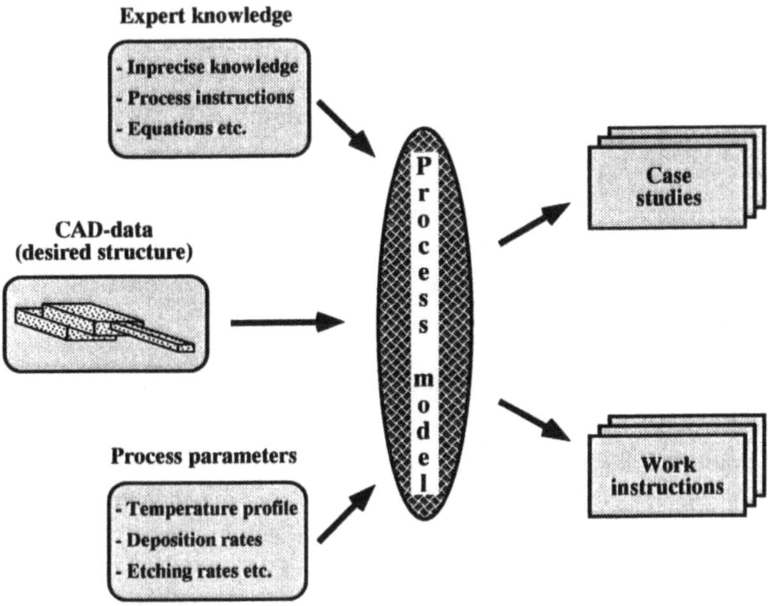

Fig. 7.4: The concept of process modelling and simulation. Courtesy of the Karlsruhe Research Center

The software tool LIMES *(Liga Manufacturing and Engineering System)* is currently being developed for simulating the LIGA process [Herp94]. Using 3D CAD data, this program can help to investigate the influence various process parameters have on a microstructure to be produced. Existing LIGA knowledge is prepared in an acquisition phase. Both knowledge in the form of functional dependencies between various parameters as well as fuzzy expert knowledge in the form of fuzzy rules are used. LIMES will be integrated in the design environment LIDES *(Liga Design and Engineering System)* for LIGA microsystems [Egge96]. LIDES contains an object-oriented data base and a framework which allows the integration of commercial design and simulation tools (mechanical/electronic CAD tools, FEM simulators and digital circuit simulators) and the test system COSMOS-2D (Section 7.2) for automatic measurement of produced LIGA structures.

- **Simulation of microsystems and their components**

Simulation methods for microsystems and their components will help to optimize the system behavior of a computer-supported design. To do this, various parameters (electrical, thermal, mechanical, chemical, etc.) of the microsystem must be gathered, calculated, and it must be shown how these quantities influence each other. As already mentioned, tools from other technical fields, especially microelectronics, are being used or adapted for the simulation of microsystems and components. There are several analog simulators for electronic components of microsystems, such as SABER, KOSIM and SPICE as well as digital logic simulators.

The most popular standard analog simulator is SPICE *(Simulation Program with Integrated Circuit Emphasis)*. It can be used to simulate all components of a control circuit and to simulate closed-loop controllers; this makes it easier to test different system configurations. Even non-electronic system components (thermal, optical, mechanical, etc.) can be modelled with SPICE if all quantities can be described via electrical models. In [Weick94], a simulation of a microlaser system using SPICE is described; the results of a thermal-mechanical cross-coupling in a piezoresistive silicon pressure sensor simulated with SPICE are shown in Figure 7.5.

The parasitical thermal effect on the functions of an integrated switching device is investigated. The cross coupling transforms a membrane displacement into an electrically measurable resistance change via a piezoresistive effect. The simulated sensor is turned on and after 100 ms (after the switching device has warmed up), a sine shaped pressure impulse is applied to the membrane. There is a temperature offset in the sensor response which must be taken into consideration to understand the functioning of this device.

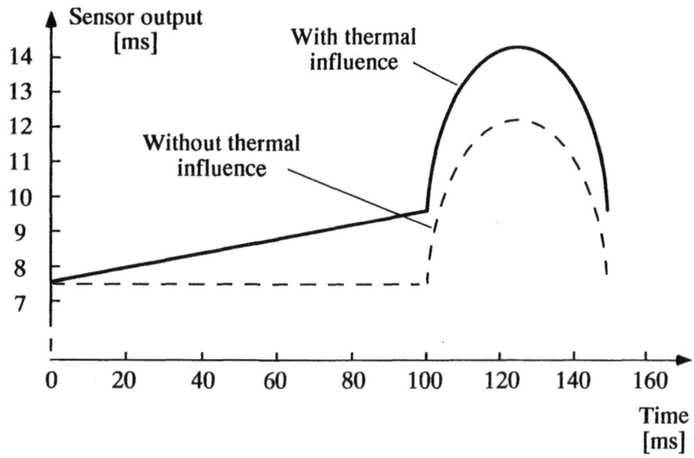

Fig. 7.5: SPICE simulation of a piezoresisitive silicon pressure sensor with an integrated switching unit. According to [Klaa94]

Besides thermal disturbances, often there is a parasitical electromagnetic cross coupling of galvanic, capacitive and inductive nature. For this reason, the electromagnetic compatibility (EMC) of the system must be considered during the design phase. In [Schra94], investigations on electromagnetic compatibility are described in detail using an example of a piezoresistive acceleration sensor simulated with SPICE.

There are various standardized electric models available in MST-specific model libraries [INFO95]. However, the progress of analog simulation is slow because electric models for micromechanical system components and for the transformation of electric quantities into non-electric ones and vice versa are not yet available. Often, the action of micromechanical components is realized with the help of resistive, thermal, electrostatic or piezoelectric effects. Here, differential equations which adequately describe the relationships between the various field parameters must be solved to simulate components. There are two types of interactions, one is unidirectional (e.g. an inhomogeneous temperature distribution can cause a mechanical tension) and the other one is bidirectional (cross-sensitivity), whereby both effects influence each other. Typical methods which are used to discretize the equations and to determine the field distributions with the help of a computer are the finite difference method, finite volume method, finite element method, etc. There are many commercial software tools available, especially for the finite element method which has been used for a long time in macroscopical elastomechanics. Often the linear and nonlinear behavior of materials can be simulated in a thermal-mechanical state. Thereby, temperature-dependent material quantities and visco-elastic behavior can be taken into account for determining temperature or mechanical stress distributions (e.g. in an elastic sensor membrane). There are also special FEM tools to calculate fluid dynamic behavior and for calculating electromagnetic fields.

A geometric model of a microstructure, including material specifications, is necessary for a successful FEM simulation. This information can also be obtained from a CAD system. The structure is then divided up into many finite elements. The nodes of the resulting net are discretized points of the continuous field quantities. There exist commercial FEM simulators, such as ANSYS, NASTRAN, COSMOS and ABAQUS, which divide a mechanical structure into finite elements. After setting the input-output relationship for the nodes, the equations are generated for the phenomenon to be investigated. This reduces the initial complex problem to numerous simpler problems (depending on the number of elements) which can easily be solved with a computer. Finally, the desired quantities are determined from the resulting field distribution. If the materials and geometries are varied, parameter studies can be carried out and the system can be optimized.

The FEM can be used to calculate mechanical stress distribution in a silicon membrane under pressure. This allows to position the piezoelements exactly on the pressure sensor. An optimal geometry for a maximum sensor sensitivity can be mathematically found this way, whereby the deviations from the actual values are about ±10%. The effects of the mechanical stresses in multi-layered systems (e.g. silicon nitride – silicon – glass structures) can also be shown. The results can be used for the optimization of manufacturing processes or for showing the temperature distributions in an integrated microchip or for determining the best placement of power electronic components on a substrate. Steady state influences are evaluated with static FEM calculations, whereby dynamic calculations are used to investigate resonant frequencies or the component's behavior at different frequencies. This for example can be used to determine a shift of the membrane resonant frequency with respect to the change of the physical parameter to be determined.

It must be stressed that experienced FEM users may need days or even weeks to find a suitable descriptive model for a finite element analysis of an object; also the actual calculation time for each structure variant can take several hours. The lack of knowledge about material properties may be a severe problem. Furthermore, the behavior of a micromechanical component may not necessarily be the same as that of a macrocomponent made from the same material (scaling effects). Surface forces and mechanical stresses play a much larger role in microcomponents than in macrocomponents.

An example of a good FEM application is the numerical analysis of the excitation of an electrostatic microcomb structure [M²S²93], [Schwa94]. The lower comb is attached to the substrate and the upper one hangs from an elastic membrane, Fig. 7.6.

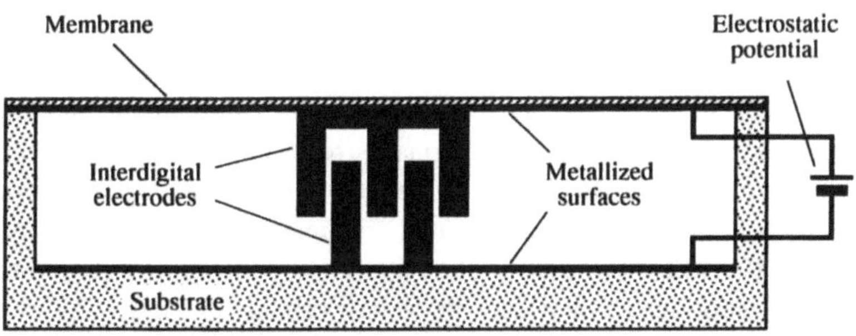

Micro comb structure

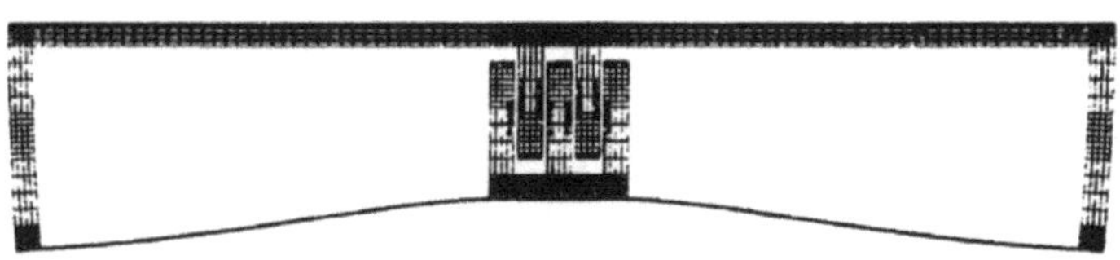

FEM simulation

Fig. 7.6: Thermal-mechanical simulation of the membrane behavior. According to [M²S²93]

By applying an AC voltage to the device the membrane is made to resonate, whereby the resonant frequency can be modified by external means. An

FEM simulation tool was used for this sensor to investigate thermal-mechanical interdependencies and their influence on the membrane distortion. The 3D simulation system SESES, which is based on the FEM, was used for this simulation; it was extended by supplementing or reformulating the model equations which describe the elastothermomechanic, magnetomechanic and piezoelectric effects on the performance of the microsystem. The numerical capability and the user-friendliness of the simulation package was drastically improved in comparison to ANSYS by a new fine-tune strategy and a functional extension of the graphical modules. Interfaces were also made with other simulators (e.g. FASTCAP) and to the process simulation environment, which should make it possible to simulate the entire design process from the layout to the finished component.

The simulation package PICUS *(Piezoelectric Systems in a Cubic Trial Function Simulation)* was developed for modelling and optimizing piezoelectric ultrasound sensors [NU-Te93]; e.g. it is possible to simulate the resonance behavior of a piezodisc by applying an AC voltage (piezoelectric effect). Thereby, the resonant frequency can be modulated by varying physical parameters. It can be shown that for every resonant frequency there is a certain form of oscillation, which is dependent on several material and geometrical parameters. PICUS helps to determine optimal frequency ranges for a desired disc deformation. The FEM was used in the simulation package for modelling the piezodisc, the coupling medium and the sensor casing. In Figure 7.7 characteristic forms of oscillation are shown which were obtained for a circular piezodisc with axial polarization by using PICUS for various resonance fre-

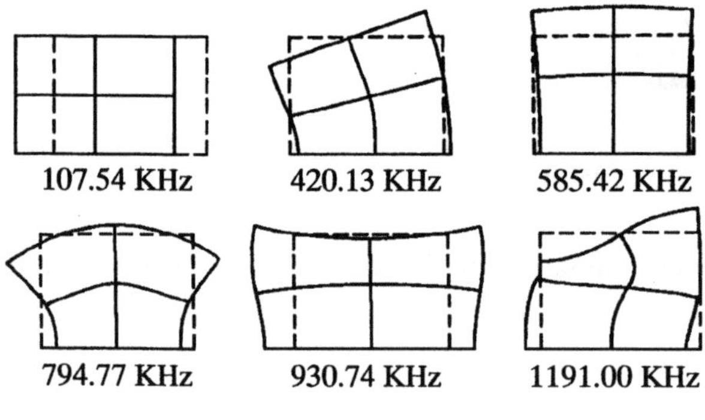

Fig. 7.7: Simulation results with PICUS package. From [NU-Te93]

quencies. These results were compared with real deformation measurements. The deviations were within a 10% tolerance limit which shows the accuracy of the PICUS' results.

The "Microsystem Technology 1990–1993" program sponsored by the German government agency BMBF (Section 1.4.1.2) included a joint partnership project to develop tools and technology for designing and realizing micromechanic (piezoresistive and capacitive) pressure sensors. Among others, the simulation tool SENSOR was developed for the construction and optimization of a pressure sensor; it is based on a commercially available FEM package [VDI94]. Since the simulation tool is modular in design, other sensor types, e.g. force or acceleration sensors, can easily be accommodated.

7.2 Test and Diagnosis of Microsystems

One important factor in developing microsystems or system components is a continuous quality control. Suitable test and diagnosis methods should be used already in the designing phase in order to be able to sufficiently verify the most important or even all functions of a microsystem. Ideally, every process step must undergo quality control in order to find errors or defects in time. Already with the first fabrication step of a microstructure, the so called layer deposition, the desired optical (refractive index) respectively mechanical properties (internal layer stresses) or the structural geometry must be examined, depending on the type of structure built. Here computer-supported methods are indispensable to produce quickly and reliably quality control data. The "Microsystem Technology Program 1994–1999" of the German BMBF (Section 1.4.1.2) is supporting a quality control project for supervising the manufacture of microsystems.

Since there are few microsystems presently available, test methods for individual micromechanical and microelectronic system components are of particular interest. While microelectronic QC methods for testing integrated circuits can be used in MST, new test methods must be developed for the quality control of micromechanical components. The quality of a micromechanical structure is usually determined by the accuracy of the geometric parameters; to realize this reliable measurement systems are needed.

A fully automatic measurement system, COSMOS-2D *(Computer System for Measurement of Optically Acquired Structure Surfaces in 2-Dimensional Space)*, is being developed for testing microstructures using image analysis [Guth93]. It uses gray-scale pictures and a computer-controlled microscope equipped with a CCD camera for digital image processing and pattern recognition. A comparison of the actual geometry with the expected geometry saved in the CAD system allows a judgement to be made on the product quality. By using pattern recognition, even an inexact positioning of a component under the microscope is not critical and permits thorough quality testing.

A laser scanning microscope can be used to acquire geometrical information about an initial 3D microstructure. Here the object surface is scanned point-by-point with a thin laser beam which is approximately 1 μm in diameter. The reflected light is detected and the single points are assembled to form a complete in-focus picture. The operation of the microscope can be controlled by an external computer, which is advantageous in industrial production. There are also various scanning probe microscopy (Section 2.5) and X-ray diffractometry methods available which can do testing of micromechanical structures. These methods can also be used in combination with numerical FEM simulation. For example, the characteristic internal stresses of a multi-layered piezoceramic wafer (Figure 5.25) which are obtained by means of X-ray diffractometry, are important input values for a numerical simulation of the mechanical-thermal behavior of a structure [Faust93].

Tests can be simplified by using defect simulations obtained in the design phase. A thorough system description (e.g. electric SPICE network model) is necessary to simulate a faulty behavior of a system. Thereby, test signals can be deduced which will detect errors or defects in the real devices.

The testability of a microsystem during its operation is also significant in MST, since ageing is much more serious here than in macrodevices. Therefore, it is tried to install internal diagnosis functions directly into the information processing components of the microsystem to be able to recognize and report a faulty function. For this purpose, internal sensors, which locally supply the real-time test algorithms with data, are of advantage. The algorithms can be run on an internal microprocessor or an external test device connected to the system. Due to the wide range of products, it is very difficult to develop standardized test and diagnosis components.

7.3 Information Processing in Microsystems

7.3.1 Introduction

An overview of information processing tasks which decisively influence the performance of a microsystem is given in Figure 7.8. The information processing component is used to generate the electrical signals for controlling the system actuators using sensor data, for self-monitoring and for communicating with external systems. An information processing cycle consists of the sensing, transformation, storing, evaluation, and generation of signals. It is obvious that an engineer working on the information processing part of MST devices must thoroughly understand the underlying computer technology.

Sensors transform chemical analysis data or physical data into electrical analog signals, which can be digitalized with an A/D converter and amplified. MST allows many sensors to be integrated onto one chip. This way, signals can be simultaneously detected and brought together in a subsequent fusion or parallel processing procedure to get the desired information. The sensor information obtained after the signal filtering and the application of pattern recognition methods is evaluated numerically and serves as a control input to the actuators. A microsystem can also be self-monitored and tested by its central processing unit, and it can communicate with other systems and microsystems, if necessary. Thereby the communication often takes place in a noisy surrounding which is far from ideal. Therefore, it is important to use error-tolerant transfer and/or intelligent information processing methods to recognize and interpret noisy signals.

Many highly integrated components for information processing in microsystems, like devices for signal amplification, transformation, postprocessing, storage, and signal processing, are already available. However, for some MST systems only insufficient results have been obtained by using conventional information processing methods, for example when the sensor signals are time invariant or contain noise. Other sources of error are cross-talk and aging, which can lead to considerable difficulties in microsystems. Under these conditions it is often very hard to find a useful system model. In classical control theory, however, it is assumed that the measured values are very exact and that a mathematically accurate model of the system is available.

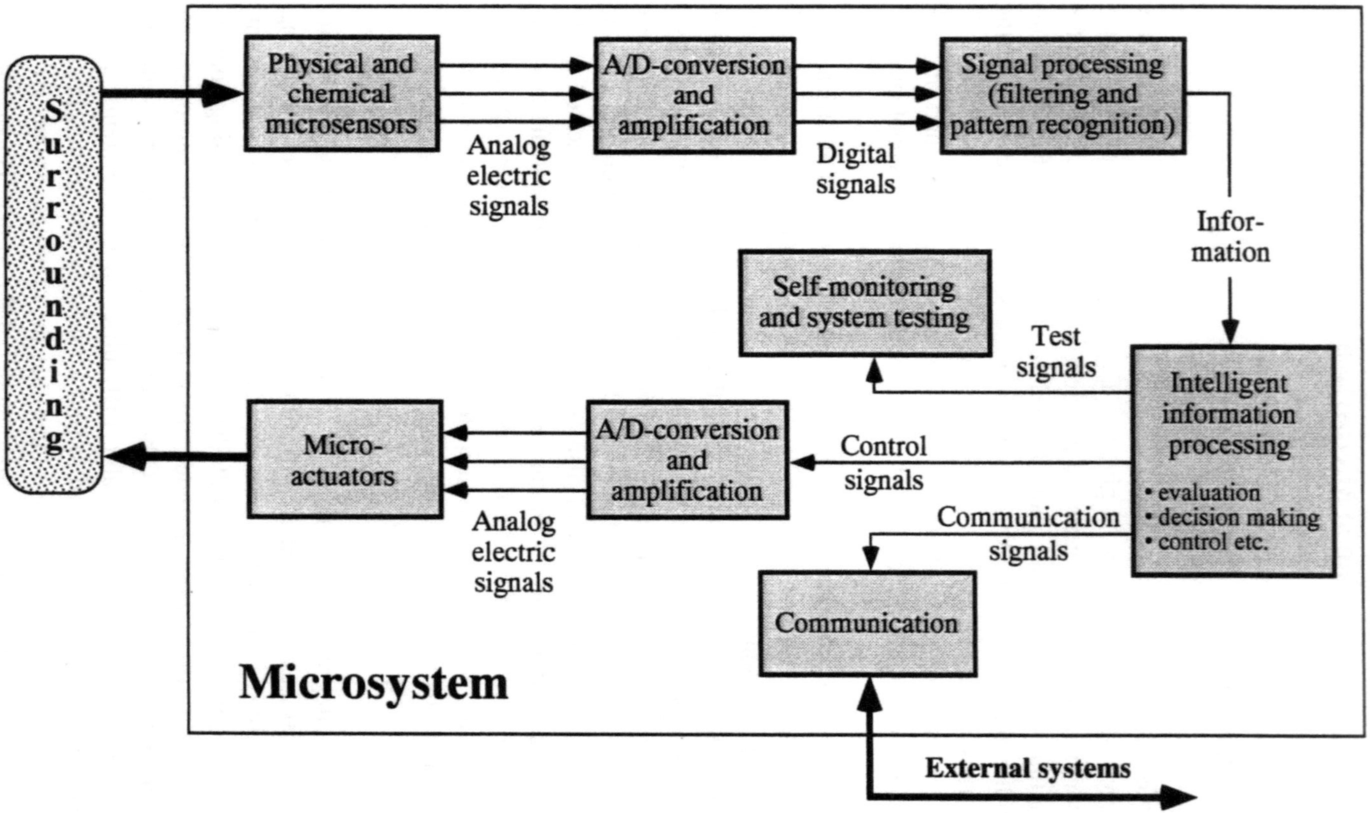

Fig. 7.8: Signal and information processing in microsystems. According to [Trau93]

Increasingly complex microsystems with their growing number of components do not satisfy this assumption. The flow of information is becoming harder and harder to follow and cannot be handled by using conventional information processing techniques.

On the other hand, by using an exact and often large model, the microelectronic realization might be complicated and there are long delay times until realization. The demands for real-time behavior and the integration on one chip make these systems difficult to design, and in the end a compromise must be found between measurement exactness, reliability, integratability and cost.

The capability of an intelligent microsystem to adapt itself to the process requirements, especially for environmental monitoring, medical *in-situ* applications, or for various types of inspection and assembly robots is very important. The microsystem should be able to operate in an undefined environment and be able to guarantee a reasonable behavior in unpredicted situations. For this new information processing methods are needed which do not require an exact system model and which allow a reasonable compromise between the real-time processing, the exactness and the amount of input data; they must make definite decisions based on vague or incomplete information. Two promising methods are fuzzy logic and neural networks. Typical examples are: the fuzzy logic supported fusion of different sensor data (often to obtain characteristic values for the system control, which cannot be measured directly), the use of neural networks, which store data associatively for classification of characteristic features of a noisy set of signals, or the use of a back propagation neural network to control a microsystem with the non-linear behavior. These promising and growing information processing methods may be able to solve the conflict to obtain a high degree of miniaturization in conjunction with high intelligence. Such methods are described e.g. in [Bekey93], [Bothe93], [Hertz91], [Kahl93], [Krat93], [Kruse94], [Nauck94] or [Zimm91] and will be introduced briefly below to give the reader the necessary basic knowledge.

7.3.2 Fuzzy Logic

The fundamentals of fuzzy logic will be described with the help of a fuzzy controller for a microrobot developed at the University of Karlsruhe. There is a more detailed discussion on this robot in Chapter 8. Figure 7.9 shows a

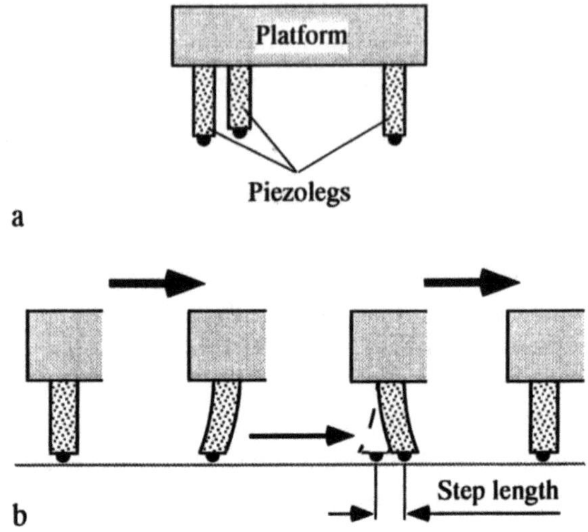

Fig. 7.9: Concept of a piezoelectric microrobot
a) sketch of the robot platform; b) motion principle of the robot

scheme of the robot platform and the principle of motion, respectively. The positioning of the platform is done by three tubular piezoceramic legs which are provided with four metal electrodes each, on their outside surface. The electrodes are parallel to the center line of the legs and equally spaced about the circumference of the legs. If a voltage is applied to an electrode the legs will bend and can assume different speeds by proper control. For moving the following sequence is maintained, Fig. 7.9b. First, the legs are slowly bent and stay in contact with the surface on which they stand. This causes the robot to walk. Second, the legs are quickly reversed and they go through a sliding motion into a new position. Finally, the legs slowly bend to reach a straight position, and the step is completed.

The fuzzy controller for the platform will serve as an example to describe the fuzzy method. Of course, there are additional parameters to be controlled with a mobile robot. The behavior of the robot motion can be described in a similar manner as that of a car. The fuzzy controller must determine the step frequency F of a robot leg, in regard to the distance D to be travelled to the robot's goal and the step length L of a leg. The latter parameter is generated by a saw tooth shaped electrical voltage supplied to the piezo legs.

The characteristics of the binary logic is that only the values of 0 and 1 (high or low) can be assumed. Contrary to this, fuzzy sets also have a degree of membership; they can assume any value between 0 and 1. This enhancement of the binary logic opens up new possibilities for the mathematical treatment of quantitative, linguistically formulated knowledge. This means that ambiguous and inaccurate information can be handled, which is very typical for complex systems. Every physical parameter can assume several quantitative or linguistic values; there is no exact border between these values. Each linguistic value is a fuzzy set and a physical parameter is defined as a linguistic variable.

Fuzzy sets are represented by so-called membership functions $\mu(x)$ which can be described either parametically by an analytic function, graphically by a transfer function $\mu(x)$ or discretely with a pair $\mu(x_i)/x_i$ parameters. Figure 7.10 shows the graphical presentation of the fuzzy sets (linguistic values) *very_small, small, middle, large* and *very_large* of the linguistic variable *Distance D*; it was defined above.

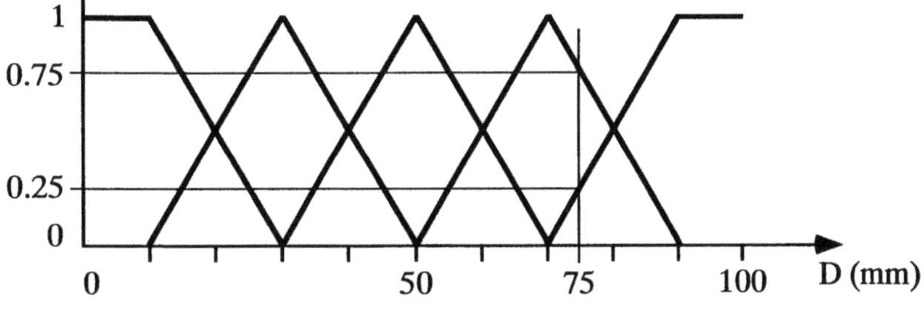

Fig. 7.10: Fuzzy sets for the linguistic variable *Distance D*

The fuzzy sets cover the entire range of the variable *Distance D*, thereby describing this physical parameter completely in the sense of the fuzzy logic. E.g. the membership function *large* represents the degree of membership with which an exact distance value will be defined as large. The distances between 50 mm and 90 mm will be assigned a degree of membership other than

0. According to the exact logic, only the value of 70 mm would be large, because only this value has the degree of membership of 1. E.g. the exact value of 75 mm has the degree of membership of 0.75 for the fuzzy set *large* and the degree of membership of 0.25 for the fuzzy set *very_large*, Fig. 7.10. This roughly corresponds to a human estimation of the distance of 75 mm between the microrobot and its goal.

In principal, various curves can be used to describe fuzzy sets. However, in practice trapezoid- or triangle-shaped membership functions with linear transistions are used. They are easy to realize both by software or hardware implementation of a fuzzy system. Figure 7.11 shows the other linguistic variables used in the example; they are *Step_Length L* and *Step_Frequency F*.

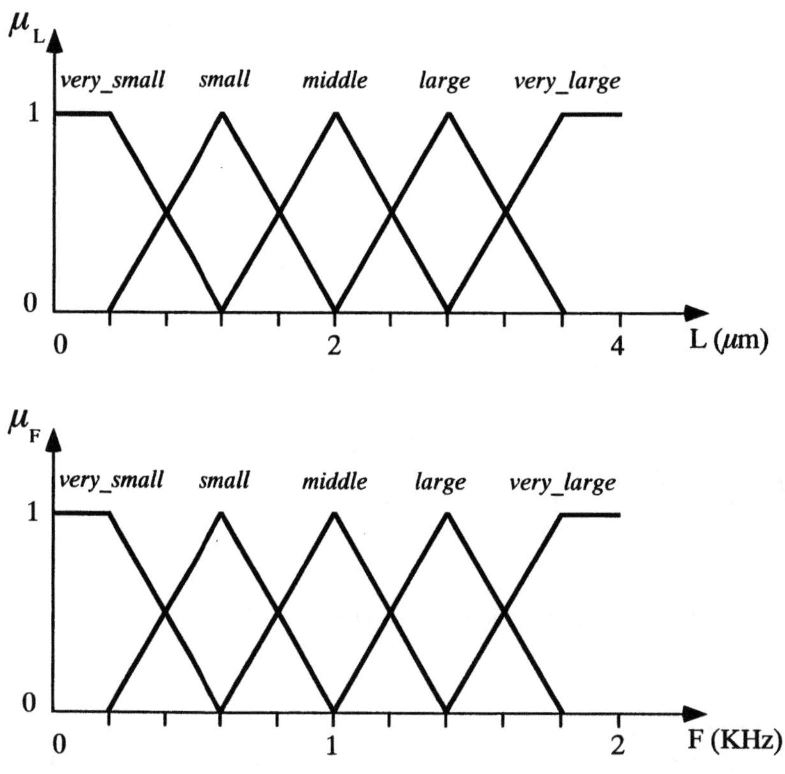

Fig. 7.11: Fuzzy sets for the linguistic variable *Step_Length L* (top) and the linguistic variable *Step_Frequency F* (bottom)

Fuzzy information can be combined using AND/OR operations. The AND operator for a fuzzy set is defined as the intersection of the areas under the membership functions (MIN operator) and the OR operator for fuzzy sets is the union of the areas under the membership functions (MAX operator). The ambiguous infomation and the AND/OR operators can be used in combination with IF-THEN rules to form fuzzy relationships (fuzzy rules). To make control desicions, an inference scheme contains specifications for processing input values to form output values using a set of fuzzy rules. This way, any problem which can be formulated with normal language can be solved with algorithmic computation methods.

To illustrate the inference scheme, we will be using the aforementioned example again. The rule base will be consisting of the following fuzzy rules:

R_1: IF D = *middle* AND L = *large* THEN F = *low*

R_2: IF D = *small* AND L = *very_large* THEN F = *very_low*

With these rules, an exact output value F has to be determined from the exact input values D = 35 mm and L = 3.1 μm. By applying the inference scheme, the following successive steps are taken:

- The transition from the exact signals D and L to the corresponding fuzzy values D^* and L^*, respectively, is called the fuzzification and consists of the computation of the degree of membership of all fuzzy sets to the linguistic variables D and L. During fuzzification, the exact values of the inputs are converted to a vector of the degree of membership values:

$$D^* = (\mathrm{m}_{\mathit{very\_small}}(D), \mathrm{m}_{\mathit{small}}(D), \mathrm{m}_{\mathit{middle}}(D), \mathrm{m}_{\mathit{large}}(D), \mathrm{m}_{\mathit{very\_large}}(D))$$
$$L^* = (\mathrm{m}_{\mathit{very\_small}}(L), \mathrm{m}_{\mathit{small}}(L), \mathrm{m}_{\mathit{middle}}(L), \mathrm{m}_{\mathit{large}}(L), \mathrm{m}_{\mathit{very\_large}}(L))$$

Accoding to Figure 7.12, we obtain:

D^*(35 mm) = (0, 0.75, 0.25, 0, 0) and
L^*(3.1 μm) = (0, 0, 0, 0.625, 0.375).

- The next step is to determine the active rules of the rule base. A rule is active if its degree of satisfaction is greater than 0. The degree of satisfaction is determined by using the MIN operator for all the membership values of the

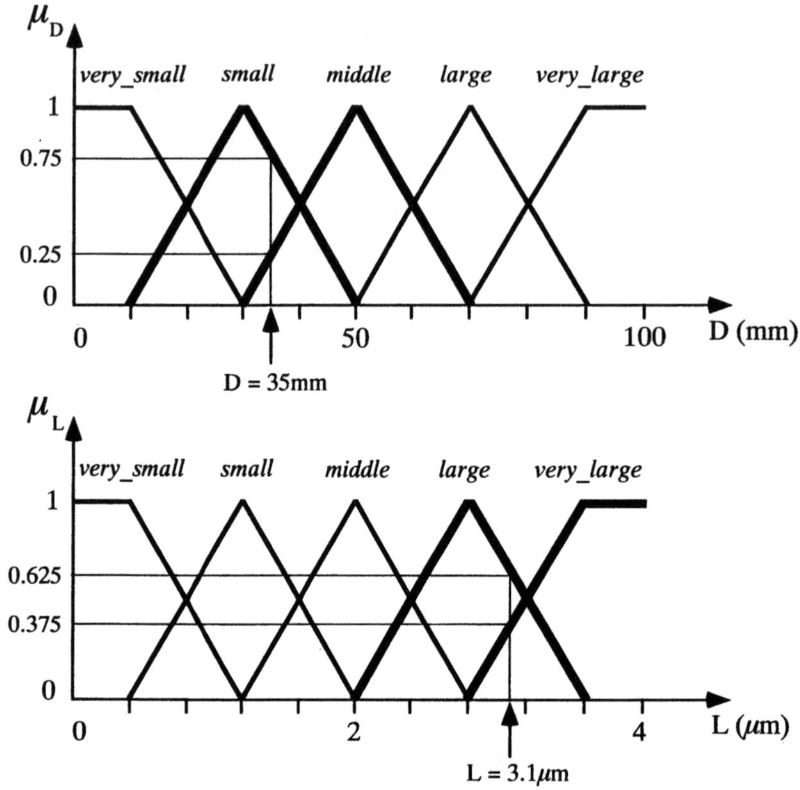

Fig. 7.12: Fuzzification of the input values *D* and *L*

control input data. In our example, both rules, R_1 and R_2, are active, since the degrees of satisfaction E_1 and E_2 of these rules are not 0, Fig. 7.13:

$$H_1 = \text{MIN}(\mu_{D\ middle}(35\text{ mm}), \mu_{L\ large}(3.1\ \mu\text{m})) = \text{MIN}(0.25, 0.625) = 0.25$$

$$H_2 = \text{MIN}(\mu_{D\ small}(35\text{ mm}), \mu_{L\ very\_large}(3.1\ \mu\text{m})) = \text{MIN}(0.75, 0.375) = 0.375$$

The degree of satisfaction of a rule is therefore determined by the smallest membership value (MIN operator) which appears in the membership function of the premises of the current event. It should be noted here that a fuzzy rule which contains OR operators in its premise can always be distributively represented as an equivalent composition of fuzzy rules which only have AND operators in their premises.

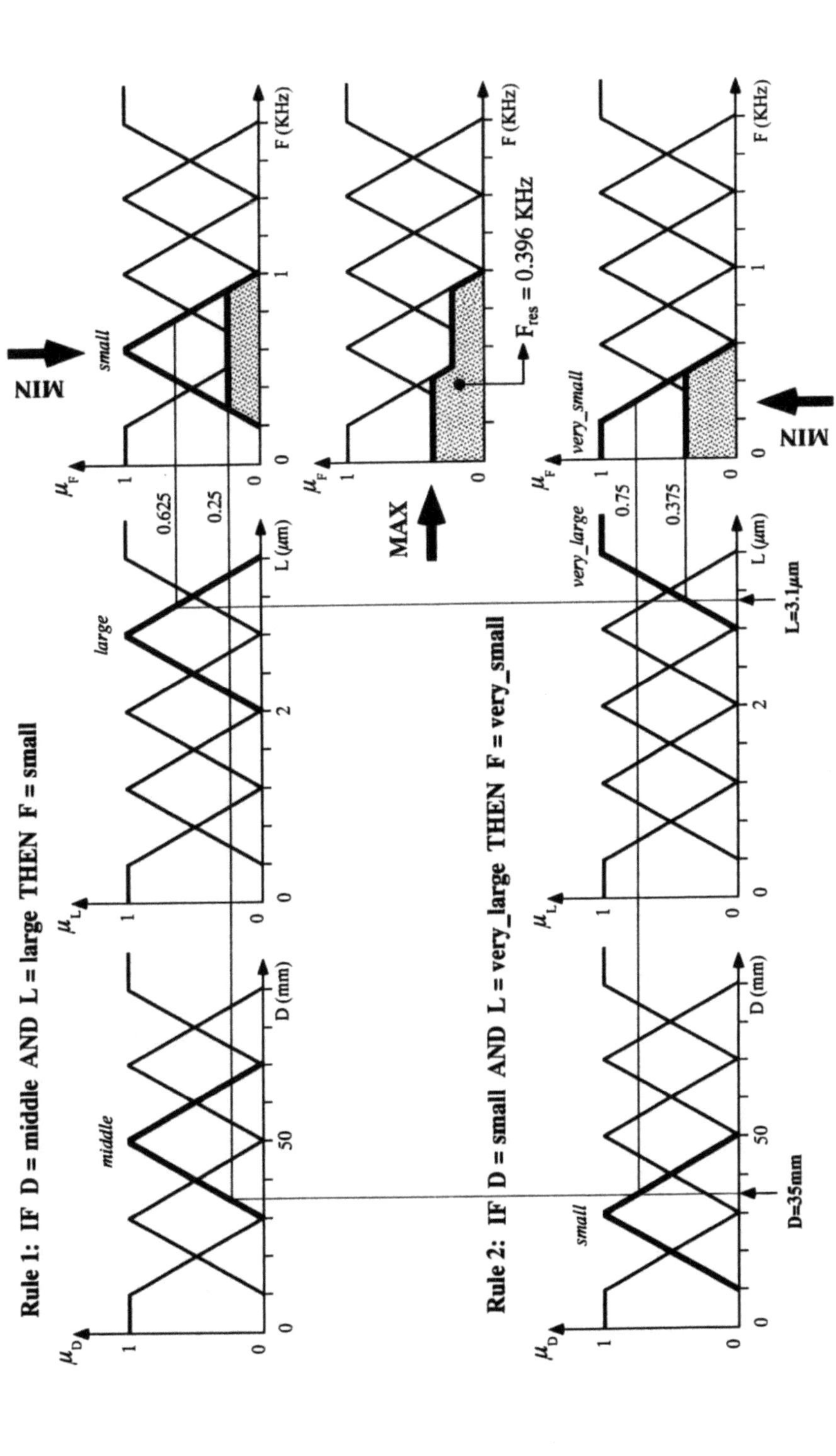

Fig. 7.13: Fuzzy information processing in the control of the microrobot

• After the active rules and their degrees of satisfaction have been determined, the fuzzy response of the system, consisting of a output fuzzy set, can be determined. In general, the MAX-MIN inference scheme is used. The resulting output fuzzy set $\mu_{F\,res}$ is determined by forming a union (MAX operator) with the output fuzzy sets of the active rules; these sets are cut off at H_i (MIN operator, see previous processing step), Fig. 7.13.

• The resulting fuzzy set $\mu_{F\,res}$ is not directly suitable for further processing. The task of the rule base system consists of finding an exact output value, therefore it is necessary to defuzzify the output fuzzy set. The transition to the exact value F_{res} in our example was made by applying the center of gravity method, Fig. 7.13, to the set $\mu_{F\,res}$:

$$F_{res} = \frac{\int_0^{\infty} F\mu_{F_{res}}(F)dF}{\int_0^{\infty} \mu_{F_{res}}(F)dF}$$

Besides the center of gravity method, there are many other ways of doing the defuzzification, such as the maximum method, the accumulation method, or the so-called F method [Kahl93]. The selction of a proper defuzzification method depends on the application and constraints.

After this short overview, it can be seen that fuzzy logic is an excellent tool for the classification of vague sensor data and control tasks and it is therefore of great interest to MST. Fuzzy approaches have their particular merit where exact mathematical modelling of the information processes in a system is impossible. Thereby, a complex system can be made manageable with the use of easy-to-understand linguistic fuzzy rules; if the latter are properly applied it is possible to shorten development times for a product and to better optimize a control system. Detailed descriptions of various applications can be found in [Altr93] and [Zimm94].

If many sensors of either identical or different type are used in a microsystem, fuzzy methods are a good tool of evaluating nonlinear sensor signals, because the tedious conception of a multi-dimensional process model is done by formulating fuzzy rules, which are based on *a priori* knowledge about the process. E.g. in [Knob96], a fuzzy classifier for chemicals was described for detecting three different gases in a gas mixture. The measurements were done with a system consisting of several temperature sensors which were integrated on a chip.

The overall information flow for the fuzzy analysis of sensor signals is shown in Fig. 7.14. This method permits the determination of quantities which cannot be directly measured. Thereby, various parameters, which can be directly measured, are placed in a fuzzy rule base which linguistically expresses the known relations between them and the constituents to be determined.

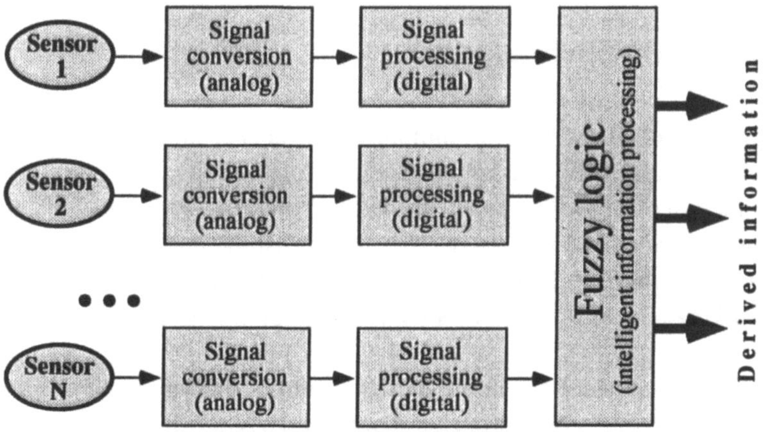

Fig. 7.14: Fuzzy logic in processing sensor data. According to [Altr93a]

The control of a microsystem, the behavior of which cannot or only insufficiently be modelled due to its complexity, is a difficult task. Wenn using a conventional control method (e.g. PID control), temporal deviations of the system parameters due to environmental influences lead to additional problems, such as unstable behavior, resonance, faulty positioning etc. In order to be able to develop robust and effective control algorithms under these constraints, the MST engineer must search for new error-tolerant control methods. The fuzzy method can handle such problems. It is also suitable for the collision-free planning of trajectories in intelligent multirobot systems (Section 8.9).

Fuzzy algorithms for MST applications can be realized both with hardware or software. Fuzzy logic hardware solutions are especially suited for real-time applications due to their high processing speed, whereby fewer circuits allow the systems to be integrated compactly onto one chip. To implement a fuzzy system, however, a software solution is usually preferred. By using fuzzy development tools, such as FuzzyTECH or TILShell [Angs93], a first solution can be found which can then be fine-tuned and optimized according to already existing methods. The optimized software version of a fuzzy system can then be implemented in hardware form. To obtain a high execution speed, there is a selection of various standard fuzzy processors available, such as OMRON FR-3000 or FC110 [Kahl93].

7.3.3 Artificial Neural Networks

Complex biological networks like the human brain consist of a multitude of neurons which exchange information over their connections. They are the backbone of intelligent behavior. With the help of neurons it is possible to learn, understand, predict and recognize. The main characteristics of a neural network are its hierarchical structure and massive parallelism; the functionality of the network can adapt to new tasks by learning or changing parameters.

In a biological neural network the nerve cell (neuron) is the information processing unit, which receives stimulating or suppressing signals by other neighboring nerve cells via synapsis connections. The part of an incoming signal that reaches the corresponding neuron is determined by the weight of the synapsis connection. In the cell body, ingoing signals are added up and the reaction of the neuron is determined. If the ingoing signal energy exceeds a certain limit, the neuron fires an output signal and transfers it to other neurons over its axon. From the summed up input values the output function of the neuron determines the magnitude of the output value. To learn a new reaction, the neuron adjusts its synaptical weight and threshold values, such that it always fires when a corresponding value combination reappears at a later time. The structure and the weighting of a neural network determine its behavior and are the degrees of freedom in optimization.

Biological networks for information transmission, processing and storing are imitated by artificial neural networks. From now on artificial neural networks will simply be described as neural networks, since we will regard only artificial neural networks. There are no instructions in a neural network to be executed, and no data which are saved in a memory, as in conventional information processing systems. The neurons or processor elements generate output values corresponding to their parallel input values. The result is the overall state of the network as soon as it has reached an equilibrium state. Consequently, the knowledge of a neural network is not located in a specific place (address), but is distributed over the whole network as a pattern. The architecture and training methods alone determine how the network works.

An abstract model of a neural network consists of connections (edges) and nodes (neurons), which evaluate the function. The edges have a direction and are usually weighted information channels, which transport the arguments for the function evaluation from one node to the other. The nodes have several independent inputs and one output, i.e. the information flows in a specified direction. The schematic representation of a neuron is shown in Figure 7.15.

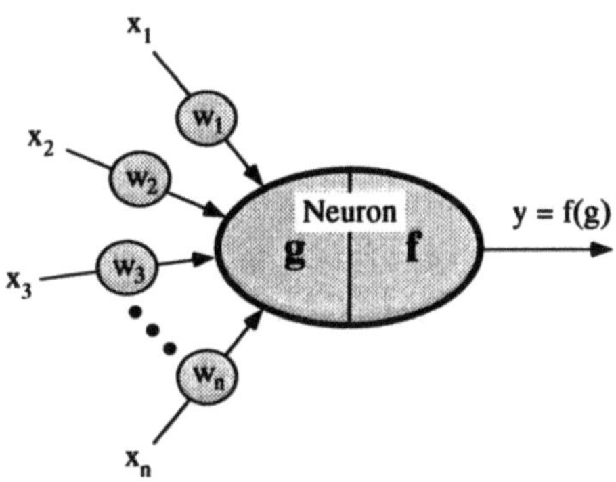

Fig. 7.15: Model of an artificial neuron

An integration function g combines the input arguments $x_1,\ldots,x_n$ of a neuron to one single argument, which is then used as an input for the activation function f. A widely used and simple neuron model, the so-called perceptron [Mins69], has a real threshold value Θ and a step function as an activation function of the neuron. If the perceptron has n input edges with the real weights $w_1,\ldots,w_n$, then it fires a 1 for the input values $x_1,\ldots,x_n$ if and only if

$$\sum_{i=1}^{n} w_i \cdot x_i \geq \Theta$$

Otherwise, its output is 0. There are also other neuron models being used having integration and activation functions different from that of the perceptron. Certain neural networks learning algorithms need, for example, an activation function that can be differentiated. In this case, a smooth function is most often used, the so-called sigmoid function:

$$s_c(x) = 1/(1+e^{-cx})$$

If elementary neurons are combined to form groups or layers, they form an artificial neural network. The individual network types distinguish themselves from one another in their architecture (number of neurons and how they are connected) and in their learning algorithms. Learning by a neural network means setting their weights to satisfy a specific task, so that a certain network input produces a desired output. The weights are gradually adjusted each time the learning rules are applied. Below three network types, which are important for practical applications, will be discussed.

- **Multilayered networks, back-propagation learning**

The back-propagation algorithm is used for learning by a neural network having a layered architecture. We talk of a layered structure when many neuron layers are connected in series, when the edges only connect neurons in consecutive layers and in a defined direction and when the neural network does not contain cycles. The input is processed and is passed on from layer to layer until a result is produced. A network having a layered architecture is shown in Figure 7.16.

The back-propagation algorithm represents one example of so-called supervised learning (learning with a "teacher") in a neural network. With supervised learning, the *a priori* knowledge about a defined problem exists in the form of input/output patterns. An input/output pattern means that a known input causes a defined output. Training is done as follows: first, the neural network will be given arbitrarily chosen weights, second, the network is provided with the pattern input. If the network output does not correspond with the pattern output, the weights are corrected to make the next output closer to the desired one. Back-propagation learning generally involves map-

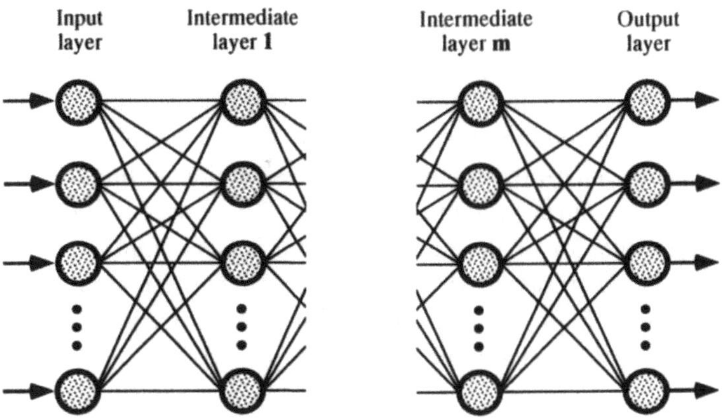

Fig. 7.16: A multilayered neural network

ping m predefined n-dimensional input vectors $o_1,\dots,o_m$ as exactly as possible onto m predefined k-dimensional output vectors $t_1,\dots,t_m$ (set values). The overall quadratic projection error of the network is minimized by following the gradient of the error function "downhill". Since the gradient of the error function must exist for all points in the weight space, a differentiable sigmoid is used as the output function of the network nodes. The combination of these network weights which minimize the quadratic error of the network

$$E = \sum_{i=1}^{m} \| t_i - y_i \|^2$$

with reference to all m training patterns is the solution to the learning problem. Here, $y_1,\dots,y_m$ are k-dimensional output vectors (actual values), which are created by inputting m pattern vectors $o_1,\dots,o_m$.

The network is first initialized with arbitrary weights. The error calculated after each pattern input is propagated back through the network from the output layer to the input layer, whereby the edge weights in the network layers are successively corrected in the opposite direction of the error function gradient. This allows a local minimum to be found for the error function E. After training with several representative examples, a back-propagation network should have good extrapolation properties, so that the network still remains functional during operation if there are small deviations from the trained range. If, for example, an input signal is close to a learned pattern input, it is expected that the output of the neural network is close to the corresponding pattern output (continuous projection). Thus, adaptive controllers can be realized with back-propagation networks, which automatically adapt to the behavior of a control system [Schod93]. This is especially practical when the transfer behavior of the control system is very nonlinear and when disturbances are likely to occur. If the deviation from the learned operating range is unexpectedly high, the controller must be externally "advised" with correctional hints, which must also be learned by the network.

- **Associative memory, Hopfield networks**

Hopfield networks are well-known recursive network models in which cycles and feedback loops are also possible, as opposed to back-propagation networks. This means that information can flow in both directions. It is possible to model with a recursive network a so-called associative memory; it associates certain input vectors with certain output vectors, thereby imitating the biological associative memory mechanism. Basic patterns are initially saved in an associative memory using certain learning or design procedures. If during operation an incomplete or faulty input appears, the network should converge after a certain amount of time to one of the stored patterns. The difference between such a network and a continuous projection network is that the neighborhood of the input vector is also projected onto its output pattern.

One distinguishes between hetero- and autoassociative neural networks. Heteroassociative networks associate m input vectors $x^1, x^2,\ldots,x^m$ with m output vectors $y^1, y^2,\ldots,y^m$ and can therefore be used for projecting patterns. Self-organizing Kohonen networks are one example which we will explain later. In an autoassociative model, such as a Hopfield network, the vectors $x^1, x^2,\ldots,x^m$ are associated with themselves. These networks can therefore be used for recovering incomplete patterns or for classifying structurally similar pattens. In a Hopfield network, all neurons are connected with each other, whereby each edge is bidirectional, Fig. 7.17. Each neuron is "responsible" for the activation of all other neurons and each plays a role in storing patterns. This mutual activation/deactivation should eventually lead to the reconstruction of (stored) patterns from the partial patterns provided to the network. The edge weights in a Hopfield network are symmetrical, i.e $w_{ij}=w_{ji}$. Each neuron has a threshold θ and fires a 1 if its overall excitement is greater than θ; otherwise, the neuron state is not changed. Possible changes in the activation state must occur asynchronously to avoid network oscillations around the solution range. The asynchronous network operation is guaranteed by various procedures which define the order of the node processing. A neuron keeps its last computed state until its excitement is reevaluated.

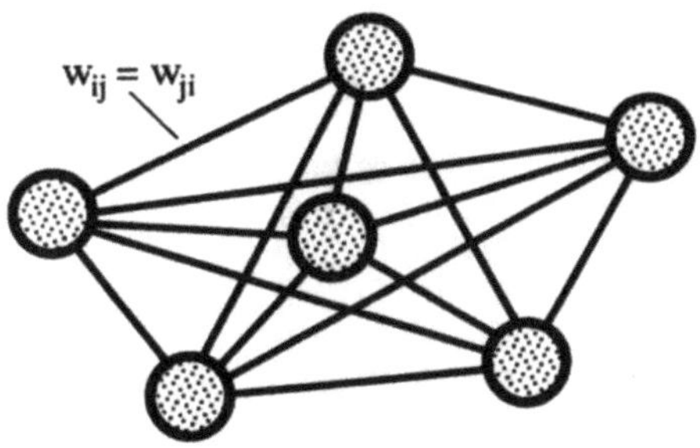

Fig. 7.17: Structure of a Hopfield network

The behavior of the Hopfield network can be described by an energy function which represents an assessment of the overall network state at a given time:

$$E = -1/2 \cdot \sum_{j=1}^{n} \sum_{i=1}^{n} w_{ij} \cdot x_i \cdot x_j + \sum_{i=1}^{n} \theta_i \cdot x_i$$

Here, x_i, x_j are activation states (0 or 1) of the corresponding neurons. The energy function defines the stable states (stored basic patterns) of a network, since they correspond to the local minima of the function E. In order to reach a stable state from an arbitrary network state, the node outputs are asynchronously changed according to the activation rule until no further change is possible. It can be shown that a Hopfield network always converges to a stable end state [Rojas93]. Geometrically, all network states which lie on peaks or on the slope of the energy function are physically unstable. The network cannot remain in such a state (i.e. there are still neurons which must be changed in the network), but must converge toward the energy minimum. Autoassociation takes place when the incomplete input of a network causes the assumption of the closest attainable stable state which is saved in memory as a basic pattern.

An application example can be seen in Figure 7.18. Here, a Hopfield network made up of 36 neurons is used to recognize noisy alphabetical characters. There are the four basic patterns A, B, C and D. The series of figures below shows the transition of the pattern B. The neuron states are asynchronously changed in pseudo-random order.

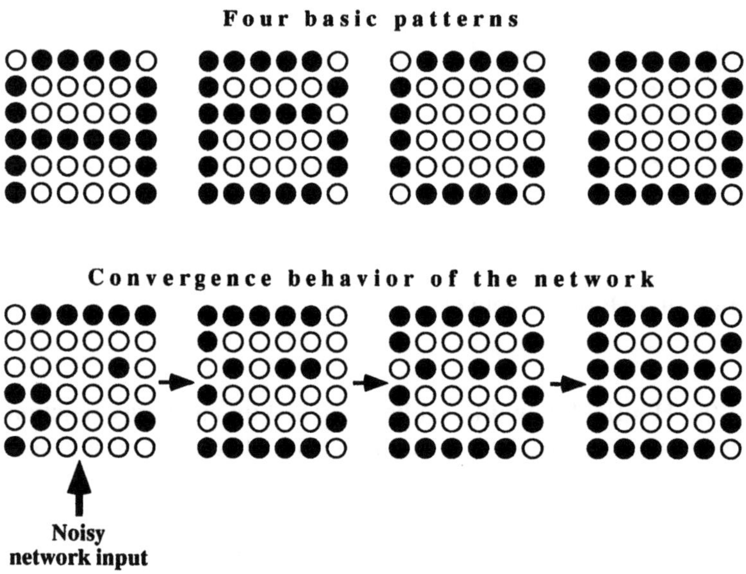

Fig. 7.18: Character recognition with a Hopfield network. According to [Krat93]

The question arises how does one save necessary basic patterns in a Hopfield network. In the simplest case, the network can be made by hand by heuristically setting the edge weights and threshold values. In more complex networks,. as in the above example, different supervised learning methods are used for saving patterns. All methods are related to the fundamental idea of D. Hebb [Hebb49], which is known as the Hebb rule; it states that the connections between two active neurons should be reinforced by the network's basic pattern.

- **Unsupervised classification, Kohonen networks**

The self-organizing network models of T. Kohonen are used as associators for solving classification tasks [Koho87]. These networks distinguish themselves from other supervised, trained network models, in that no explicit output is defined; this means that no error function can be defined for a Kohonen network. The learning process in self-organizing networks is therefore an unsupervised, self-controlled process, whereby each neuron of the network should be specialized for a certain category of input patterns. The classification task is illustrated in Figure 7.19. The input space of the network should be mapped such that only one neuron i is responsible for one input, for example from the subspace a_i; this means its excitement is maximal for an input from this subspace, under this condition the neuron fires. All other neurons are impeded. The 2-dimensional topology presented here is usually used in practice; topologies of higher dimension are possible. There are also other associators which can solve such classification tasks [Carp86], [Krat91].

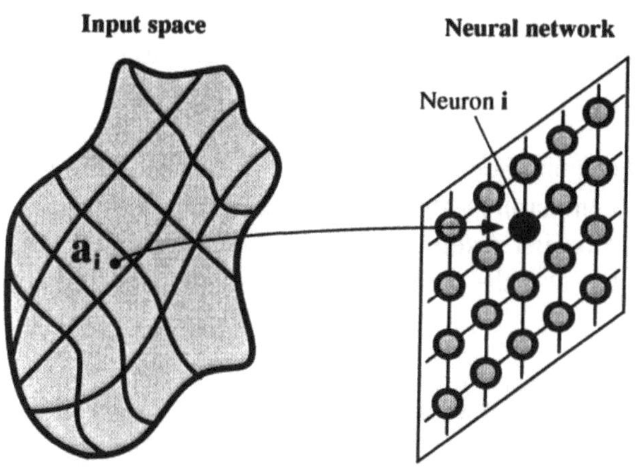

Fig. 7.19: Heteroassociative mapping of an input space

In self-organizing feature maps, the topology is conserved during the mapping. A topology of the input space is defined *a priori* for the neurons of the network, and it is conserved during the entire unsupervised learning process. It is known that the human cortex processes the multitude of sensory impressions it receives in exactly this way [Rojas93]. The topological constraints are taken into account during the learning process of the Kohonen network by defining the neighborhood of a neuron. The neighborhood definition allocates to a neuron a set of other neurons as neighbors in the network. The neighborhood of planar topologies can easily be defined based on linear numbering, whereby all neurons which are shifted r positions left or right, up or down, form the neighborhood within a radius r of a neuron, Fig. 7.20.

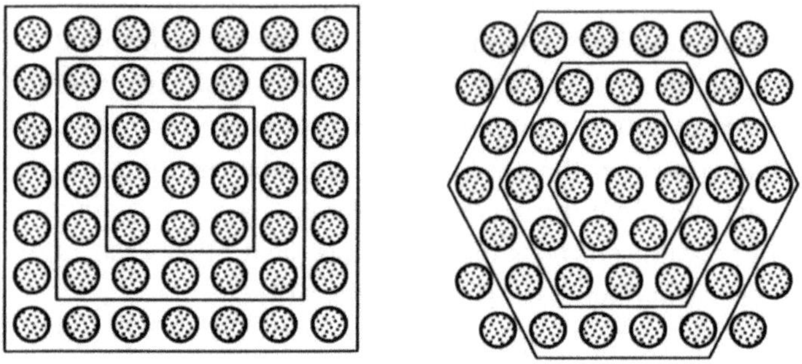

Fig. 7.20: Possible neighborhoods of a neuron. According to [Krat93]

In the Kohonen model, the neighborhood of a neuron is influenced during the learning phase. In the Kohonen learning algorithm, for example, the neuron k with a maximum excitement is first determined for an input vector ξ. Afterwards, not only its weight, but also the weight vector w_i of neighboring neurons is updated:

$$w_i := w_i + \lambda \cdot \varphi\,(i, k) \cdot (\xi - w_i)$$

Here, λ is a learning constant and $\varphi\,(i, k)$ is the neighborhood function which determines how strongly a neuron i is coupled to the center of neighborhood k during the learning process. The learning strategy is to start with a wide neighborhood whereby each neuron tries to influence its neighborhood. The latter is then successively decreased as the algorithm progresses (i.e. the influence of each neuron on its neighbors gets smaller and smaller) until only single neurons are processed. The result of the learning process is a uniform distribution of the weight vectors in the input space, whereby each neuron is "responsible" for one subarea. Figure 7.21 shows the spread of a 2-dimensional Kohonen network after 0, 20, 100, 1000, 5000 and 100,000 training cycles.

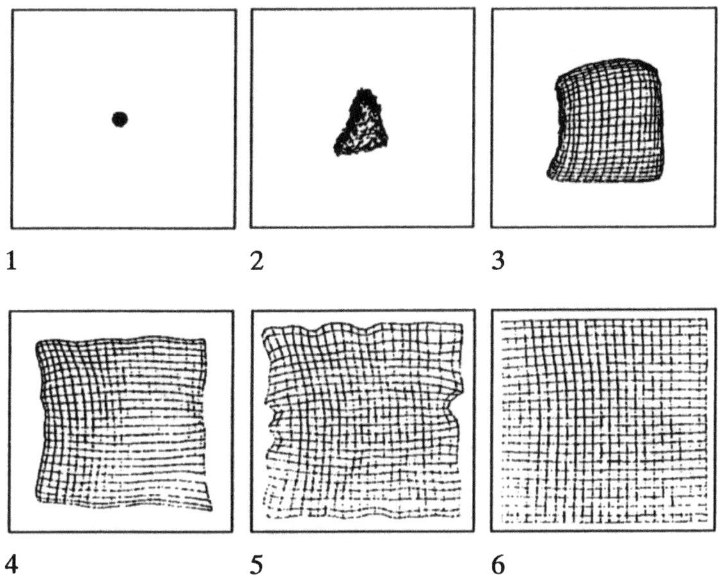

Fig. 7.21: Distribution of the weight vectors in a quadratic input space. From [Koho87]

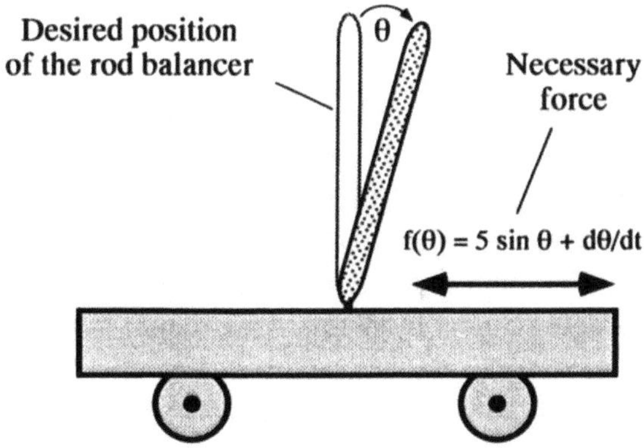

Fig. 7.22: Control of a rod balancer. According to [Rojas93]

As a result of the learning algorithm, Kohonen networks can adapt to input spaces of any form. This is illustrated by the following example of a neural control system of a rod balancer, Fig. 7.22.

A rod, which can rotate on one end, should be kept perpendicular by shifting it either right or left. The force necessary for this is given by the equation: $f(\theta) = 5 \sin \theta + d\theta/dt$. This function was learned by a 2-dimensional Kohonen network, Fig. 7.23. Thereby, the input space (θ, $d\theta/dt$) was mapped onto the application-relevant area by the network, Fig. 7.23, left.

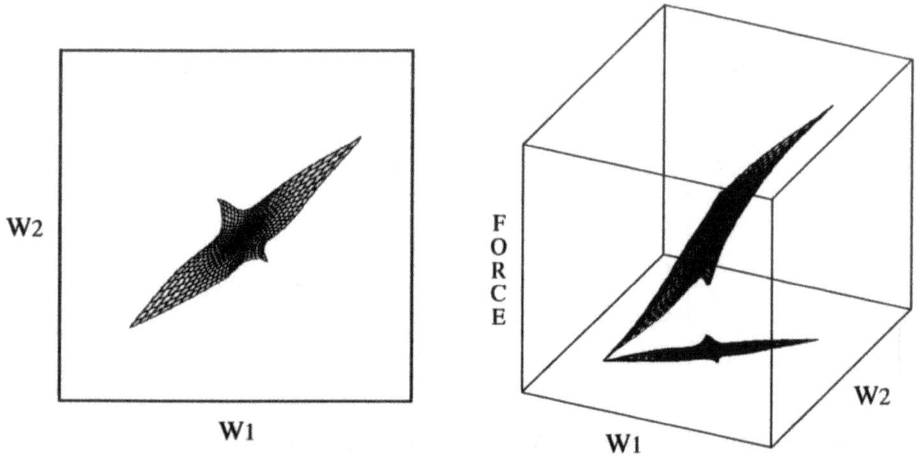

Fig. 7.23: The mapped input space (θ, $d\theta/dt = w_1, w_2$) of the learning problem (left) and the resulting control area (right). From [Ritt91]

The trained network can be used as a real-time controller of the rod balancer. If an updated combination of θ and $d\theta/dt$ are fed into the network, then the neuron responsible for this subspace of the input fires and provides the corresponding force value, Fig. 7.23, right. Therefore, the network behaves as an adaptive table of function values. Kohonen networks can be used for positioning endeffectors of a robot at a defined goal in its operating range [Ritt91].

The importance and the application potential of neural networks should have become clear after this short introduction to the most widely used network classes. They are especially of interest where modelling is very difficult and when there is only a set of training examples available instead of linguistically described expert knowledge. Neural networks, like fuzzy systems, can be used to detect multidimensional functional relationships between different types of information. E.g. nonmeasurable values of a process can be determined from many different sensor values or the necessary control parameters can be computed from a multidimensional input space, independent of how complicated the functional relationships are. The neural networks are especially good for handling incomplete information, and therefore for solving classification problems. Today, most neural network applications are used for pattern recognition and classification.

Since pattern recognition is very important to MST, neural networks will play an ever increasing role in information processing in microsystems. A characteristic task is the processing of incoming signals from many different sensors, so that even noisy and partially distorted signals will lead to reasonable results. A very good application is the so-called "artificial nose", an on-chip sensor system made of several gas sensors [Kohl96], [Gemm96]. Pattern recognition is necessary for the quality control of MST products. Here, neural classifiers which are suitable for processing visual sensor information can form the heart of automatic non-destructive tests and diagnoses. Very specific MST applications such as classifying acoustical sensor signals in a cochlea implant are also conceivable.

We only gave a short insight into practical applications of neural networks, however, there are many other problems that can be handled. Presently, many design steps have only heuristic solutions and there is a need to obtain an intuitive understanding of the behavior of a network. The design of back-propagation networks, which can be transferred to other types of networks, are discussed in [Lawr92]. There are also difficulties in technically realizing neural networks in microsystems. This problem is being widely researched today, both hardware and software realizations are considered. Software solutions are easier, especially with the help of the many software tools available on the market. An overview of them is presented in [Berns94], but they are not suitable to be used for autonomous microrobots. Also, data busses and interfaces to external computers are often indispensable, which strongly influence the application of microsystems.

Hardware solutions open up many possibilities, the most promising solution are application-specific neural chips. Usually, they are made up of many interconnected elementary processors (e.g. for addition and multiplication) and special memory banks (e.g. for vector and matrix computations). Entering the network weights is done with an external processor on which the learning algorithm is implemented. Neurocomputers built from standard VLSI components are also of interest for MST applications since they can be implemented by connecting resistors and electrical circuits on silicon circuits. However, a major disadvantage is their lack of learning capabilities and the accompanying inflexibility of this method. Recent improvements in microoptics make optical neurocomputers a tempting possibility for MST applications. There are no wires, so the neural networks can be made from microoptical elements only, such as lenses, diodes, light waveguides or solid state lasers. An overview of commercial hardware realizations can be found in [Rück93].

7.3.4 Combined Neuro-Fuzzy Approaches

Fuzzy systems are relatively easy to conceive if a knowledge base exists in the form of IF-THEN rules, but they are not able to learn. On the other hand, neural networks are able to learn even without previous knowledge. However, the learned knowledge is stored in the form of network weights, and is therefore due to the lack of a structure difficult to understand. One promising approach which has become of interest within the past few years is a combination of these two techniques, whereby the advantages of each are combined and the disadvantages eliminated.

A classification of various combinations of both these information processing technologies and an introduction to this innovative topic can be found in [Nauck94]. The method mostly investigated today is using the learning capability of a neural network for optimizing of the characteristic parameters of a fuzzy system (linguistic rules and membership functions). This capability decisively influences the system behavior and is defined by a heuristic approach. The logical structure of the fuzzy rules is transferred to the neural network. In a next step, the network is trained with the existing example data, whereby the parameters of the fuzzy system are reoptimized. A fuzzy system having this new learning capability can then be better adapted to the problem to be solved.

8 Microrobotics

8.1 Introduction

In the previous chapters we have discussed the most important components of a complete microsystem, which are the microactuator, microsensor and information processing unit. Although there have been enormous advancements on the component level there are very few complete microsystems available presently. The spectrum of potential microsystems is very broad and was already introduced in the introductory chapters. This chapter will be dedicated to an important area of MST, the microrobotics.

We are all familiar with the definition of a macrorobot. The concepts of a robot originate from the human desire to do away with hard labor and dangerous work and have such jobs done by mechanical means. The word "robot" originates from the Czech word for worker, "robota". Robots have many capabilities with regard to force, speed, reproducibility and endurance, superior to those of human operators. Robots have improved many production processes and have led to their rationalization. However, there are also other robots which, compared to industrial robots, provide advanced functions and take on the tasks of a servant. These robots are known as service robots. The concept of "service" comprises a multitude of activities, such as service in medicine, inspection, maintenance, household, entertainment, etc. In order to meet future demands and to offer "services", robots should be mobile, easy to handle by the user, and be adaptable; all this requires a certain amount of intelligence. The long-term goal is to develop a robot with some human capabilities.

Like conventional robots, microrobots are very complex systems using various types of microactuators and microsensors. They are provided with algorithms for intelligent signal and information processing. Microrobots also move, apply forces, manipulate objects, etc. Some drive principles from the macroworld can be applied to the microworld, but a possible scale down effect must be taken into account; otherwise, performance data may be calculated for a scaled-down micromachine which are not realistic. With the exception of the specific microworld problems and the size difference, the design criteria and steps to build micro- or macrorobots can in many cases be identical. Analogous to the design and making of a macromachine, first the functional components of a microrobot with the desired dimensions and internal structures must be produced and then assembled and fine tuned. Finally, tests are run to make sure the robot is functioning correctly.

Usually, microrobots have specific requirements to make them useful for handling workpieces with very small dimensions. That is why microrobots pose a great challenge to MST researchers; there is a continuous increase in research activities directed to new application fields in which conventional robots can not be used due to conceptual limitations. Great importance is attached to the capability of microrobots to move over long distances at adequate speed, to manipulate different types of objects, to be robust and able to operate in a hazardous environment, and to be functional over long periods of time without maintenance. Studies performed on current MST issues in Europe (NEXUS), the USA (National Science Foundation) and Japan (MITI) have shown that in the near future, microrobots will play an important role in industry and that there will be a growing market for such systems.

A major breakthrough is expected by devices that can handle many practical applications with the help of flexible, powerful microrobots (Chapter 2). Important criteria for using microrobots are high reliability and low cost. The reason for the high cost of the existing microrobots are the special development efforts needed to design the components such as actuators, sensors and signal processing units. High cost is a general MST problem, which can only be eliminated by developing hardware and software standards that can be used by many microrobot systems.

Due to the advancements of conventional robotics and the good microproduction techniques, mobile microrobots with dimensions less than a cubic centimeter can be developed today. For many applications a microrobot must be mobile and have excellent micromanipulation capabilities to handle tiny objects with μm or nm dimensions in a work space. Although the human hand is a very versatile instrument and has an almost unsurpassed dexterity, it does have limitations when working in the microworld. E.g. manipulating biological cells or assembling microsystems pose a big problem without suitable aids.

8.2 Microrobotic Applications

It is not yet possible to predict how much influence microrobots will have on our lives in the future, but in many areas, we can already notice a big influence, Fig. 8.1. Many applications will be in medicine, production technology, biology, and metrology. In the following we will be discussing promising applications.

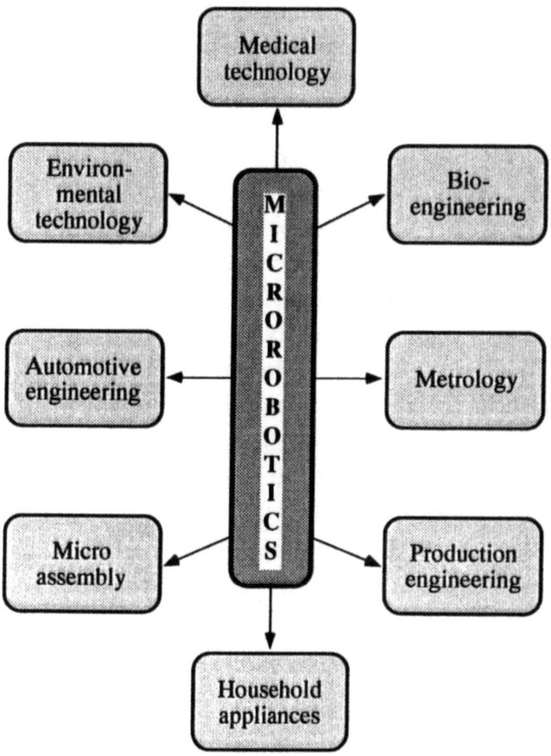

Fig. 8.1: Potential microrobot applications

In industry, especially in production and metrology, highly sensitive test systems with microdimensions are becoming increasingly of interest. One important task is the testing of microchips, where several contact points on a wafer have to be touched with temperature or voltage sensors. In order to automate this difficult and often manually performed operation, robot systems are needed as wafer probers which can precisely manipulate and move over certain distances. The same applies to intelligent maintenance and inspection robots, which are intended to reach otherwise inaccessible areas (e.g. pipe systems, heat exchangers, or airplane turbines) or even a dangerous work environment to detect leaks or defective parts, and possibly carry out the repairs while they are there. Such flexible robot systems are currently being investigated in Japan within the framework of a 10-year research project.

Microassembly is one of the most important microrobot applications of the future. There are particular problems with the mass-production of microsystems. Batch processes are rarely applicable for the manufacture of complex microsystems. Such systems usually consist of several microcomponents made of different materials and manufactured by different microtechniques. These components must be very exactly assembled in one or more steps to form the desired microsystem. Often it is necessary to combine conventional components and microcomponents, which requires very accurate fine tuning and high flexibility on the part of the assembly system. The microassembly facilities which exist today are rather large, are usually tailored to a specific task and depend on the manual skill of the operator. In conventional systems there are drives with mechanical transfer elements, which are subject to frequent mechanical wear and maintenance, making the systems expensive. Considering the rapid progress being made in MST, it becomes obvious that the assembly of microsystems (i.e. non-destructive transportation, precise manipulation, and exact positioning of microcomponents) is one of the most important microrobot applications. Microrobots are needed that can carry out micromanipulations and can move over relatively long distances. Such applications, which are important for the industrialization of MST, will be described in more detail below.

At the same time, it is expected that the advancements in microrobotics not only will affect various industrial segments, but also our daily lives. Minimal-invasive surgery has become a very important medical application. Smaller and more flexible active endoscopes are needed which assist the surgeon and can react to instructions in realtime. They may enter into the body through small incisions or through natural pathways, proceed through blood vessels and enter various cavities (angioplasty) by remote control, where they carry out complex *in-situ* measurements and manipulations (gripping, applying tourniquets, incisions, suction and rinsing operations etc.). In order to meet these demands, an intelligent endoscope must have a microprocessor, several sensors and actuators, a light source and possibly an image processing unit integrated in it. These microrobot devices will revolutionize classical surgery, but their realization is still a problem because of friction, poor navigability, biocompatibility etc.; they are also not small enough yet. In biotechnology, microstructured tools are needed which allow the performance of various micromanipulations, such as sorting and combining cells, measuring profiles in tissues, or the injection of foreign substances in a cell using a microscope. For example, it is often necessary to search for certain cells in a sample, to separate them from tissue and to transport them to a test location, or to precisely insert a microprobe equipped with biological sensors (e.g. O_2 or glucose) in a small area of a tissue. In genetic research and environmental technology where cells are used to indicate presence of dangerous substances, the precise, non-destructive manipulation of individual cells is also necessary.

Before we explain various microrobotic devices, a definition and classification will first be done.

8.3 Classification of Microrobots

So far, the term microrobot has not been defined. The first effort to do research on microrobots comes from Japan. During the preparation phase of the aforementioned MITI research program, various experts were requested to look closely at potential applications, to screen existing knowledge and special technologies and also to come up with definitions. While the expression "microsystem" is appropriate for all types of microdevices, there is a conceptual distinction between "micromachines" and "microrobots". A micromachine is a device which can generate or perform mechanical work without needing direct "on-board" control. E.g. a micromotor or valve is a micromachine which can consist, however, of several components in order to be capable of doing a complex movement. Basically, a micromachine is a "passive" system which is an operating part of a more complex system.

The definition of a microrobot derives from the macroworld: a microrobot is characterized by its programming features, task-specific sensors and actuators, and, in general, by its unrestricted mobility. Like conventional robots, microrobots consist of two generally autonomous subsystems which determine the overall agility of the microrobot; they are the actuators for manipulating objects (robot arms and hands) and actuators for moving the robot platform (robot drive). The first subsystem determines the manipulation capabilities and the second one the mobility of the microrobot. Manipulators can be successfully integrated with a drive unit into a flexible microrobot system. Several prototypes of such systems will be introduced below.

- **Classification with regard to size**

When classifying microrobots, it is important to distinguish between minimechanisms, micromechanisms and nanomechanisms. Minimechanisms are being more and more often used in practice. Micromechanisms are standing on the threshold between theoretical concepts and the first prototypes whereas nanomechanisms will not be realized in the near future. Usually, miniature robots are a few cm^3 in size and are made up of conventional miniaturized components and micromachines. They can generate forces which are comparable with those the human operator exerts when carrying out fine manipulations. They may be remotely controlled, or, in order to be able to work autonomously, have a certain degree of intelligence and an on-board energy source. A miniature robot is actually a complete robot system which is very small in size and tailored to a specific MST application.

Microrobots are a few μm^3 in size. They are structured on a chip consisting of a microactuator, microsensor and signal processing unit and can be produced by micromechanical manufacturing technologies such as the substrate or surface micromechanics or the LIGA method. A microrobot must also be programmable and should be able to react to unpredictable events. If necessary, it should be remotely controllable.

According to this classification scheme, the only relevant difference between robots of the macroworld and microworld is the size of their work place in which they operate. A small work place may lead to difficulties, since the operations of the microrobot can often only be observed with a light-optical microscope or a scanning electron microscope. This substantially limits the robot design, especially its grippers. Frequently, a miniaturization changes significantly the physical properties of the construction material. This may have an effect on the forces and torques that can be transmitted and will lead to mechanical actuators with a specific behavior. This problem has to be taken into account by the researcher; he may have to use new MST-specific motion principles for microrobots. Here actuators are required which have extremely small dimensions, a simple mechanical design, direct driving gears and a high reliability; such actuators were described in Chapter 5.

The nanotechnology is concerned with objects or mechanisms in the μm-range or even smaller [Drex92]. Conventional mechanical principles cannot be applied here, as opposed to microrobots. There are many biological organisms produced by nature that serve as examples for nanorobots; they are efficient and simple and may serve as excellent models for electrochemical "drives". It was shown during the development of nanomanufacturing techniques that solid state technology is not suitable for manipulating molecules and atoms. Hopes are being set on the new polymer techniques, such as the Langmuir-Blodgett method [Gard94]. However, these techniques are still in the basic research phase.

- **Functional classification**

A microrobot as described here and throughout the rest of the chapter can belong to any of the above-mentioned classes. It usually has sensors and actuators, a control unit and an energy source, which determine the capabilities of the microrobot and its performance. By choosing different combinations of these components, various types of microrobots can be built, Fig. 8.2. Here the abbreviations are as follows: CU – control unit, PS – power source, AP – actuators for positioning and AO – actuators for operation.

Wenn using this classification method, there are three important criteria to be considered for microrobots. They are: mobility (yes/no), autonomy (energy source on-board/not on-board) and control type (with cable/without cable). The microrobots on the left side of Figure 8.2 are all remotely controlled. The separation of the power supply and control from the manipulation unit simplifies the construction and realization of a system. Type (a) is similar to conventional miniaturized industrial robots. The microrobots of types (b) and (c) have integrated actuators and they are connected to the controls and power

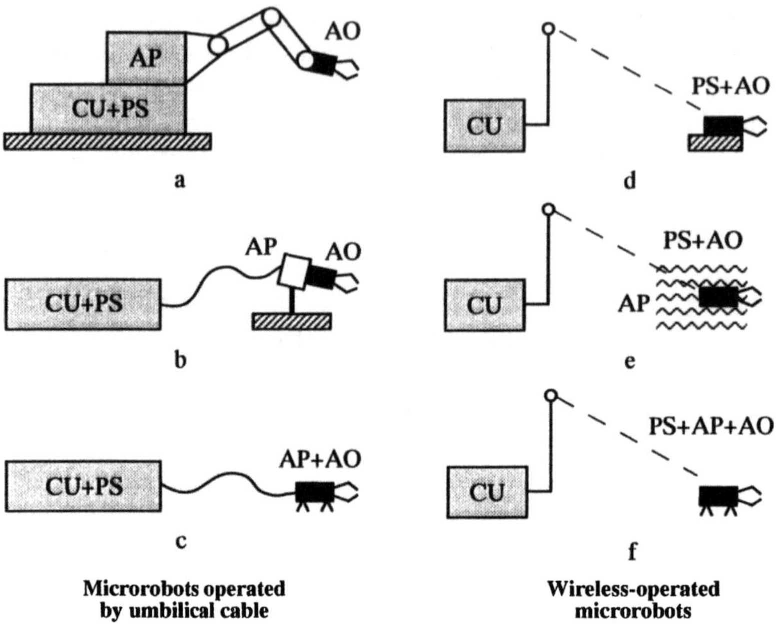

Fig. 8.2: Functional classification of microrobots. According to [Dario94a]

supply by electric, hydraulic or pneumatic means. Type (c) is also mobile, as opposed to (b), which increases the application potential, but at the same time imposes a number of difficulties. As opposed to conventional robots, the mobility of a microrobot can be achieved by transporting it with the help of external means, e.g. having it carried through an artery by the blood stream. Several researchers are working on methods to navigate micromechanisms through human blood vessels; these microrobots, however, are difficult to control. There are no on-board energy sources available to power the microrobots, despite of great efforts to design such devices. This is a typical problem for autonomous microrobots: efficient actuators usually have such a high energy requirement that an on-board power supply cannot provide enough electricity over a long time. The information transmission for a robot controller can easily be realized cableless with acoustic, optic, electromagnetic or thermal interfaces. New solutions are being sought to use such interfaces for the energy transfer as well. It is obvious that the development and control of the microrobots (c), (e) and (f) will be very time consuming and expensive, but a breakthrough with these devices would open up many new applications.

- **Task-specific classification**

A task-specific microrobot classification is based on the ratio C between the physical dimensions of the robot and its achievable range of operation [Remb95]. On one end of the classification scale (C>>1) are the stationary micromanipulation systems, which are a few decimeters in size, but can carry out very precise manipulations (in the μm or even nm range) [Oliv93], [Joha95], [Sato95]. On the other end of the scale (C<<1) are microscopically small mobile robot systems, which can be used, for example, for transporting, inspection or assembly. In between these extremes are miniaturized industrial robots of which the dimensions are similar to those of their range of operation (C≈1).

A flexible universal microrobot should have the features of the first and third classes, i.e. it should be able to manipulate microscopic objects very precisely with its effectors and should also be able to move over long distances. In order to combine these conflicting features in one robot system, suitable actuator principles must be developed. Various new robot principles are now being investigated; however, to date only few prototypes exist (Section 8.8). In the next section, several drive principles and practical solutions will be presented which can operate a mobile platform of a microrobot.

8.4 Drive Principles for Microrobots: Ideas and Examples

The most interesting drive concepts for mobile microrobots originated in Japan. The Japanese research program "Micromachine Technology", which started in 1991 and which involves many prestigious universities, research institutes and industrial firms, has already produced many promising microrobotic devices (Chapter 1). The goal of the project is to develop very small, intelligent robot systems for medical and industrial applications. A concept of a multi-agent robot system, for inspecting and maintaining the inside of inaccessible or dangerous objects, was introduced in [Idog93]. It consists of four subsystems: a microcapsule, a mothership, an operation module and a wireless inspection module. The mothership is responsible for the transportation of the modules, supplying them with energy and transferring data between the modules and an external control unit. The microcapsule having its own energy source, will be missioned to survey inaccessible places which might be damaged and then will report defects to an external control unit. After that, the inspection module will perform a precise analysis of the defective locations and also transmit the gathered data to the control unit. Finally, the operation module, which is connected to the mothership over a communication and power cable, will carry out the necessary repairs based on the inspection results. This multi-robot system may be used for diagnostic and therapeutic purposes in medicine. Several of the examples which will be introduced below were developed within the framework of this concept.

Tiny Silent Linear Cybernetic microactuator

The so-called Tiny Silent Linear Cybernetic microactuator was discussed in [Ikuta92] and [Ikuta92a]. The device uses a new type of drive principle, taking advantage of the excellent real-time response properties of piezoelectric actuators, Fig. 8.3.

The schematic design of the microactuator is shown in Figure 8.3a. The device consists of a piezoelectric multilayered stack element. A weight is attached to one end to serve as the inertial force drive. An electromagnetic positioning element with an integrated magnetic coil is fastened to the other end of the stack by which the actuator can be fixed to a guide rail made of steel. The multiphase operating principle is depicted in Figure 8.3b. In the first phase, the piezoelement is energized and expands quickly. The weight is then at its maximum distance away from the positioning element. Then, the positioning element is fixed to the guide rail by energizing the electromagnet. Thereafter, the piezoelement is quickly contracted by applying a voltage to it; this action accelerates the weight toward the positioning element. When the piezoelement has contracted completely, the electromagnet is switched off and the inertia of the weight moves the device in direction of the acceleration. The distance covered during one step depends on the inertia force and the sliding friction forces present. The parameters of the smallest actuator prototype which has been built are $5 \times 5 \times 12$ mm^3; it has a weight of 1 g. The piezodrive element is $1 \times 2 \times 5$ mm^3. Under an operating frequency of 37 kHz, the microactuator can move at a speed of 35 mm/s. This actuator can be used for many linear displacement functions.

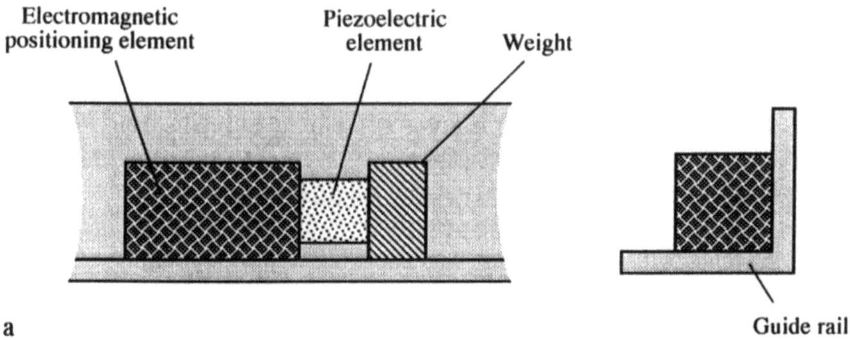

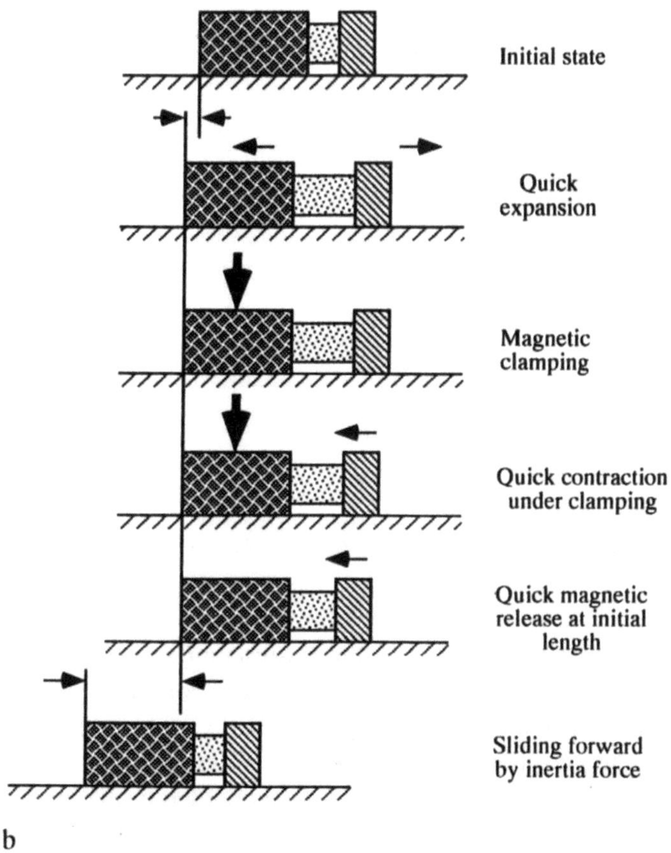

Fig. 8.3: Tiny Silent Linear Cybernetic microactuator
a) schematic design; b) motion principle. According to [Ikuta92]

Piezoelectric micropositioning device

A similar drive principle was used in a micropositioning system shown in Figure 8.4. Here, a piezoelectric multilayer actuator is moved forwards using the inertia and friction principle. The piezostack is located between the actual positioning element (the heavier mass) and a smaller weight.

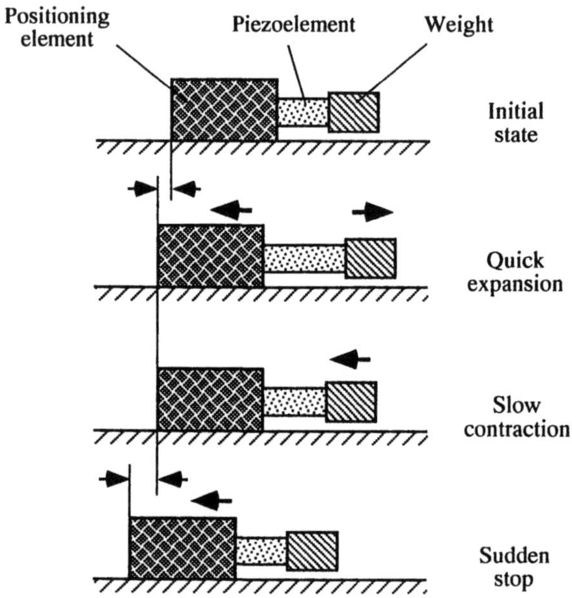

Fig. 8.4: Motion principle of the piezopositioner. According to [Yama95]

First, the piezobody expands quickly, moving the positioning element in the desired direction to the left. Then, the piezostack slowly contracts with a speed that does not counteract the started motion. Finally, this motion is suddenly stopped when the piezobody has completely contracted. The heavy positioning element is pushed further in the desired direction due to the inertia of the weight and comes to a stop. With this event a tiny step (nanometer range) is completed. Thus, the device can be used for micropositioning. An actuator system which uses this principle is commercially available; it can reach a maximum speed of 5 mm/s, a positioning accuracy of 0.1 μm and can transfer a force of up to 13 N.

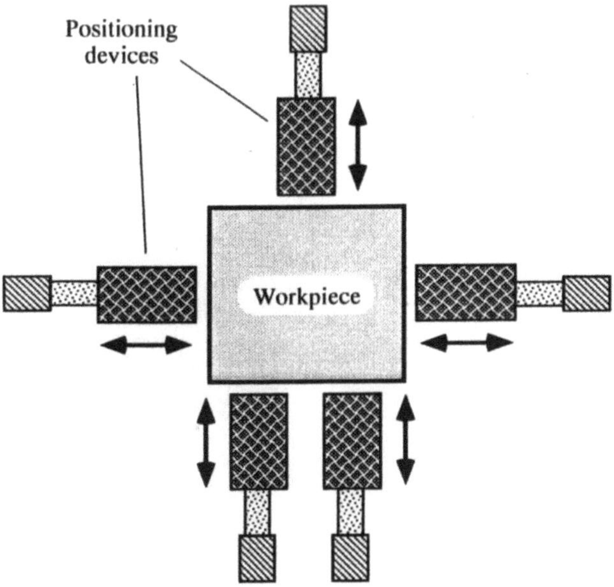

Fig. 8.5: Arrangement of piezoactuators for the micropositioning of a workpiece. According to [Yama95]

Figure 8.5 shows a possible arrangement of several piezopositioners to precisely align objects for microassembly. Objects can be pushed and rotated. The piezoactuators can be placed as shown to form a micropositioning unit; it may be used as an endeffector of a conventional robot arm. The robot arm performs a rough positioning by moving the micropositioning unit over the workpiece to be aligned, and then the workpiece is micropositioned by the endeffector. This operating principle can be used in mass production.

Micro-crawling machine

A few years ago, a research program was started in Switzerland for developing new devices and aids for microassembly. Within the framework of this program, a so-called micro-crawling machine was developed, which can make precise inchworm-like movements, and can be used as a driver for a positioning table in a microassembly station or as a small mobile robot platform [Codo95]. The device is about 60 mm × 60 mm and consists of two triangular legs, Fig. 8.6. The inner leg is connected to the outer one via three piezoelectric stack elements. There are electromagnets embedded in all three corners of the outer triangle of the device and one in the center of the inner leg.

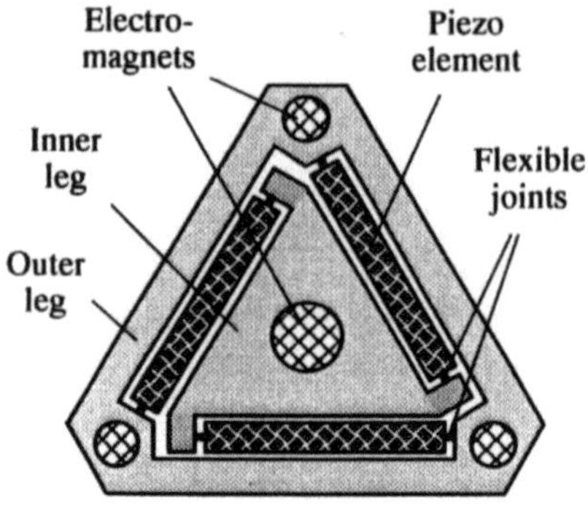

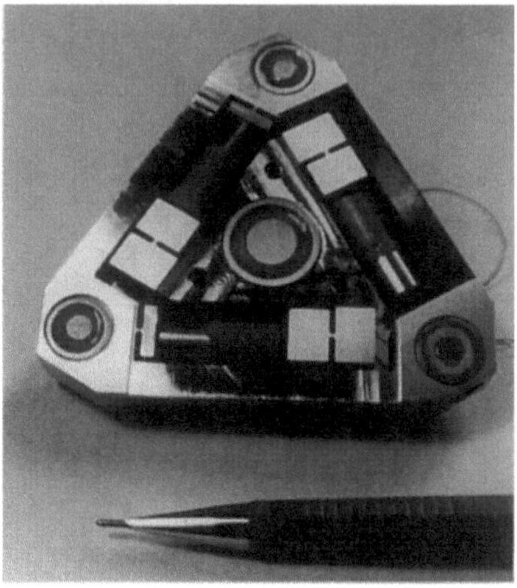

a b

Fig. 8.6: Micro-crawling machine using magnets
a) schematic design; b) a prototype of the device. Courtesy of the Eidgenössische Technische Hochschule, Zürich

The device can crawl along by holding on to a ferromagnetic base with one leg and moving the other one. The motion sequence is as follows: in the initial state, the outer leg is fixed to the base by the three magnets (current on) and the inner leg is free (the current is turned off from the inner magnet coil). If the piezoelements are then energized by an electric voltage, the inner leg moves into a new position within the x-y plane depending on the interaction of the piezoactuators. At this time, the inner leg is fixed to its new position by turning on the current in its magnet coil. The outer leg is freed at the same time by turning the current off the electromagnets and the device takes another step, etc. The platform can be precisely moved over a long distance in an omnidirectional mode. The piezoactuators reach their maximum expansion or contraction of 5 μm with a voltage of 150 V. The resolution of the motion is about 10 nm.

With this principle, a good grip of the base can be obtained and therefore can be used for pushing loads, climbing walls or even wandering around on a ferromagnetic ceiling. However, it is not suitable for use within a scanning electron microscope due to the magnetic fields produced. For this reason, another mobile platform was designed, which is based on the inertia motion principle and on the non-linearity of the Coulomb friction [Zesch95]. The schematic design and a prototype of the device are shown in Figure 8.7.

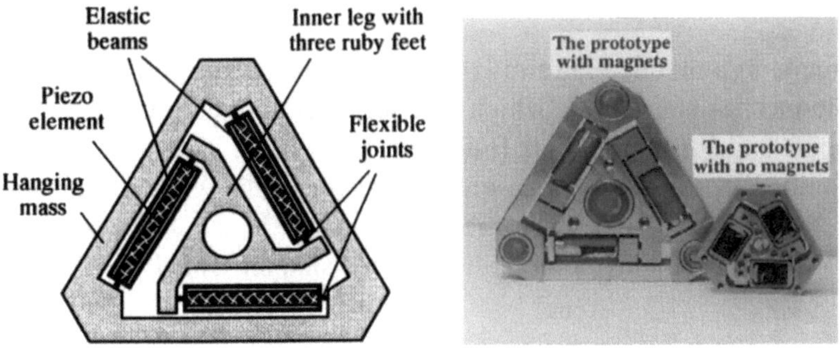

a b

Abb. 8.7: Micro-crawling machine with no magnets
a) schematic design; b) a prototype (as compared to the first micro-crawling machine). Courtesy of the Eidgenössische Technische Hochschule, Zürich

The device is similar to the first micro-crawling machine, but has no magnets; therefore it was possible to reduce its dimensions to 38 × 35 × 9 mm^3. The photo clearly shows the difference in size. Since the outer component does not have any contact with the ground, it no longer acts as a leg but as a suspended mass to generate the inertia force and moment of inertia needed for motion. Three ruby spheres are fixed to the inner component; they are in contact with the work base, and act as feet. The two platform components are interconnected by three piezoelectric stack actuators. The device is actuated by expanding or contracting the piezoelectric stacks according to a defined strategy. E.g. if the piezoelements expand quickly, the platform slides away from the center of gravity due to the inertial force created and the low friction between the rubies and the base. The piezoelements are then slowly contracted so that their inertial forces are small and the friction forces are high enough to keep the platform from sliding back. For a maximum operating voltage of 100 V, the resulting displacement is 2 μm with one step. For an operating frequency of 400 Hz, the platform speed can reach up to 1 mm/s. The platform can reach the same positioning accuracy of 10 nm, as the first micro-crawling machine, but for this motion principle the navigation control problems are more difficult.

Bristle-based motion principle

Another example of a mobile platform is a bristle actuated device (Fur Driven Micro-mobile-mechanism, FDM), which can serve as an electromagnetically driven transport unit for microrobots [Fuku92], [Fuku93a]. Two different designs were built, Fig. 8.8. Unit 1 can only move forwards, while Unit 2 can additionally turn left and right. The motion principle is the same for both devices and therefore will only be described for the simpler Unit 1.

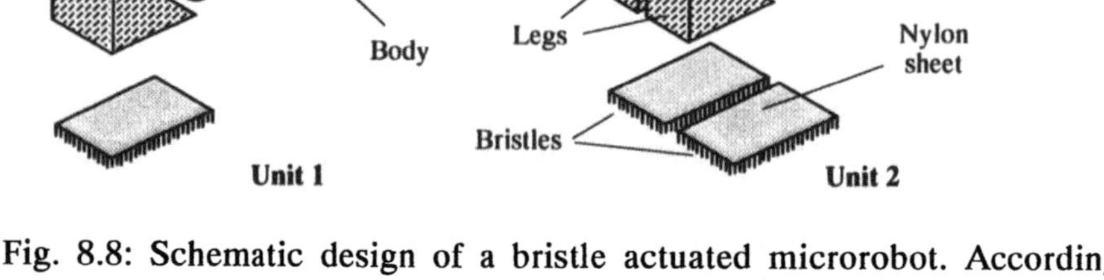

Fig. 8.8: Schematic design of a bristle actuated microrobot. According to [Fuku93a]

The robot consists of a body made from aluminum, a solenoid coil, a leg part with a permanent magnet, an L-shaped sheet metal leg to the bottom of which a nylon cloth with bristles is attached, and a stopper made from copper. The magnetic coil is made from an iron core and an enamel coated wire coil having 600 windings. The coil and the permanent magnet make up the electromagnetic actuator, Fig. 8.9.

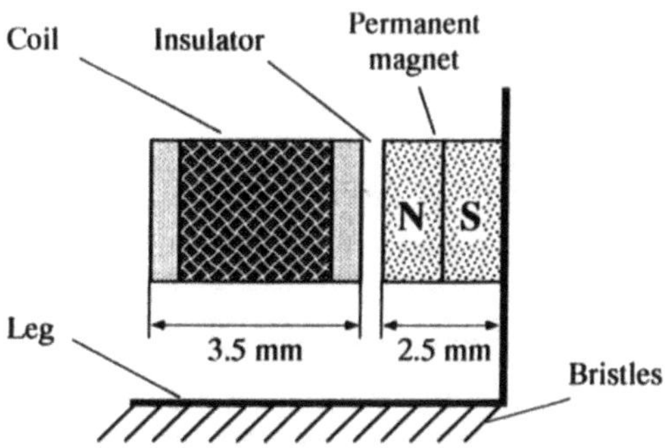

Fig. 8.9: Structure of the electromagnetic actuator of the microrobot. According to [Fuku93a]

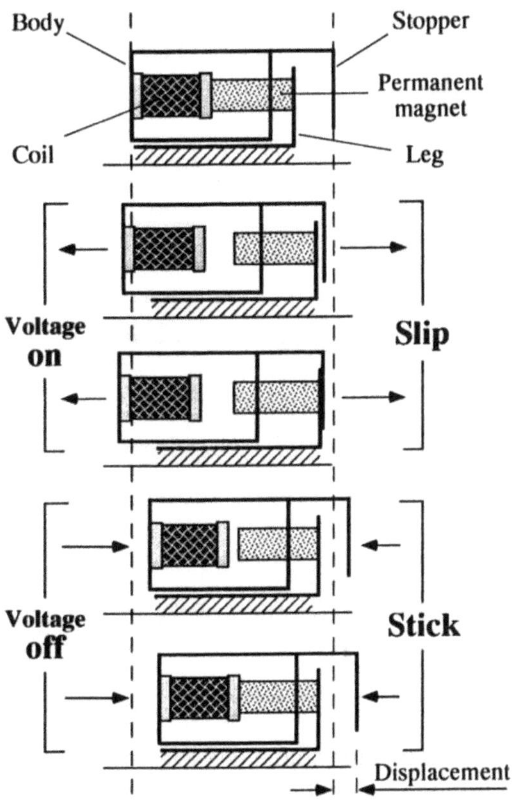

Fig. 8.10: Motion principle of the microrobot. According to [Fuku93a]

An insulation layer is attached to the core, which guarantees a fixed distance of 0.5 mm between the permanent magnet and the iron. When no current flows through the coil, an attractive force of 0.018 N is present between the permanent magnet and the iron. When a voltage is applied to the coil, a repulsive force of 0.054 N is exerted on the upper part of the leg and the leg starts to move, Fig. 8.10. By successively switching on and off the current a sliding motion in discrete steps is obtained. The exerted impact force and the generated inertia take advantage of the different friction forces with which the nylon bristles engage the floor. Since the bristles are tilted at an angle of 45°, the actuator can move better forward than backward. Thus, the forward and backward motions of the coil cause the device to walk. The microrobot can climb a 25° inclined floor carrying a load of 5 g.

Another microtransport unit, the Micro-Line Trace Robot, also uses a motion principle based on tilted bristles and inertia. This microrobot can follow a marked line on the floor by means of two infrared sensors. The structure of the robot platform is shown in Figure 8.11.

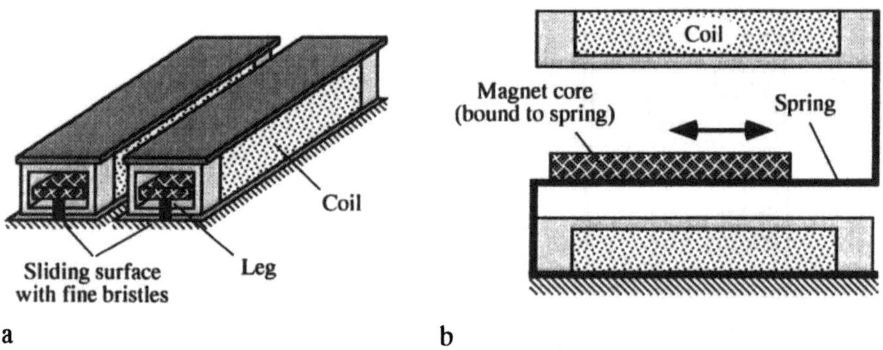

Fig. 8.11: Schematic design of the Micro-Line Trace Robot
a) schematic design of the robot platform; b) cross-section of the leg actuator. According to [Ishi95]

The robot moves by means of two sheets of fabric covered with fine bristles which are affixed to its legs. The latter are mounted on springs and vibrate by applying an alternating field to the magnetic drive coil. The electromagnetic drive allows quick motions, however, they have a relatively high power consumption. When the coil operates the spring at its resonant frequency, the power consumption can be reduced. A programmable logic device is placed onto the platform. In this device, the control algorithms which move the robot are executed. The complete robot system is 16 × 10 × 13 mm^3 in size and weighs 2.3 g. The functionality of the motion principle was shown with experiments. E.g. the device could successfully follow a spiral-like path; the path had a maximum width of 13 mm and the smallest curve radius was 30 mm. The navigation was supported with the help of two infrared photosensors integrated into the platform. The driving current was 70 mA and the operating frequency of the electromagnets 120 Hz.

Another development was a Micro-Autonomous Robotic System, which, in addition to a programmable logic device, carries three silver oxide 4.65 V batteries on its platform. Therefore, this robot does not need an external supply line, and it can be operated autonomously. In order to be able to keep the robot's dimensions within 1 cm^3 including the battery, the bulky electromagnetic coil had to be eliminated. Therefore, the motion principle introduced above was replaced by cantilever shaped piezoactuators, Fig. 8.12. A prototype of this microrobot could move at a speed of 2 mm/s carrying a load of 2 g.

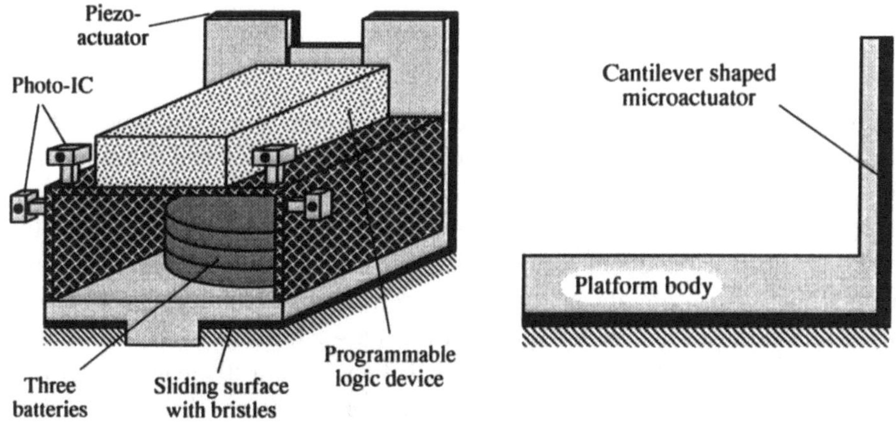

a b

Fig. 8.12: Schematic design of the Micro-Autonomous Robotic System
a) schematic design of the robot; b) cross-section of the leg actuator. According to [Ishi95]

Distributed magnet actuator

In [Inoue95], a new drive principle was introduced, using several cylindrical permanent magnets to move an object about a work table. Under the work table electromagnets were mounted in a matrix arrangement. Objects can be moved and rotated by coordinating the action of the permanent magnets, e.g. they can be arranged to perform gripping operations, Fig. 8.13.

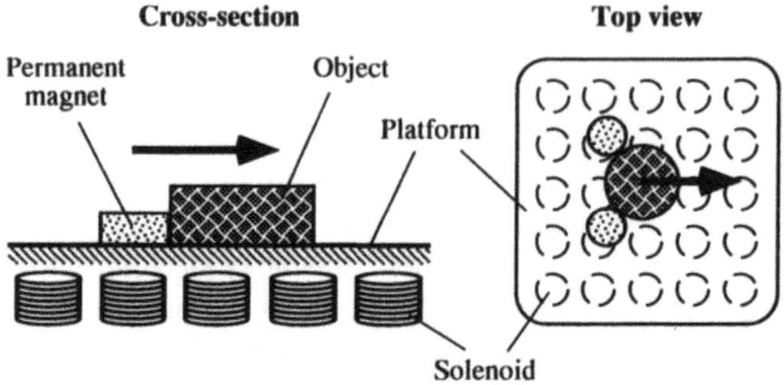

Fig. 8.13: Manipulation principle with distributed magnet actuators. According to [Inoue95]

The power of motion is provided by energizing selectively the solenoids to produce the desired magnetic field. The magnetic actuators can move various objects and can perform complex manipulations since they are not interconnected, Fig. 8.14. By using this principle, objects can be moved across the work place or rotated about themselves. This manipulator does not have any singularities, which are always present in conventional robot arms. The prototype of the device shown in Figure 8.14 has a 4.5 cm × 4.5 cm work place, under which 36 electromagnets are located; they have a diameter of 3.5 mm and are 17 mm high). The maximum force that could be applied was 7.4 mN. A computer system controls the electromagnets and can supervise the manipulation process with a camera.

This principle has an unlimited motion potential and also allows performance of 3D manipulations. For a practical application, a four legged robot was built which holds a rod as an endeffector; each leg is fastened to the rod through a

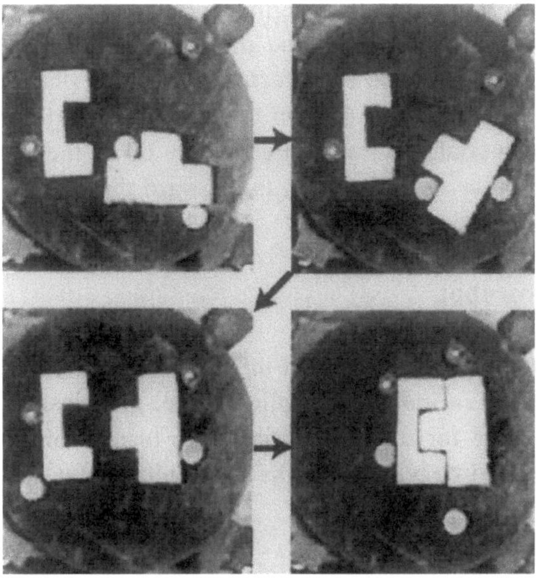

Fig. 8.14: Microassembly example. Courtesy of the University of Tokyo (Department of Mechano-Informatics)

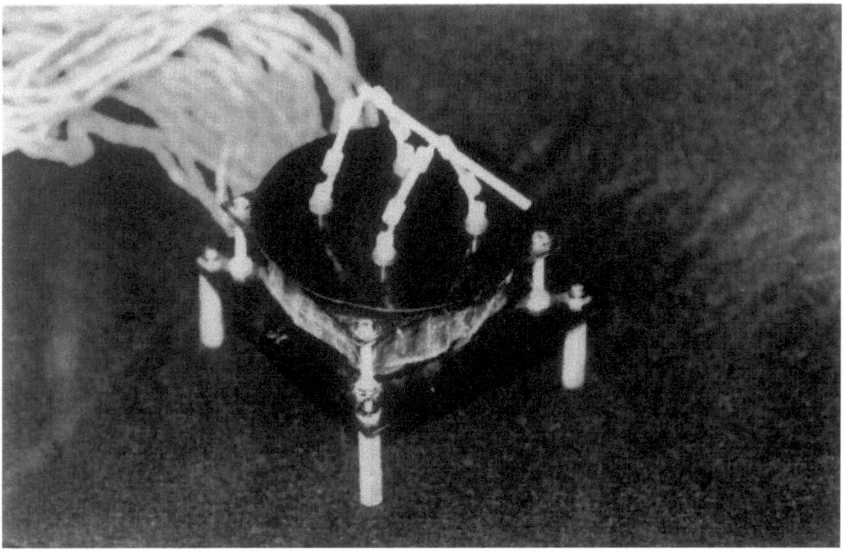

Fig. 8.15: Prototype of a micromanipulator. Courtesy of the University of Tokyo (Department of Mechano-Informatics)

flexible joint, Fig. 8.15. The legs are provided with permanent magnets, which allow the robot to position the endeffector using the motion principle discussed in Figure 8.13. This robot can carry the endeffector to different places, can rotate it and tilt it. The rod of the robot is 60 mm long and the length of the legs is 35 mm. The force which could be transmitted by the tip of the rod was 6.8 mN in the downward direction and 3.2 mN in the upward direction.

Piezoelectric swimming microrobot

A concept of a swimming microrobot was introduced in [Fuku95]. The robot can be used as a mobile platform for an industrial application, e. g. to inspect pipelines or as a miniaturized device for medical purposes to inspect blood vessels. The robot motion is obtained by vibrating fins using piezoactuators, Fig. 8.16.

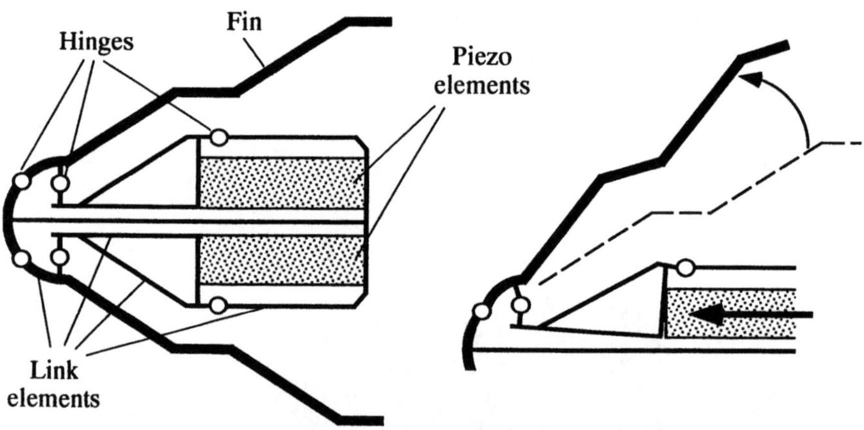

Fig. 8.16: Operating principle of the swimming robot: robot design (left) and motion principle of the fins (right). According to [Fuku95]

The multilayer stacked piezoelements exert high forces but their displacement is small. However, they can easily operate a lever mechanism. In this microrobot, a small 8 μm stroke of the piezoelements is magnified 250 times to generate a secondary motion of about 2 mm. The swim motion is accomplished with two 32 mm long fins and a lever mechanism, as shown in Figure 8.16. The resulting propulsion force was 10^{-4} N for forward motions and 2×10^{-5} N for backward motions. The robot has a length of 34 mm and a width of 19 mm. The fins provide the robot with one rotational and one translational degree of freedom, enabling it to move forward and evade obstacles in its path. The robot was able to move at a speed of 30 mm/s at an operating frequency of the piezoelements of 100–350 Hz and a voltage of 150 V.

Levitated mobile unit with optical power supply

Usually, miniaturized mechanical actuator systems have problems with high friction. Also, when miniaturizing a robot, the electric supply lines limit the action radius of the device since batteries or other energy cells cannot be integrated due to their size. For this reason attempts were made to realize a cable-less mobile microplatform which levitates on an air cushion and can be supplied with energy by using UV radiation. The work place of the mobile actuator system was an air cushion table made of steel, which had very thin air slits (180 μm wide) evenly spaced over the work area to produce the air bearing surface, Fig. 8.17, top. The air table has an overall surface area of 80×80 mm^2 on which the robot platform could move about. The platform is $15 \times 15 \times 2$ mm^3 in size, and consists of a pyroelectrical element, two electrodes and a base, Fig. 8.17, bottom.

The air table made from steel serves as one electrode in this positioning system, and the other electrode is the robot platform itself. The motion of the microplatform is obtained by the electrostatic force which is created when a voltage is applied to the electrodes (Figure 5.4b). The horizontal component of the force activates the motion; it brings the linear electrode system into a state of maximum potential energy. The electric energy needed for actuation is generated by UV light. When light falls on the platform it is converted into a voltage by the pyroelectric element. Figure 8.18 shows the operation sequence of the robot platform. The upper figure depicts the initial state in which the platform is lifted from the table by the air pressure and stays in a stable condition. When UV radiation with a wavelength of 365 nm falls on the pyroelectric element, an electric voltage is generated, resulting in an electrostatic force. Due to the horizontal force component, the robot platform begins to slide, Fig. 8.18, middle. The pyroelectric element has a dimension of $5 \times 6 \times$

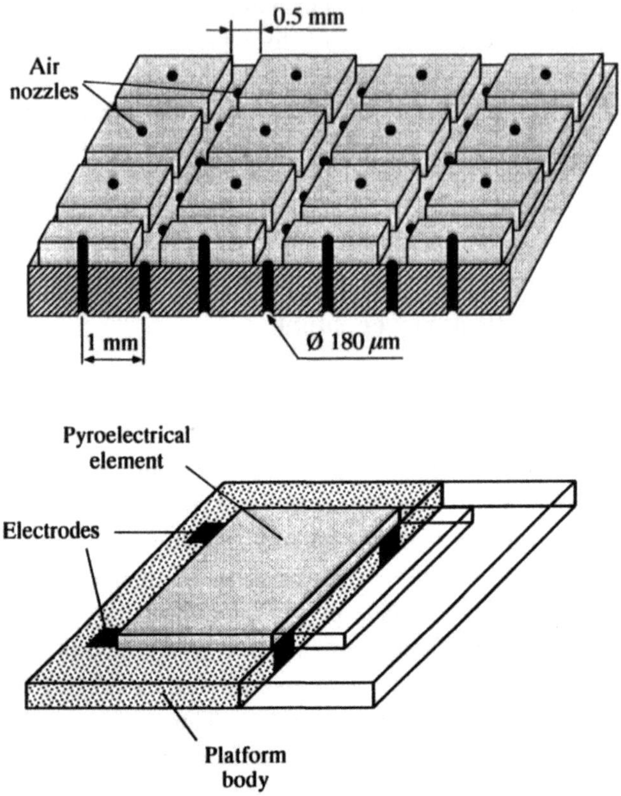

Fig. 8.17: Air table (top) and the robot platform (bottom). According to [Fuku93b]

0.2 mm³, which is sufficiently large to generate the voltage needed for the propulsion of the platform. Since the robot does not have the capability of controlling or changing its direction, it is confined on two sides by two thin guide rails with a thickness of 100 μm. The rails also increase the lift force of the air table. If the radiation is stopped, the platform continues its movement with a nearly constant speed of 7 cm/s due to its inertia and the very low friction of the air bearing, Fig. 8.18, bottom. The control of the freely sliding microrobot is very difficult, e. g. the direction of motion depends on the initial position of the platform. A platform operation under constant UV radiation with the platform's motion controlled by modulating the UV light still has to be investigated. It is also difficult to stop the robot, since the braking forces are

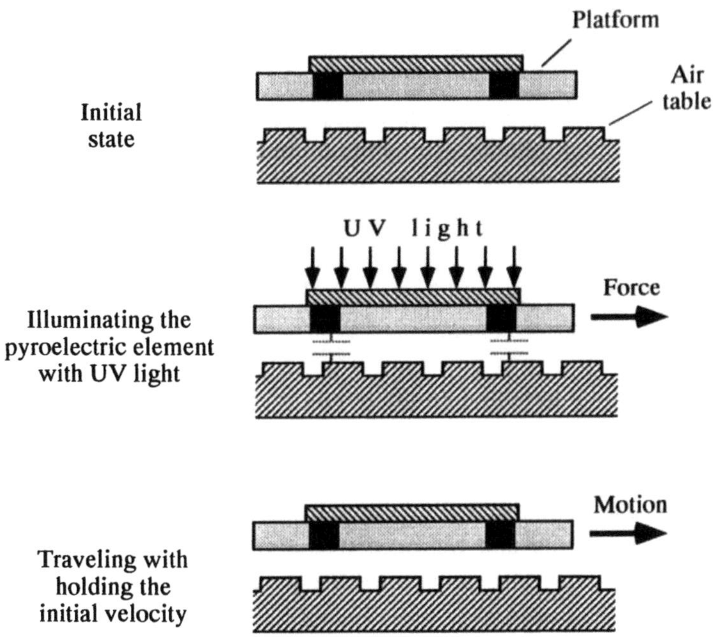

Fig. 8.18: Microrobot motion sequence. According to [Fuku93b]

too small compared to the robot's inertia. The shown platform has only one degree of freedom. To design a robot with more degrees of freedom, several pyroelectric elements must be fastened to the platform. However, one problem with this is the selective illumination of each pyroelectric element.

After the presentation of numerous promising ideas of how to move a microrobot, we want to discuss the possibilities for realizing micromanipulators in the next section.

8.5 Manipulation of Microobjects

Microassembly in production, cell manipulations in biomedical research and eye and plastic surgery is an important objective of MST. The topic "flexible micromanipulation systems" has become of interest, especially for microassembly. If a microsystem requires a hybrid design, assembly-specific problems must be considered in addition to task-specific requirements. Questions of importance are the availability of suitable interconnection technologies, the possibility of automating the assembly steps and the handling of assembly related costs. The availability of highly precise assembly processes will make it easier to economically realize operable microsystems. In order to efficiently produce microsystems and components in lot sizes or by mass-production techniques, it is absolutely necessary to introduce flexible, automated, highly precise and fast microassembly systems.

Different concepts are being followed to do micromanipulation for specific classes of application:

- Purely manual micromanipulation is the most often used method today. It is a common practice in medicine and biological research. Even in industry, microassembly tasks are very often carried out by specially trained technicians, who, for example, align assembly parts using screws and springs, then position the parts with tiny hammers and tweezers, and finally fasten them in the desired position. However, with increasing component miniaturization the tolerances become smaller and smaller, and the capabilities of the human hand are no longer adequate.

- The application of teleoperated, partially automatic micromanipulation systems of conventional size, which transform by means of a joystick or mouse the human operator's hand motions into the finer 3D motions of the manipulators of the micromanipulation system. Here, special effort is devoted to the development of methods which allow the transmission of different types of signals to the operator from the microworld (images, forces, noises) to provide the operator with a better feedback from the manipulation task. The dexterity of the human hand is supported by sophisticated assembly techniques. However, the fundamental problem of resolution of the fine motion and of speed remains since the motion of the tool is a direct imitation of that of the operator's hand.

- The use of automated multifunctional micromanipulation "desktop stations" supported by miniaturized flexible robots which employ MST-specific direct-drive principles (Chapter 5). These robots could be mobile and able to perform manipulations in different work areas. The transport and micromanipulation units performing the assembly may be integrated by hybrid techniques onto one chip; this may include microelectronic and power electronic components. The flexibility of such a microrobot can further be enhanced by simple and automated tool changers. As opposed to the aforementioned micromanipulation technique, there is no direct connection between the operator's hands and the robot. The assembly steps are carried out with the help of open-loop control algorithms, or, if the assembly tasks are more complex, with closed-loop control. The human operator assigns all tasks to miniaturized assembly mechanism and, by doing so, tries to compensate for his limited micromanipulation capabilities. The operator's commands are given to an input unit and are passed along to the robot actuators in a suitable form with the help of the control system. The degree of abstraction of the commands is determined by the capabilities of the control system. Many microrobots can be active at the same time in a multifunctional manipulation desktop station.

- The use of several flexible nanorobot systems which carry out the manipulation tasks in close cooperation. Here, the robot size is comparable to that of the manipulated object. The realization of this concept, which is based on human behavior, is in the distant future.

Although in general manipulations vary from an application to an other, the following operational sequences are always taking place: grip, transport, position, release, adjust, fix in place and processing steps like cutting, soldering, gluing, removal of impurities, etc. Figure 8.19 shows some characteristic microworld manipulation methods.

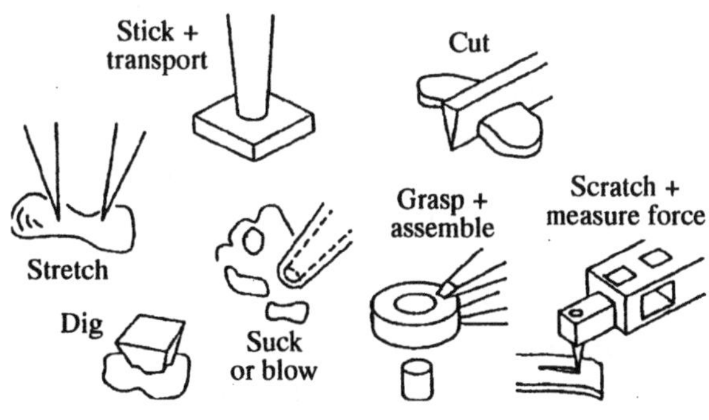

Fig. 8.19: Microworld manipulations. According to [Hata95]

In order to be able to carry out these operations, corresponding tools are needed, such as microknives, microneedles to affix microobjects, microdosing jets for gluing, microlaser devices for soldering, welding or cutting, different types of microgrippers, microscrapers, adjustment tools, etc. Microgrippers play a special role, since they considerably influence the manipulation capabilities of a robot. Microgrippers can clamp, make frictional connections or adhere to material, depending on the physical and geometrical properties of an object. Adapting a gripper to the shape of the object to be gripped is often the best solution in the microworld, even if it is at the cost of flexibility. This allows handling of a workpiece with a complex shape, such as a gear; the gripper securely attaches to the contour of the part. For small, smooth parts, a suction pipette might be a practical tool. If the upper surface of a workpiece must not be touched/gripped because of technological reasons, it can be protected by a corresponding form-fit of the pipette hole. For contour clamping and frictional connections in manipulations involving fragile parts, elastic grippers made of soft plastics are preferred to metal grippers. Due to the variety of task-specific gripping tools in automated micromanipulation systems, a suitable gripper exchanger system might be necessary.

It should be mentioned that it is not always possible to adapt conventional manipulation methods to the demands of the microworld. Often, additional capabilities are needed which are specific to a micromanipulation station. Microscopically small objects cannot be easily handled as bulk material. They entangle or jam and can be damaged during transport or isolation of single components. Especially microcomponents which are manufactured by the surface micromachining are often adversely affected by parasitical stresses in the thin layers. This may lead to a distortion of the microstructure. Microobjects may be sensitive to dust particles, humidity, vibration or temperature changes. Usually, special clean rooms are necessary which are air-conditioned, have air purifiers and vibration dampers. Depending on the application, different purity concepts may be applied. In a simple application, local shielding of a dust-prone process may suffice; in an extreme case a well-equipped cleanroom might be necessary, but the latter is a very expensive solution.

One major problem is the effect of various forces, which is completely different from the macroworld. Gravity only plays a minor role in the microworld, but attractive forces, such as electrostatic forces or Van-der-Waals forces, are significant. Liquid surface tension can also act as an attractive force in a micromanipulation if humidity is high or if a manipulator is wet. This can be eliminated by the use of an air-conditioned room where the humidity is below 9% [Arai95]. This unusual sensitivity to forces can be very irritating in a micromanipulation station. E.g. it can be easier for the micromanipulator to grip and manipulate an object than to release it afterwards. On the other hand, such an adhesion force can be used to develop new gripping methods which can fundamentally differ from the familiar mechanical (gripping with pliers or tweezers or affixing with needles) and pneumatic methods (vacuum-based gripping using suction pipettes). In [Arai95], a few interesting ideas were shown for adhesive gripping, such as electrically charging a manipulator (electrostatic force) or wetting a gripper surface by special micromachined orifices. In the latter case, a moisture removal process by means of an installed microheating element may be employed to release the microobject.

A very important aspect is the transmission of information from the microworld to the macroworld to facilitate the control of the manipulation processes and the necessary feedback. With the present technology it is very difficult to obtain force information from the microworld. One must depend on visual process monitoring, for which stereo light-optical microscopes are often used. However, since the distance between the objective and the probe table is usually too small (10–20 mm), this method often encounters several problems. Certain manipulations cannot be carried out, or only with difficulty, because the manipulator simply does not fit between the objective and the object. Also, the resolution of the microscope is limited by the wavelength of the visible light (up to 400 nm) and may not be sufficient to obtain the accuracy needed for the manipulation.

To solve this problem, the manipulation station can be placed in a vacuum chamber of a scanning electron microscope. The latter has an ample working space, high resolution and a large depth of field. A scanning electron microscope is the heart of the integrated microproduction system presented by [Hata95a]. This system, called nano-manufacturing world, consists of three vacuum chambers in which a complete microassembly process can be placed. In the first chamber, microparts are produced under optical microscope control. The resulting parts are transported to the vacuum chamber of the scanning electron microscope over a special valve with a small transport robot. Here, they are connected to other parts and further processed by other manipulators. The microscope can provide images from many view points in order to make it easier to observe the object's position. The finished components are then brought into a buffer chamber by means of another valve, from which they are later removed. The entire nano-manufacturing world is 1.8 m long, 0.9 m wide and 1.6 m high; it is a genuine "desktop plant".

Using a scanning electron microscope also means that the manipulators must be operable in a vacuum and withstand the electron radiation. For this reason, all microelectronic system components must be outside the microscope chamber to avoid damage or interference from the electron beam. The traditional size limitation for the manipulators is not so much a problem any more, since there are large-chamber microscopes available [Klein95]. With the help of movable electron optics, these large-chamber microscopes greatly improved the flexibility to manufacture and assemble, which had been a severe limitation with conventional devices. Objects can now be visualized and worked at from various views. The vacuum chamber has a dimension of about 2 m^3. Several processing and assembly stations can be set up in it in a sequential order, thereby forming a micro-production line. With the microscope a resolution of up to 4 nm is obtained using a tungsten cathode. The magnification can be continuously adjusted from 15 to 300,000.

The productivity of a micromanipulation system is low for a manual operation; it improves by going to a teleoperation and further on to an automation; this is similar with conventional robots. Most micromanipulation investigations today focus on improvements obtained by progressing from a purely manual to a teleoperated system. As the overview of the micromanipulation systems discussed below will show, most systems are teleoperated. The operator executes all motions as if the manipulator were his own arm. However, there are two main constraints: first, the operation must be perceptible to the operator and second, the processing information must be correctly conveyed to him. As previously mentioned, attempts are being made to make the transmission of effects from the microworld to the operator as realistic as possible. It is important that the operator can observe the entire scene in his field of view and that he can see the workspace from different angles. Besides visual information, the operator should also be able to receive acoustic and force signals if possible, to increase the accuracy of his movements and to avoid destroying the microobjects. For this, force sensors are needed which are implemented into the microtools (e.g. a microgripper). Suitable solutions are now being sought after to realize such sensors [Mori93], [Horie95].

8.6 Micromanipulators: Ideas and Examples

Teleoperated microhandling system

Several micromanipulation systems were developed based on the above considerations. Figure 8.20 shows the schematic structure of a teleoperated microhandling system [Mits93].

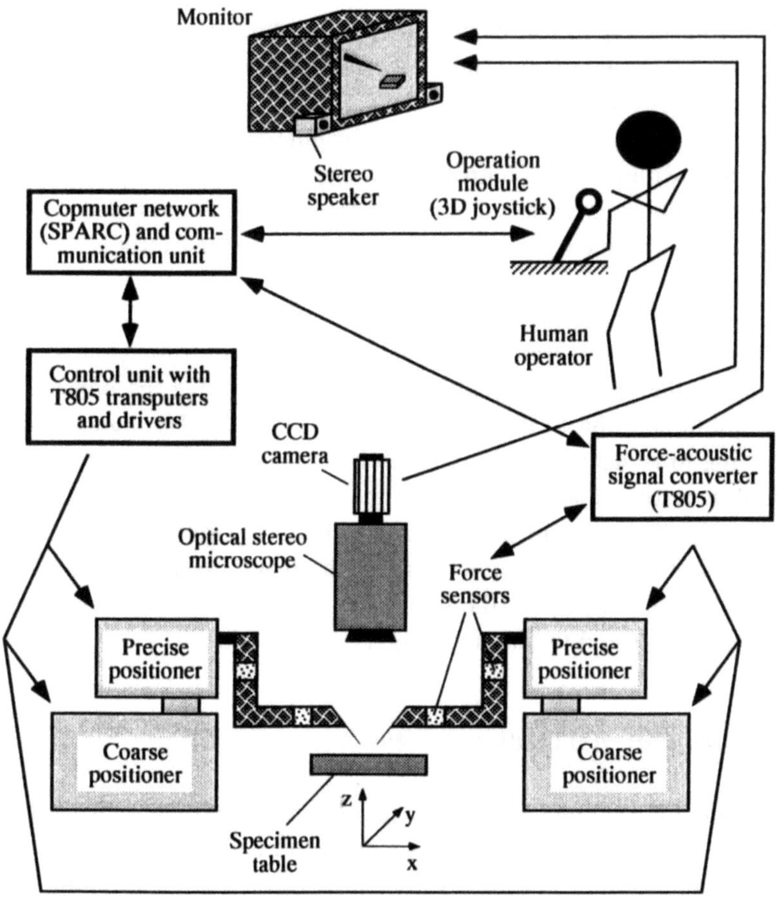

Fig. 8.20: Schematic structure of the teleoperated microhandling system. According to [Mits93]

The system contains five components. These are: the actual manipulation unit, the operation module for the operator, the optical stereo microscope with a CCD camera, the control unit and the module for transforming force signals into acoustical ones. The manipulation unit represents the right and left hands of the operator. Each mechanism consists of rough and fine positioners, a multi-axis force sensor and an endeffector. The rough positioner is driven by step motors with a resolution of 3.5 mm. The fine positioner uses an electromagnetic principle and has a guaranteed resolution of 600 μm in both the x- and y-direction and 800 μm in the z-direction. In addition, each manipulator is equipped with a multi-axis force sensor, whose signals are converted into corresponding acoustic signals after amplification, which in turn are sent to the operator. The CCD camera and the stereo microscope are responsible for transmitting the visual information and for displaying it on a TV screen. The computers of the control system are T805 transputers; each the rough positioner and fine positioner are controlled by one transputer. The operation module is a special joystick with 3 degrees of freedom. It is equipped with a multi-axis force sensor which detects the forces the operator exerts.

Several experiments with the system (for example, the removal of a 100 μm-thick paint layer from a substrate) have shown that the transmission of the acoustic signals to the operator considerably helps to improve the control of the manipulation in addition to the tactile feedback signals. The human capability of reacting to acoustic signals is much better than reacting to force by touch. Without acoustic feedback, the force the operator puts on the joystick varies considerably. This system can be used to handle parts in a range of a few μm to several mm. Possible applications of the manipulator are testing of LSI circuit connections, inspecting the surface of harddisks and optical disks, repairing printed circuits and to do microsurgery.

Micropositioning unit AUTOPLACE 400

An assembly unit for micromechanically manufactured airbag sensors has been successfully developed [Meiss94]. Its task is to bond a 1 mm^3 large cubical part on a chip with an accuracy of ±3 μm. Here, the positioning was controlled with visual sensors; the accuracy of the chip was about 0.2 mm in all three main axes, and the production system could adjust to an offset of the parts by ±50 μm. The assembly unit is a flexible positioning table with 5 degrees of freedom, Figure 8.21.

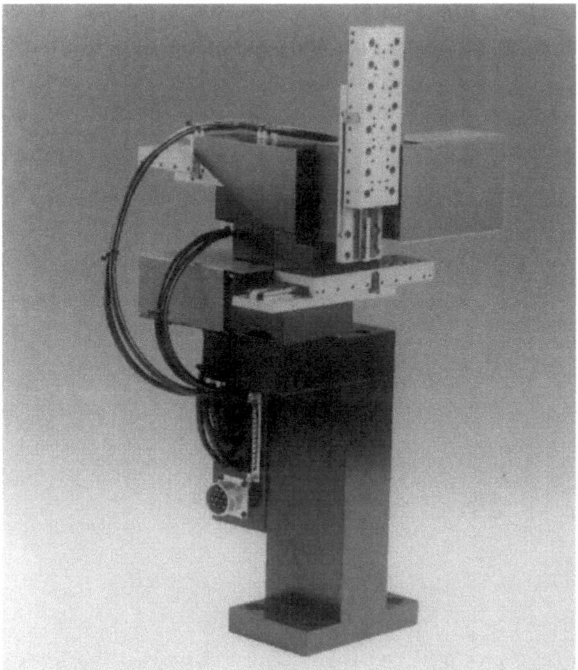

Fig. 8.21: Micropositioning unit AUTOPLACE 400. Courtesy of SYSMELEC SA, Switzerland

The reproducibility is around ±1 μm, there is a high static and dynamic rigidity at a maximum speed of 0.8 m/s; accelerations of up to 15 m/s^2 can be handled. The high rigidity is especially important in this application, since during the hardening stage of glue used in assembly, the parts may not move. The sensor system consists of three CCD cameras and a laser measuring unit. The device is controlled by a multi-processor unit using a multi-tasking operating system. The assembly system can process 520 sensors per hour.

A micromanipulation system with a shape memory alloy gripper

Attempts are now being made to use shape memory alloy actuators to perform micromanipulations. One example is a robot system for handling objects in the submillimeter range [Russ94]. The system consists of a manipulator with 6 degrees of freedom, which is controlled by special tendons and has a two-finger gripper at its end. There are a total of six kevlar tendons, which are held under mechanical tension by a spring; they position and align the gripper platform, Fig. 8.22.

Fig. 8.22: Structure of the micromanipulation system. Courtesy of the Monash University (Department of Electrical and Computer Systems Engineering)

Each tendon is driven separately by a pulley and stepper motor. The gripper has a cylindrical workspace with a radius of 2 cm and a height of 5 cm, Fig. 8.23. The gripper can move to any point in its work space with a speed of 7 mm/s and an accuracy of up to 0.1 mm. It consists of two 0.13 mm thick, flat steel fingers, which are 15 mm long and 4 mm wide, tapering off to 1 mm at the fingertips. There is a U-like shape memory alloy wire between the fingers; the wire has a diameter of 0.3 mm. It gives the gripper a linear degree of freedom, to open and close the fingers. If the shape memory alloy wire is heated with the help of an external power supply (0.8 ampere), it changes from the martensitic state to the austenitic state (Section 5.6) and "remembers" its original straight form. When the wire tries to take this form on again, a force is created which bends the fingers apart. If the current is turned off, the wire cools and changes back to the martensitic state and the gripper

Fig. 8.23: Gripper actuated by a shape memory alloy wire. Courtesy of the Monash University (Department of Electrical and Computer Systems Engineering)

closes. The gripper system is controlled by a 68HC11 microcontroller, which in turn is operated by a PC; it also controls the motions of the fingers. A phototransitor is used as a sensor, which measures the opening angle between the fingers by means of an LED light beam.

The systems's structure is very simple and robust; it has the advantage that the gripper system may collide with stationary objects without causing a serious damage. If this gripper were miniaturized and a vision system were installed to recognize the positions of parts on a work surface, this system could also be used for microassembly.

A micromanipulator for a scanning electron microscope

A manipulation system is currently being developed, which can be used for assembly of hybrid microsystems [Joha93]. The unit is installed inside a scanning electron microscope, which is suited for highly precise assembly; it has a

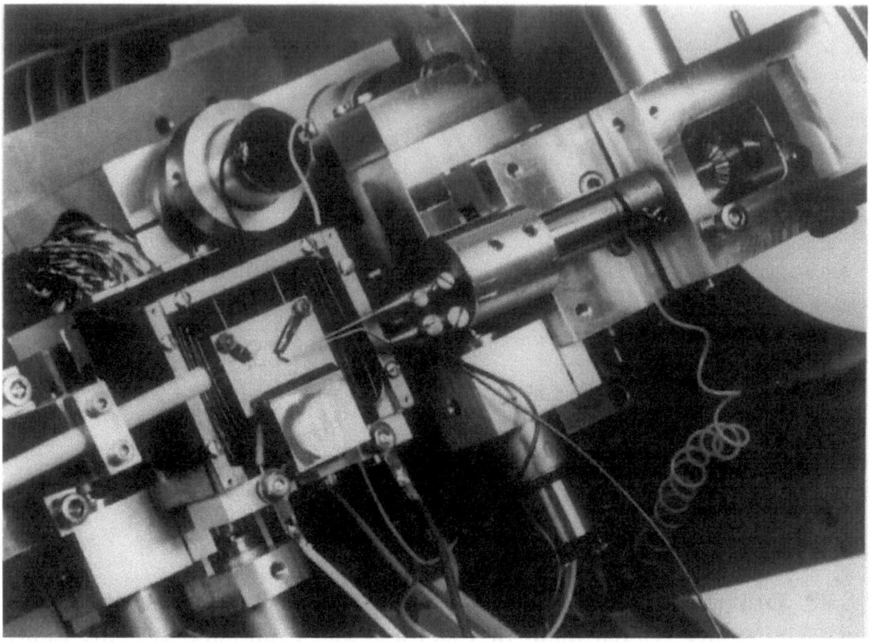

Fig. 8.24: Micromanipulator for a scanning electron microscope. Courtesy of the Uppsala University (Department of Technology)

large working space and a large depth of field and is free of dust, Fig. 8.24. The work place of this system is a table which can be heated up to reach 1000°C for carrying out various microassembly steps requiring high temperatures, such as bonding. A supply storage is also incorporated for holding workpieces until they are needed. The microscope table can be moved with an accuracy of about 1 μm both in the x- and y-directions. The microscope's vacuum chamber is 200 × 150 × 150 mm^3. Microtweezers using piezoactuators do the fine positioning of the microobjects, Fig. 8.25.

The tweezers have an opening clearance of 200 μm and can apply a force of 0.3 N. They are attached to a positioning unit which has 4 degrees of freedom and which does the rough positioning of the gripper; this unit is driven by d.c. motors. For the next prototype of this system, another positioning unit will be installed for the gripper, which has 6 degrees of freedom and a positioning accuracy in the submicrometer range. This will make it more suitable for the assembly of three-dimensional microsystems. An additional manipulator can be used for performing measurements (like temperature) or as another tool for supporting the assembly process (left side in Figure 8.24).

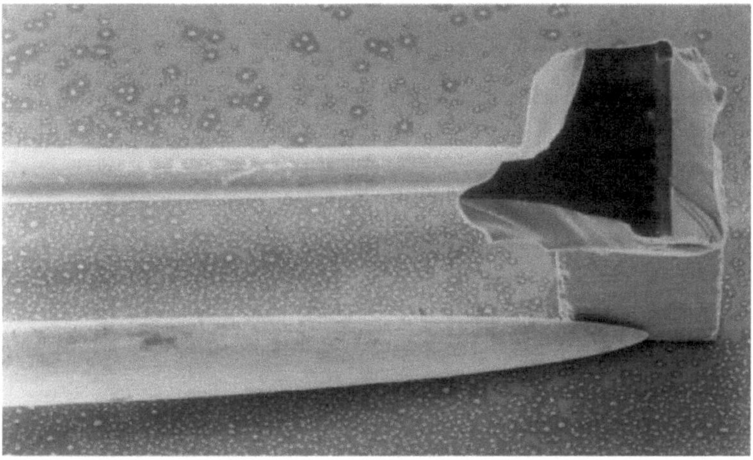

Fig. 8.25: Photo of the piezoelectric endeffector. Courtesy of the Uppsala University (Department of Technology)

A manipulator system for a scanning electron microscope

Another example of a teleoperated microhandling system operating in connection with a scanning electron microscope was introduced in [Mori93]. The concept was shown in Figure 8.20. The operator receives visual, acoustic and force information from the work area when handling microobjects. The movements of the operator's hands are transmitted by two joysticks and a control unit into the work space, which is the vacuum chamber of a stereo scanning electron microscope, and are then transformed by a manipulator system to the fine motions needed in the work space. The manipulator system consists of two piezoelectrically driven robots located in the chamber of the microscope, Fig. 8.26. The robot controlled by the operator's left hand holds the object and changes its position, taking on the function of a controllable positioning table. This positioning table can move within the range of $20 \times 20 \times 20$ mm^3 with an accuracy of 0.01 μm. The right-hand robot is the actual manipulator. It is able to move within an area of 15 μm^3 with the same accuracy and to hold different tools. The tools currently used are tungsten or diamond needles which can be exchanged during manipulation.

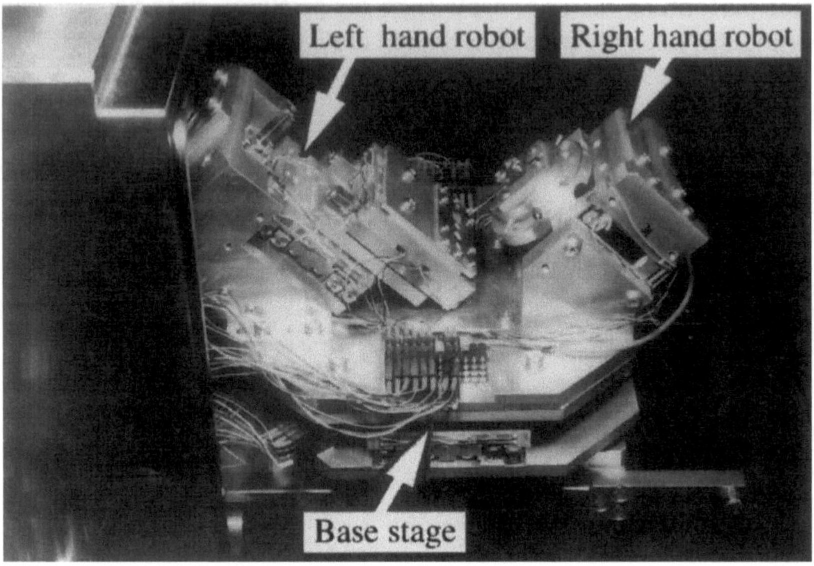

Fig. 8.26: A manipulator system for a scanning electron microscope. Courtesy of the University of Tokyo (Research Center for Advanced Science and Technology)

It is also planned to use a highly sensitive multi-axis force sensor for the manipulator, which will give the operator a feeling for the reaction forces via the joystick. In order to duplicate the operator's movements as realistic as possible, both robots have 6 degrees of freedom. A series of teleoperated experiments was done with a prototype of this system. A tungsten needle was used with a diameter of about 1 μm to separate aluminum connections on an LSI chip at a defined location. The system component which transforms information on the forces encountered by the manipulator into acoustic signals was also very helpful to the operator; he was able to carry out the given manipulation task without any problems. Removing aluminum scrapings which had accumulated on the tip of the needle, however, was difficult.

The system was enhanced to control the manipulators automatically with a real-time image processing module; the operating modes considered were either "open loop" control or a "visual feedback". Figure 8.27 shows the system structure. A host computer controls all system components. They are: an image processing computer, a transputer system for controlling the two mani-

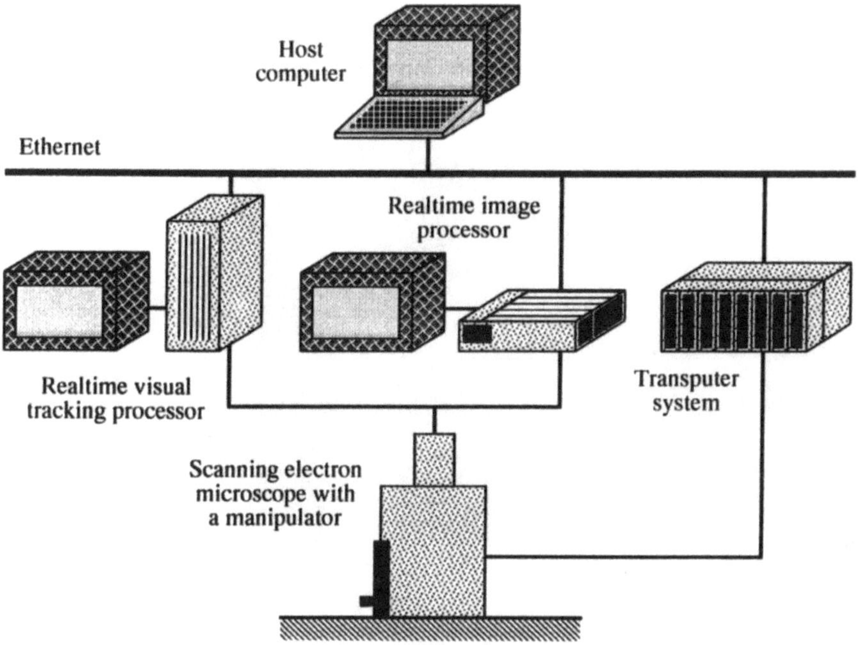

Fig. 8.27: Control structure of the manipulator system installed in a scanning electron microscope. According to [Sato95]

pulators and a so-called visual tracking processor which allows keeping track of a moving object. For the used scanning electron microscope a magnification range from 15 to 200,000 can be automatically selected.

Initial experiments have shown that the manipulator system can automatically carry out micromanipulations under sensor support. The tool was a tungsten or diamond needle, which can be sharpened electrochemically and has a tip radius of about 100 nm. In an automatic "open loop" mode (no visual feedback), it was possible to draw up with the needle Chinese characters of 15 μm × 15 μm size. In order to prevent the needle tip from being plastically deformed, the drawing was performed very slowly; 20 minutes were needed to draw one character.

This system was also the first to demonstrate controlled micromanipulation. A groove of 3 μm width was made in the substrate using a visual feedback mode. Figure 8.28 shows the test results, whereby the upper line was engraved without visual feedback, and the lower one with visual feedback.

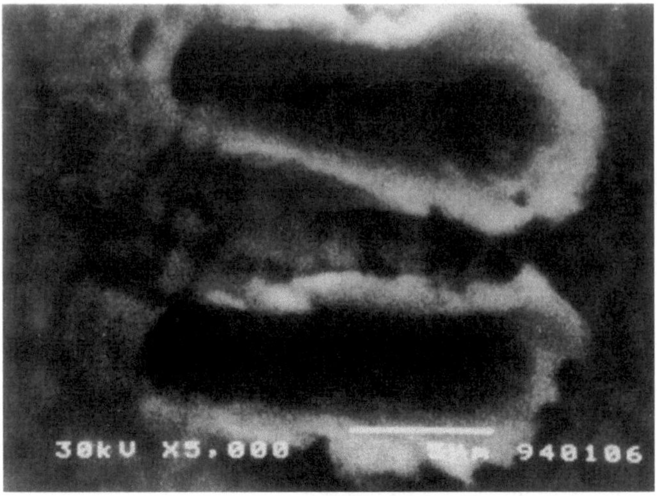

Fig. 8.28: Engraving a microgroove with a needle: open loop control (upper line); visual feedback mode (lower line). Courtesy of the University of Tokyo (Research Center for Advanced Science and Technology)

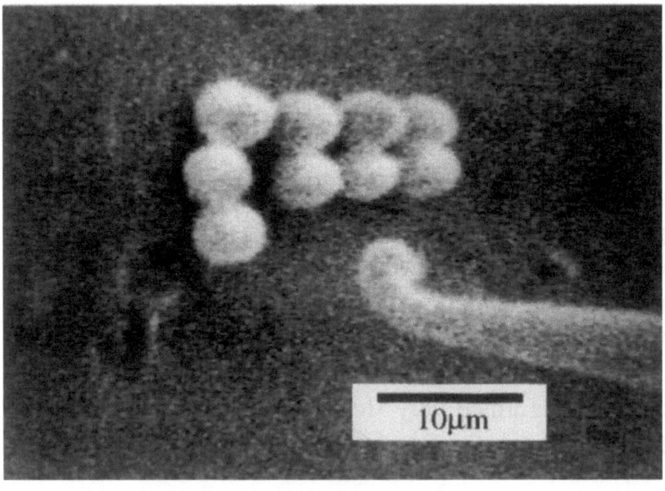

Fig. 8.29: Teleoperated manipulations of iron particles. Courtesy of the University of Tokyo (Research Center for Advanced Science and Technology)

The closed-loop mode was significantly more accurate, it made the desired horizontal groove of uniform width. The maximum deviation of width was only 300 nm; this is a very good value considering the minimum resolution of 200 nm which the scanning electron microscope had in the video mode.

As other experiments have shown, the unfamiliar force conditions in the microworld often lead to problems with a microassembly. Figure 8.29 shows an experiment in which several 5 μm small iron particles are telemanipulated by using a needle as a tool; they are positioned as shown. The particle, which is on the tip of the needle, sticks to it by adhesive force. The operator had major problems trying to remove the particles from the manipulator. It took 30 minutes to arrange the nine iron particles shown on the photograph. In the end, the entire arrangement was destroyed when trying to remove the last particle from the manipulator.

8.7 Microassembly with the Help of Microrobots

The examples introduced above point out many problems connected with microrobotics. As we have seen in the last section, modern micromanipulation systems depend on the skill of the operator. However, he is limited by his own skills to handle small objects and the tools available to him. With the growing variety of design elements there is a definite trend toward the development of hybrid microsystems. Therefore, the further development of MST depends on the availability of flexible manipulation stations, which allow components to be automatically assembled. This will reduce production costs and simultaneously create high quality. For this reason, the manufacturing engineer is looking for flexible micromanipulation robots which have both transportation and manipulation skills. These robots must be automated with the help of visual and force sensors, in order to free humans from the tedious task of having to manipulate miniscule objects directly. This automation is necessary to make the production of microsystems more efficient. An automated robot-based station can take over precise processing and assembly tasks and thereby further the growth of MST. They should be quickly adaptable to different types of products with the support of sensors and suitable controllers. Problems involved in this promising field will be considered below.

The spectrum of tasks in microassembly ranges from simple preparatory operations like applying adhesives, drawing adjustment marks, cleaning objects, etc. to the performance of the final assembly and inspection of the finished microsystem. A well conceived microassembly station must be able to automatically accomplish the following steps with its microrobots:

- preparation of the parts to be assembled
- transportation of parts
- positioning and fixing of parts
- connecting the parts
- testing and measuring the finished microsystem.

Typically, in a conventional automatic or semiautomatic assembly station, standardized mechanical parts are assembled in well-defined work positions. The robots performing the work are usually of multi-axis arm design or they are gantry systems, usually driven by d.c. motors. Today, it is being attempted to use these type of familiar systems for handling and assembling of miniaturized components with dimensions in the millimeter range. E.g. a modular microassembly system with 4 degrees of freedom is currently being developed [Geng95]. It has an x-y positioning table, above which a overhead manipulator and a stereo microscope with a CCD camera are located, Fig. 8.30.

The overhead manipulator can move along the z-axis and rotate around this axis; it can be equipped with different grippers or applicators by an automated revolver tool exchanger. The positioning accuracy of the microassembly system is 2 μm. Its task is to assemble microoptic duplexers, which consist of two 0.9 mm spherical lenses, a $3 \times 3 \times 1$ mm^3 wavelength filter and a glass fiber cable. These components are assembled onto a $4 \times 14 \times 1$ mm^3 base plate made by the LIGA method. The required positioning accuracy is within the range of ±20 μm. All assembly parts are manually inserted into magazines and then placed in exactly defined positions on the positioning table. The actual assembly steps, such as gripping the parts, applying adhesives and positioning or fixing the parts, are carried out semiautomatically. The assembly process is interactively controlled by a personal computer. The operator obtains the visual information from the work space with the help of the stereo microscope and the CCD-camera (left side in Figure 8.30a), and initiates the necessary assembly steps with a mouse click on a graphical user interface. It is planned to add a force control system for inserting the lenses and the wavelength filter in their guide fixture.

a b

Fig. 8.30: A microassembly system
a) overall view; b) the system's work table. Courtesy of the Karlsruhe Research Center

Another microassembly system with a 6-axis vertical manipulator arm called μ-KRoS 316 is shown in Figure 8.31. The robot has a repetition accuracy of 3 μm in a spherical work space with a diameter of 1 m and can be used for joining, adjusting and testing mechano-optical components. Presently, various microobject handling techniques are being developed with this device within the framework of a German BMBF research project [Frick95].

With increasing workpiece miniaturization, however, it becomes more and more difficult to use conventional manipulation robots for assembling microsystems. The manipulation accuracy is mechanically limited for conventional robots, since disturbing influences which are often negligible in the macroworld, such as fabrication defects, friction, thermal expansion or computation-

Fig. 8.31: The high precision robot μ-KRoS 316. Courtesy of JENOPTIK Automatisierungstechnik GmbH, Jena

al errors, play an important role in the microworld. Due to the mechanical drives needed for the actuator's motions, these robot systems are subject to mechanical wear and must undergo regular maintenance, which makes them expensive. The assembly process is influenced by the dynamics of the robots and the object being handled.

Different processing conditions exist for manipulating microscopically small components (Section 8.5). The positioning accuracy and the tolerances of the microcomponents are in the nm range, a few orders of magnitude smaller than in conventional assembly. These accuracy requirements can only be obtained with manipulators having highly accurately drives utilizing MST and advanced closed-loop control. In general, robots which only execute preprogrammed motions, are not suitable for microassembly. The microrobot-based "desktop station", which was introduced in Section 8.5, is of particular interest, and we will be discussing it in more detail.

The operations of a microassembly station may be described as follows:

- The parts are first separated and placed into magazines in order to have them correctly positioned for automated assembly; this is necessary since microcomponents are often delivered as bulk material. This step can also be automated in a versatile microassembly station, to avoid the expensive manual handling.

- A microrobot removes a micromechanical component from the magazine and transfers it to a processing cell, where the component can then be prepared for microassembly by other microrobots. In this step, adhesives or solder are applied, adjustment marks are provided, or other simple operations are carried out.

- After the part has been processed, it is gripped by a robot and brought to a microassembly cell.

- If necessary, the same operations are repeated many times in order to fetch other necessary components from a supply container and prepare them for assembly.

- All components are positioned correctly, affixed to each other and adjusted; thereafter they are joined together by various interconnection techniques, e.g. laser spot welding, gluing, insertion, wire bonding, etc.

- After assembly, a robot brings the finished component either to another work station or a microassembly cell for further processing or to an inspection cell, where all functions of the microsystem are being tested. Finally the finished system is transported to a storage.

The advantages offered by a highly precise microrobot can only be used if the automated microassembly station also has high-resolution sensors; in addition there must be a good control strategy. Since the handling forces encountered when manipulating microobjects are often very small (within the range of 1 nN to 1 μN), the required force sensors are very expensive. For this reason, often only optical sensors offer a feasible solution to obtain a position feedback in the assembly station [Sato95], [Fati96]. The entire assembly process is done either under a light-optical microscope equipped with a high-resolution CCD-camera or in a chamber of a scanning electron microscope. Laser measurement instruments are also suited for determining the position and orientation of a microrobot or the parts. Sometimes the applied forces can be inferred by optically examining the deformation of a tool tip. In the future, new types of force or contact sensors will considerably improve the capabilities of a microassembly station.

The above description of microassembly was very general and perhaps makes the assembly process sound very simple. Specific difficulties encountered with micromanipulation were addressed in Section 8.5. In order to obtain a breakthrough in the production of a microsystem using automated microrobots, many additional problems must be solved. The design rules, which describe the product to be produced, must also consider its functions, the manufacturing constraints and must be able to handle assembly aspects. A thorough planning is fundamental for the preparation of the production and for obtaining a low price for the microsystem. To name an example: a system component to be assembled must also be provided with surfaces to which the assembly tool can hold on.

After the completion of the design of an "assembly-friendly" microsystem all tools and techniques necessary for its automated assembly must be determined so that the microassembly station can be set up for a task-specific operation sequence. The specified techniques and tools must take the geometry of the components of the microsystem into consideration, as well as their physical properties, such as rigidity, texture and temperature stability. Therefore, the planning phase of an automated microassembly involves many problems and requires a high degree of competence at the higher levels of the planning system. Often experiences gained from conventional automated robot-based production can be used [Ramp94].

Pure top-down planning in a microassembly station seems to be impossible, since the selected robots and their manipulation tools determine both the flexibility and the degree of automation of a microassembly station, i.e. they determine the performance limits of the station. One possible planning strategy is the so-called *meet-in-the-middle* strategy, which allows the intermediate interface to be on the tool level. Indeed, the main functions of assembly planning are the determination of the task-specific sequences of the elementary operations and the selection of necessary tools for carrying out the work (top-down planning). On the other hand, the tools and the elementary operations needed for the assembly of a microsystem also require that the microrobots have specific functional properties which may influence the robot design (bottom-up planning).

In order to achieve the desired performance of the microassembly system, the microrobots should provide mobility and should also be able to carry out task-specific manipulations with high accuracy. The mobility of an assembly robot may solve the problem of material feeding, exchanging of tool pallets and of transportation; these are also typical problems in conventional assembly systems. The robot's flexibility should be improved through simple and automated tool exchange; thus, it is possible to quickly adapt to different types of workpieces. In the following section, two possible approaches of designing such robot systems will be introduced.

8.8 Flexible Microrobots

Microrobot using a magnetic actuation principle

First, a microrobot will be introduced which uses the electromagnetic principle for the robot legs, similar to the Tiny Silent Linear Cybernetic microactuator described in Section 8.4. This microrobot moves on two metal legs, each of which has an electromagnetic coil, which can adhere to a metal surface by a magnetic field [Aoya95]. A multilayered piezoelement, located between the magnetic legs, allows the robot to move forwards. A leaf spring is fastened in parallel to the piezoelement and transforms the piezoelement's expansion into a robot leg motion by a simple mechanism. If one of the magnetic legs is clamped down to the metal floor the other free leg is moved by the expanding piezoelement. During this motion, the other leg is held in place by switching on the current to its coil and generating a magnetic grip with the floor. Then the first leg is freed by turning off its current and the process is repeated. By timing the interaction between the leg coils and the piezoelement, the robot can even walk upside-down on a magnetic ceiling. The robot has an omnidirectional movement potential and is only confined by the metal base plate. It is able to move at a speed of 20 mm/s with a control frequency of 100 Hz and a step length of 20 μm.

Attempts are being made to build a flexible microrobot system based on this concept. Different types of micromanipulation tools are placed on the robot. One robot, for example, was supplied with a microhammer, whose tip was made of a hardened steel ball of 2 mm in diameter. The tool is driven by an electromagnet and has two degrees of freedom, one lateral, one in the direction of stroke. With this tool it was possible to make a precise, 500 μm long impression on a wall under teleoperation control. In another experiment, a diamond cutting tool with a tip radius of only 0.5 μm was attached to the platform and was operated by a micro stepping motor. This tool could be used to make parallel grooves less than 1 μm apart on the surface of a glass substrate. A special feature of this experiment was the use of an automatic closed-loop control system which was provided with optical control information by another mobile unit equipped with sensors. This monitoring robot was moved close to the manipulation robot and recorded its movements optically with an LED and a photodetector. This is one of the first examples where the activities of two mobile autonomous microrobots were coordinated and controlled. At the moment, an attempt is being made to pursue the problem of robot coordination, which is needed to operate several flexible microrobots. This topic will be discussed further in Section 8.9.

Microrobot using a piezoelectric actuation principle

The next example discusses two microrobots which were reported in [Magn94], [Magn95] and [Fati95]. The first one, a one-armed microrobot is named PROHAM (Piezoelectric Robot for Handling Microobjects); it is equipped with a piezoelectrically driven base platform and a manipulator which can hold different tools. The second robot is named MINIMAN (MINIaturized MANipulation robot); it uses a similar platform as PROHAM, but has a more versatile two-armed manipulator. Both models will be discussed below.

The platforms of both robots move with the help of three piezoceramic legs, Fig. 8.32. The 13 mm long legs are made of VIBRIT 420 ceramics; they are sintered in a tubular form and can change their length when a voltage is applied (Section 7.3.2). The outer diameter of the legs is 2.2 mm and the inner one 1 mm. They are readily available on the market and are inexpensive. Each leg is covered with two metal electrodes, an outer and an inner one. They are used to force the ceramic leg to change its length when a voltage is applied to it. The applied electric field either causes the ceramic to grow or shrink, depending on the polarisation of the electric field. In order to be able to bend a leg, the outer electrode is divided into four segments which are equally spaced about the circumference of the ceramics and which are isolated from one another and are parallel to the center line, Fig. 8.33.

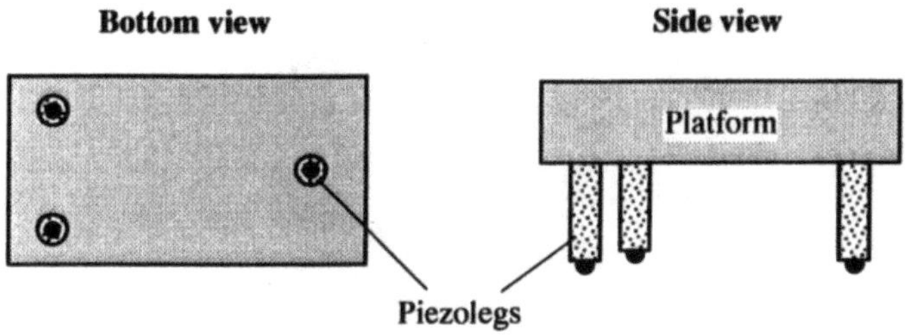

Fig. 8.32: Leg configuration

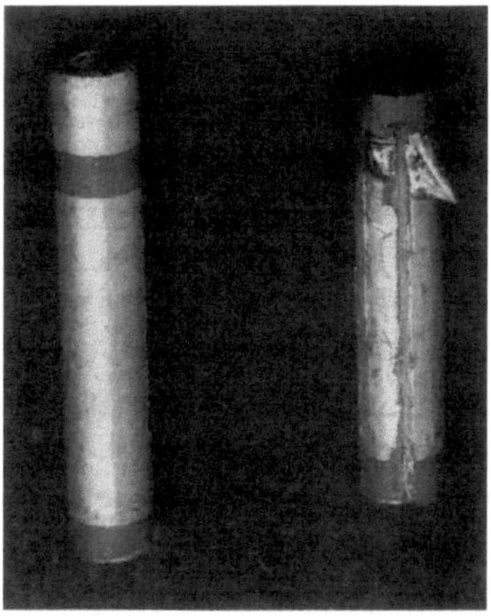

Fig. 8.33: Piezoceramic tubular legs with electrodes

If two electrodes lying opposite to one another are energized by a voltage having an opposite polarization, the ceramic will expand on one side and contract on the other, making the pipe bend. If a voltage of the same polarity is applied to the other two electrodes, the leg also changes its length. This method of supplying energy makes it possible to exactly control the motion of the robot leg, just by varying the strength and direction of the electric field.

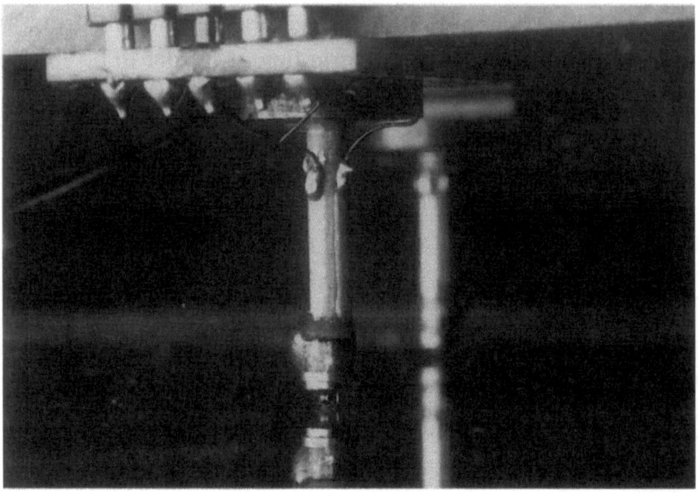

Fig. 8.34: Piezolegs of the robot platform contacting the work surface

The mechanical design of the robot leg is shown in Figure 8.34. The legs are bonded to metal receptacles which are screwed onto a platform. The receptacles also contain a small connector board which interfaces the piezoelectrodes with the on-board control electronics. Small ruby spheres with a 1 mm diameter are attached to the legs, which serve as feet and guarantee precise motions and a constant friction on the glass base. The glass base was chosen due to its smooth and flat surface; this is important for the motion principle, since the piezoceramics can only take tiny steps. Also, in some applications, the lighting for the light-optical microscope must come from under the microscope table, e.g. for manipulating biological cells; for this purpose the robot base must be transparent. The glue for attaching the receptacle and rubies is very hard; it allows the transmission of the smallest piezoceramic displacement into motions of the robot platform.

Two possible motion principles of the robot platform are shown in Fig. 8.35. The motion in Fig. 8.35a is basically divided into 4 phases. At the beginning, no voltage is applied to the piezoceramics. To move the robot the legs are bent relatively slowly at first so that they stay in contact with the base. The platform is thereby displaced into the desired direction of motion. Second, the polarity of the voltage energizing the actuator electrodes is suddenly reversed so that all three legs bend quickly in the opposite direction. Simultaneously,

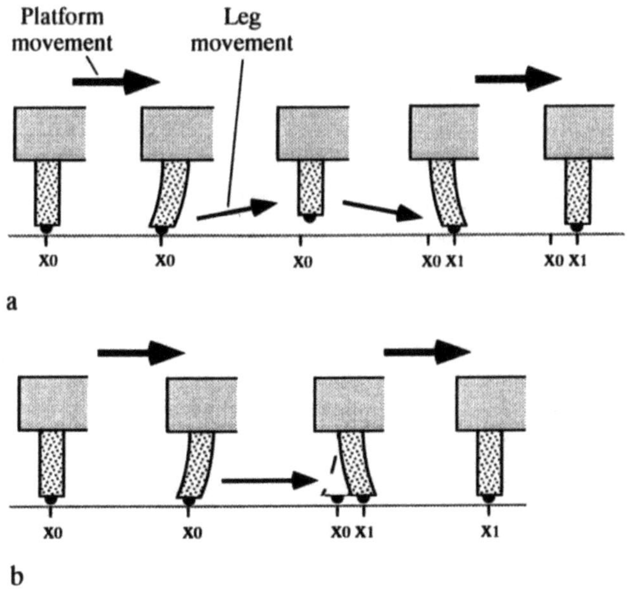

Fig. 8.35: Two types of robot platform motion
a) lifting of legs; b) sliding the legs along the base

the robot lifts all three legs at once and performs a "microjump", which prevents sliding friction between legs and base. The bent legs are then lodged in their new position; the platform does not change its position. Third, the legs are straightened out and the platform moves a little more in the desired direction, taking the step $\Delta X = X_1 - X_0$, Fig. 8.35. The last leg actuation is again carried out relatively slowly to keep the legs from sliding on the glass base. As this sequence is repeated, the robot platform performs a continuous "gallopping" motion. The motions can be modified in different ways, since the legs can be controlled and moved separately. Thereby it is possible that each leg influences the position of the platform in a controlled way. This makes the positioning of the robot more flexible, but requires a sophisticated control algorithm. The motion possibilities of the piezoelements may be fully used for this type of walking, whereby the legs alternate between two oppositely bent states when taking a step. The maximum leg stroke (at ±150 V) is about ±3 μm, and the largest possible step of the platform is 6 μm. One disadvantage of this principle is the non-linear trajectory of the leg motion. This requires very exact control and puts excessive strain on the ceramics.

With the simplified motion principle shown in Figure 8.35b, the gait of a leg is controlled by moving the leg in both directions at different speeds. First it is bent slowly, then it takes a very quick "step". Contrary to the first method, a leg is not lifted but slides on the surface. The legs slide on the glass surface due to the inertia of the robot platform and their high speed, since the sliding friction between the ruby spheres and the glass is relatively low with a lightweight robot. The platform jerks back a little, but this is usually negligible compared with the entire step length. The legs stretch out again when the new position is reached and the step is completed. The disadvantage of the simpler principle is a more unstable platform motion, which is very dependent on the robot's mass. The heavier the robot, the greater the sliding friction and the smaller the steps. If the robot is made lighter, the stretch of the legs may cause parasitic jerklike platform displacements and this effects the smoothness of the platform motion; thereby the speed and positioning accuracy are adversely affected. This effect can only be compensated with a very complex control algorithm placing more demand on the control computer.

The robot platform is able to move and turn in any direction by controlling the drive voltages for the legs. Since there are no mechanical restrictions, the legs can move in any direction, i.e. the robot platform can be positioned very exactly. By controlling the level of the supply voltage in 256 incremental steps (using an 8-bit D/A converter), the platform can be moved with a very fine resolution of 10 nm. A further increase in resolution is limited by the hysteresis of the ceramic material.

The first prototype of the robot is shown in Figure 8.36. The control and power electronics are directly installed on the platform in order to make it as autonomous as possible. The processor board with a Siemens C167 microcontroller is located above the base plate, which is where the leg control algorithms run. Above the processor board are three driver boards equipped with D/A converters and amplifiers to generate the control voltages for the legs. Each driver board is responsible for one leg. A needle-like metal arm is used as the manipulation tool (endeffector), which is fixed to the platform.

The endeffector carries out its manipulations through the fine motions of the robot's legs. The third degree of freedom to lift the needle is realized with the help of a tilting system in the microrobot. This tilting system is driven by a d.c. electric motor and spindle actuator.

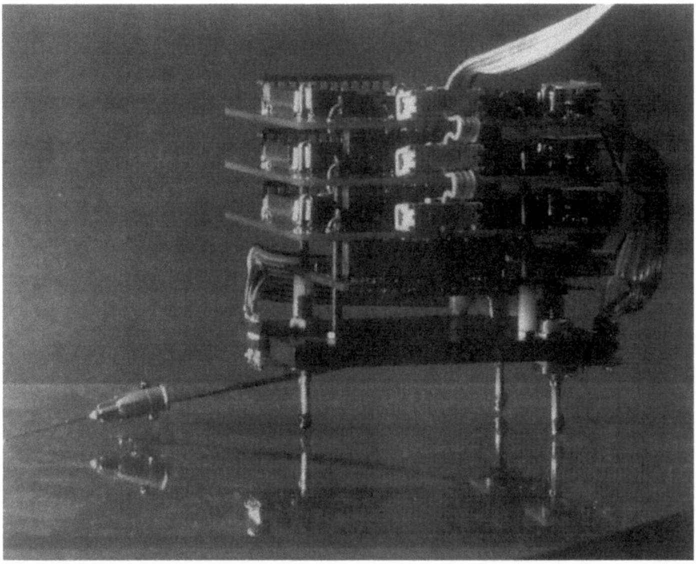

Fig. 8.36: One-armed piezoelectric robot PROHAM

The overall size of the robot is about $8 \times 8 \times 5$ cm^3; the device can carry a load of about 500 g. The omnidirectional movement of the legs allow to move the robot on a smooth surface like glass in any direction or about itself. For a step frequency of 5 kHz, the maximum speed of about 30 mm/s was obtained ("gallopping"), which is adequate for most applications, it depends on the control frequency of applying the drive voltage. The robot is easy to produce and can be produced at low cost, since it is made of few mechanical components. There is little maintenance and the probability of a defect is small.

The possibilities of the robot to do work under a microscope were tested. The task of the robot was to remove dyed biologic cells from a tissue sample by scratching them off with a needle-like arm and then transporting these cells to another place for further analysis. The manipulations were done in a tele-operated mode by manual control with a computer mouse. The mouse had three regular tracking balls which were integrated, so that the rotational and displacement motions of the operator's hand could be followed. For this operation different speeds could be selected. This one-armed robot can also be used for testing microelectronic chips using a probe as an endeffector. A disadvantage of the robot is that it has to be operated with ±150 volts.

However, by using other piezoelectric drives it is possible to reduce the voltage to ±20 V. At the moment, studies are performed in this direction, thus allowing a fututre miniaturization of the robot.

For more intricate assembly operations in the microworld, two endeffectors are necessary. For this purpose, another robot was developed with two controllable piezoelectric manipulators. The mechanical structure of the manipulator is shown in Figure 8.37.

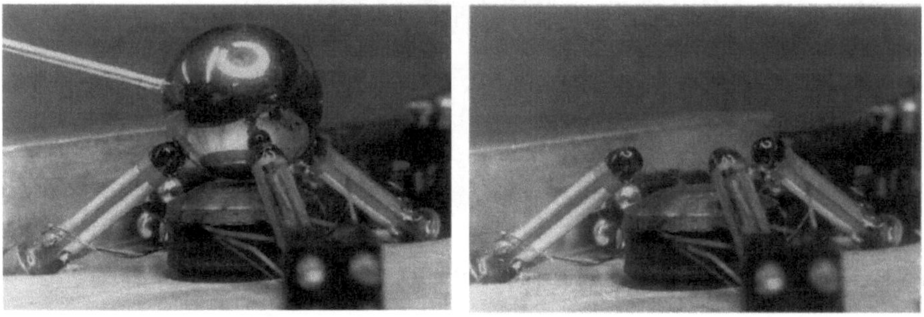

Fig. 8.37: Manipulation unit of a two arm microrobot using piezoelectric drives (right: without toolholder)

The manipulator base is made up of a steel ball to which the end effector is fastened. Three piezoceramic actuators like those used for the robot's legs serve as drive mechanisms to rotate the ball and with this to move the endeffector. The metal ball is pulled against the three piezoceramic actuators by a permanent magnet and held in position, Fig. 8.37, right. If the actuators are controlled in a similar manner as the legs of the robot discussed above (Figure 8.35b), the metal ball can be precisely rotated in any direction due to frictional forces. The force which the endeffector can exert depends on the friction and the pull of the magnet. This actuation mechanism gives the robot arm three rotational degrees of freedom. Thereby the tool attached to the metal ball can be lifted, lowered, slued or turned. The precision of positioning the tool only depends on the motion resolution of the actuators, it is about 10 nm. The speed of the micromanipulation arm also depends on the control frequency of supplying the drive voltage to the piezoactuators. This manipulator design is simple and it is easy to exchange tools by using another tool holder with the desired tool attached to it.

A prototype of a robot using the above discussed motion principle with three walking legs is shown in Figure 8.38. It consists of a platform with the legs and two micromanipulator arms, attached underneath the platform. In the picture, the steel balls of the arms are equipped with simple, needle-like effectors. By coordinating the motion of the robot platform via its legs and the motion of the arms via the piezoactuators the robot can travel and the tip of the tool can reach any point in the workspace. Different gripping and manipulation principles can be realized. For example, one manipulator can hold a

a

b

Fig. 8.38: Two-armed micro robot using piezodrives
a) frontal view; b) bottom view

microshovel and the other can place objects on it with a microscraper. The objects can then be transported. Another possibility is clamping of an object between two special tools. Microobjects can also be speared up by pointed, needle-like effectors. With this robot it is possible to carry out manipulation tasks in a vacuum chamber of a scanning electron microscope since the microelectronic components are separated from the robot's platform.

Since the robot has three autonomous multiactuator systems (platform and two arms), its control system is very complex. A powerful computer system is required to obtain a fast control speed of the piezoactuators. The coordination of the operation of the platform drivers and the manipulator drivers is done under hard real-time conditions, otherwise the control system does not respond quickly. The system will be further complicated if sensors and special actuators are incorporated. For this reason it is necessary that the resulting system structure remains manageable and maintainable, despite of its complexity. In order to be able to control this robot, a parallel computer system with C167 microcontrollers was developed which can be adapted to an application [Magn96]. This system provides the necessary high sampling rates and computational speed for the robot control. Each independent subtask, in this case, the control of one of the actuator systems, is processed by its own microprocessor. The processors are arranged on a base plate and the units which must exchange data are adjacent to each other. Thus very fast and inexpensive dual port RAM communication is possible. With this architecture there is much freedom to adapt the computer structure to an application (number of processors, task subdivision, interface structure, topology). Also actuators and sensors can be added to the parallel computer system, just by supplying each one with its own processor.

A typical computer structure is shown in Figure 8.39. Here a central processing unit takes over the coordination of all three system processors. This unit is placed in the center of the parallel computer system. Dual port RAMs and a high speed communication module perform communication between the processors of the manipulators and the processor for controlling the platform's legs. A multifunction communication card connects other micromanipulation devices to the system (Section 8.9). The processors for controlling the actuators are arranged around the central unit, they are equipped with a communication board and three high-voltage signal generator boards, each of which is used to control one piezoelement.

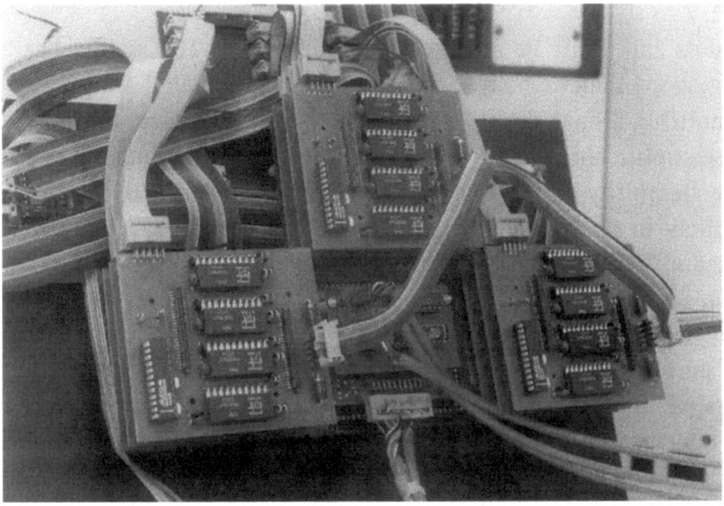

Fig. 8.39: Parallel computer system for a microrobot. According to [Magn96]

The microrobots introduced in this section are very versatile, as opposed to conventional micromanipulation systems. Despite of their ability to perform exact positioning, the robots do not lack mobility and they allow the attachment and exchange of different grippers. Also microobjects can be transported over long distances. It is possible to connect several workcells in a microassembly station to form a production line; this is a common practice in robotic macroproduction. In the following section, we will discuss how such a microassembly station works and how flexible microrobots can be included. In addition problems are pointed out which may result when automating assembly operations for microparts.

8.9 Automated Desktop Station Using Micromanipulation Robots

The design of a micromanipulation station is an interdisciplinary task and is a major challenge for MST, it requires thorough knowledge in mechatronics and computer sciences. A new model of an automated micromanipulation "desktop station", which uses piezodriven microrobots is reported in [Fati96] and [Fati96a]. The robots of this station have a micromanipulation unit integrated in their platform, which makes them capable of moving and manipulating. Various tools can easily be attached and exchanged. The station has a light-optical microscope under which sensor supported automated assembly is possible. This multifunctional desktop station can also be used for online testing of microelectronic chips or for other operations like manipulating biological cells. The flexibility of the system can also be enhanced by accomodating several robots in the station, which can cooperate and carry out manipulations as a team. The schematic design of the micromanipulation desktop station is shown in Figure 8.40.

It is possible to perform an entire assembly process on the desktop station under the automated light-optical microscope; the latter is equipped with a RS232-standard interface. The microscope also has an automatic positioning table with two translational degrees of freedom (x-y plane) and a glass plate fixed on top of the table. By controlling the movement of the table, each assembly cell can be brought in the field of view of the microscope.

The station has a central computer (Pentium PC) which is used for task-specific assembly planning. With it the operational steps are defined and carried out successively. The commands of the central computer are further processed on a lower control level, by the parallel computer system of a type mentioned above; they are resolved into a defined command sequence to activate the system components, such as the robots, microscope and positioning table. The central computer is coupled with the parallel computer system of the robots over serial and parallel interfaces. In the multiprocessor system, the generated commands are executed in parallel, which makes the realtime operation of the microassembly station possible. This way the 2D movements of the positioning table, the microscope functions (changing of objectives, focussing and adjusting of the light) and every piezoactuator are controlled.

In order to automatically control the manipulation processes in the microassembly station, there is sensor feedback from a CCD camera. The camera and the microscope form a local sensor system, which supplies visual information about the manipulation with the microobjects to the central computer; for this purpose the position of the microobjects and that of the robot tools must be determined. The coarse positioning of the robot on the glass base is supervised by a global sensor system, consisting of another CCD camera and a laser measurement system. The visual sensor information of both sensor systems is used by the control algorithms to generate the commands for the robot, microscope and the positioning table. Vision is supported by a frame

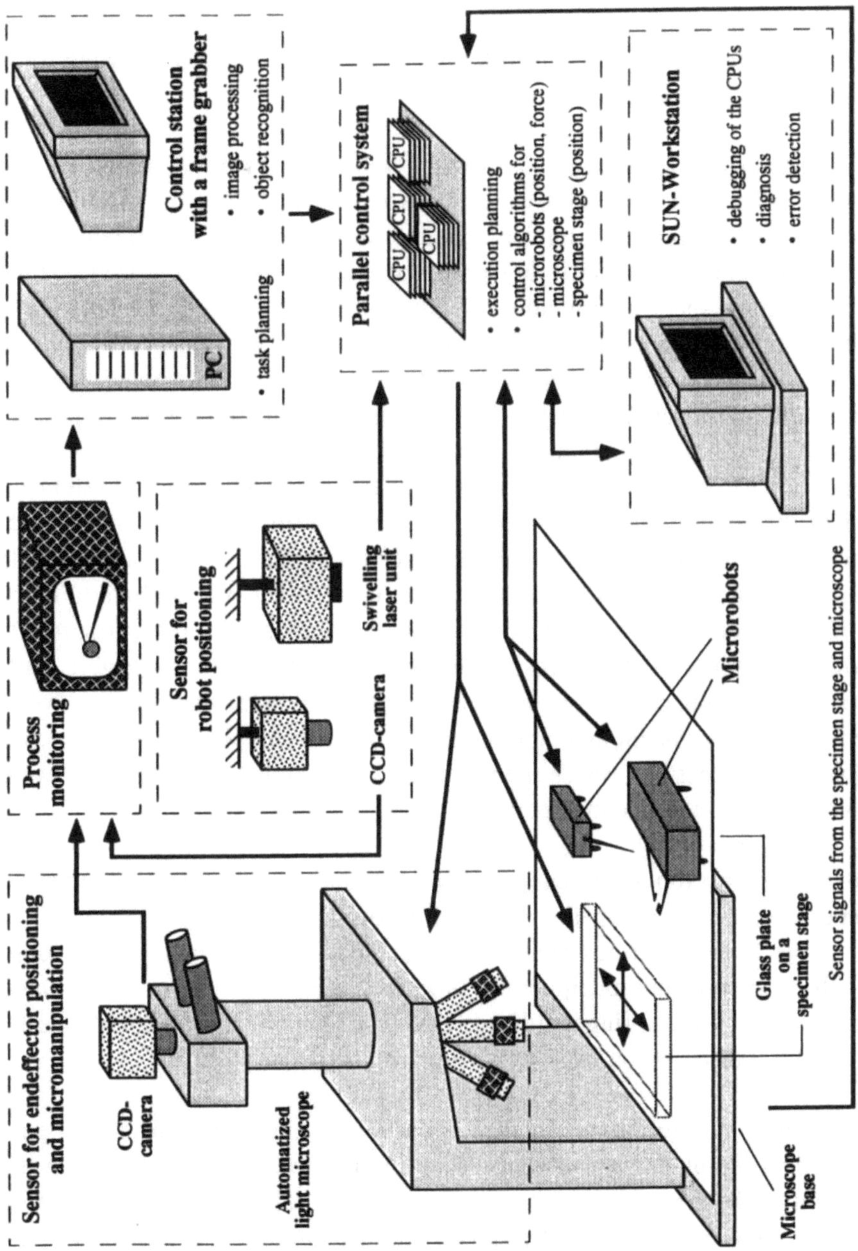

Fig. 8.40: Concept of the micromanipulation desktop station

grabber in connection with a fast real-time image processing system. The vision pamareters are passed on to the parallel computer system, where they are used as set points for the control loop.

For more complex assembly tasks several robots can work together in this desktop station. This brings the advantages of sharing work by parallel execution, the use of small and light robots and an improvement in flexibility of the assembly cell. Individual robots can, for example, be specialized to take on one or more specific assembly operations. In this case, the robots carry out their manipulation tasks in a fixed sequence which was defined during the planning phase. For more complex operations robots can be pooled together to do simultaneous actions using different tools, e.g. transferring or gripping of objects. The operator's commands are no longer transmitted sequentially to the manipulator arms, but are distributed to the entire multirobot system by means of the "one-by-multiple" method [Hirai93]. With this method, one microrobot acts as a leader of the group, receives micromanipulation assignments from the operator and coordinates the work of the other microrobots to complete the planned task. By automatically communicating with its fellow robots it provides them with the corresponding commands. Another possible multirobot cooperation can be accomplished with a hybrid macro-micro robot system. With this system assembly concepts are conceivable in which a conventional robot does the coarse manipulation of an object while the miniature or microrobots do the fine manipulation.

If the cooperating robots are equipped with sensors, new object manipulation procedures can be developed, which use information based on the observation of objects from various positions. Force sensors in a manipulator would also help to control the manipulation process. Such methods were developed for gripping by a multi-fingered hand [Dörs94], [Fati94a]; they could also be used for the transportation of objects with a multirobot system.

An intelligent multirobot system requires a very large amount of expert data for coordinating, planning and controlling. One of the biggest problem in the realization of the micromanipulation desktop station is the intelligent assembly planning on the uppermost control level, the distribution of the specific tasks to individual robots and the control of the movements and forces. It is necessary that the assembly process works error- and collision-free. The information which the operator gives to the multirobot system must be control information for the moving and handling of objects and not for the individual tasks the robots have to perform. The monitoring of the assembly is another important element of a multirobot system. For conventional telemanipulation systems, monitoring is relatively simple; both the manipulator and the object are observed by the operator. In multirobot systems, however, it is extremely difficult even for the most experienced operator. A manipulation can only be sufficiently monitored by using self-checking strategies. An independent monitoring robot may offer a possible solution, with the unit serving as an integrated sensor carrier and passing on the relevant information to the operator.

It is also very difficult to control several robots at the same time. Only advanced information processing systems can handle this job. An adaptable and hierarchically distributed control system must be used, which makes it possible to quickly initiate a cooperation at the request of an operator or a robot or several robots. A centralized system is not suitable here due to the high communication requirements. It is also nearly impossible to explicitly program every possible scenario that can happen in the surrounding of a complex operation. In a flexible system, the individual robots should be able to adapt their behavior to the immediate requirements and react flexibly to a new situation. Decentralizing of a control system is also of advantage to facilitate the robot's ability to learn. Machine learning by single robots has been investigated for a long time, but the learning process of a group of robots is not known. An important concept in multirobot systems is learning by observation. This should happen on the lower robot control level, where the momentary process situation is recognized by the individual robots, which then automatically form their own work groups and define their action sequences accordingly. On the other hand, it may be necessary for the operator to intervene and give the robots commands directly, for example when the handling of the task is likely to fail due to a mispositioned robot or a missing component.

- **Other activities and future aspects**

Many research groups around the world are investigating various aspects of multi-robot problems, such as motion planning and navigation, human-robot and robot-robot communication, task distribution and allocation, and distributed intelligence in a large group of robots, [Hirai93], [Fleu94], [Aoya95], [Ishi95]. A control algorithm must be simple enough so it can operate a flexible microrobot. Today, it is still not possible to implement complex control algorithms on a microrobot, and it cannot be equipped with complicated communication hardware. A multi-robot system presented in [Ishi95] contains control algorithms which allow the robot to react "reflexively"; thereby the individual robots cooperate on the solution of a complex task by using simple control algorithms.

Another control concept of a miniaturized factory model similar to the one introduced above is presented in [Aoya95]. It consists of an operator, a network of processor modules with several control levels, microrobots and monitoring instruments. The system is organized in a partially distributed and partially centralized manner. The microrobots are connected over a shared memory in which a blackboard is located; by this method the participants can communicate via messages which are accessible to all other microrobots. Thereby data is quickly and asynchronously processed between the microprocessors. The control system has three hierarchical levels. The highest level serves as a user-friendly interface, via which the operator can input the production plan in a general form. Global assembly planning is done here. The second level is the group planning and management tier. The lowest level serves as the numerical device control tier which can be accessed over an I/O interface. In assembly planning, the central processor needs different status information from the lower levels which it gets via vertical communication channels. This information may include the performance specification of an individual robot or a status report of a device. With this information the assembly task is divided into elementary motions and distributed among the available robots. It can happen that one or more robots remain idle and simply have to wait. The active participants approach the work area autonomously with the help of the visual monitoring unit.

Different simulation tools can be very helpful in assembly planning. In order to be able to evaluate the assembly task and choose the best strategy, various operating sequences can be tried out by using CAD models of the microrobots and workpieces. Evaluation criteria could be freedom of collision, best use of resources, duration of assembly, compliance with technological parameters, etc. In [Aoya95], a simulation system was implemented, which aids the prevention of collisions as several microrobots are simultaneously moving around. It is assumed here that the participating robots are randomly distributed on the surface of the table at the beginning of the simulation. If the operator is not satisfied with the results, he can choose other parameters or navigation strategies, depending on the performance limitations of the devices. In tests, many robots were given tasks of different priority and a simple collision prevention algorithm was used. A robot pursues its goal as long as it is not blocked by another robot; if the latter happens it tries either to go to the right or left, as long as there are no obstacles in the way. If there is a danger of many robots colliding, the units which have lower priority stop. This system succeeded in coordinating 24 microrobots, which were divided into two groups and which had two different work areas and various priority levels.

However, it is not always possible to get help from a simulation module. As long as there is only path planning involved, like in the above example, there are many ways of finding a solution. When no trajectory restrictions are given other than that a goal has to be reached, collision avoidance methods based on fuzzy-logic seem to be suitable [Reig94]. But it is very difficult to simulate the force distribution between several robots which cooperate in a manipulation task. Another unsolved problem is simulating the task-specific manipulation of microobjects, for example, for the purpose of testing the assembly task of a given robot and a tool, or to determine how much time the process will take. To do this, it is absolutely necessary to know the forces occurring in the microworld and the dynamic behavior of microobjects, for neither of these problems there are models available. Dynamic assembly planning cannot be carried out without such simulation tools.

Coming to a conclusion, it can be stated that presently no easy solutions exist for assembling microparts, especially when taking into consideration the hardware and software and the costs involved. It can be clearly seen, however, that the availability of very versatile automated microassembly stations will greatly contribute to the long-awaited industrial breakthrough of MST. A current industrial success that encourages the development of MST is the rapid progress of microelectronics, where automated production methods led to the birth of an entire new industry. This development is a motivating example for MST researchers, who are the innovators in the interdisciplinary field of microassembly. The development of suitable microrobots, which are capable of operating efficiently in the microworld, is becoming a major goal for many new applications. These new applications might lead to new technological advances and make microrobots an important part of industrial production.

Bibliography

[Ache91] Ache, H.J.: "Entwicklung von Analytischen Mikrosonden und Chemischen Mikrosensoren im Institut für Radiochemie", *KfK-Nachrichten, Jahrgang 23*, Forschungszentrum Karlsruhe, 1991 (2–3), S.148–157

[Ahn93] Ahn, Ch.H., Kim, Y.J. and Allen M.G.: "A Planar Variable Reluctance Magnetic Micromotor with Fully Integrated Stator and Wrapped Coils", *Proc. of the IEEE Int. Conf. on Micro Electro Mechanical Systems (MEMS),* Fort Lauderdale, Florida, 1993, pp.1–6

[Ajlu95] Ajluni, Ch.: "Accelerometers: Not Just For Airbags Anymore", *Electronic Design*, 12.06.95, pp.93–106

[Akin94] Akin, T. et al.: "A Micromachined Silicon Sieve Electrode for Nerve Regeneration Applications", *IEEE Transactions on Biomedical Engineering*, 1994, Vol. 41(4), pp.305–313

[Albr93] Albrecht, A. et al.: "Mikrobewegungssysteme", *Mikroelektronik*, 1993 (5), Fachbeilage "Mikrosystemtechnik"

[Altr93] von Altrock, C.: "Fuzzy Logic: Technologie", Oldenbourg, München/Wien, 1993

[Altr93a] von Altrock, K.: "Signalverarbeitung für Sensorsysteme", *Mikroelektronik*, 1993 (3), Fachbeilage "Mikrosystemtechnik"

[Ande94] Andersen, B. and Millar, C. E.: "Performance of Multilayer Actuators Based on Piezoelectric and Electrostrictive Materials", *Proc. of Int. Conf. on New Actuators,* Bremen, 1994, pp.167–170

[Ange92] Angerer, G. und Hiessl, H.: "Sensoren für den Umweltschutz", *Mikroelektronik*, 1992 (4), Fachbeilage "Mikroperipherik"

[Angs93] Angstenberger, J.: "atp-Marktanalyse: Software-Werkzeuge zur Entwicklung von Fuzzy-Reglern", *Automatisierungstechnische Praxis*, 1993 (2), S. 112–124

[Aosh92] Aoshima, Sh., Yoshizawa, N. and Yabuta, T.: "Micro Mass Axis Alignment Device with Piezo Elements for Optical Fibers", *Proc. of the 3. Int. Symp. on Micro Machine and Human Science*, Nagoya, 1992, pp.89–96

[Aoya95] Aoyama, H., Iwata, F. and Sasaki, A.: "Desktop Flexible Manufacturing System by Movable Miniature Robots", *Proc. of Int. Conf. on Robotics and Automation*, Nagoya, 1995, pp.660–665

[Arai92] Arai, T., Stoughton, R. and Jaya, Y.: "Micro Hand Module Using Parallel Link Mechanism", *Proc. of the Japan U.S.A. Symp. on Flexible Automation*, San Fransisco, 1992, pp. 163–168

[Arai93] Arai, T., Larsonneur, R. and Jaya, Y.: "Basic motion of a micro hand module", *Proc. of Int. Conf. on Advanced Mechatronics (ICAM)*, Tokyo, 1993, pp.92–97

[Arai95] Arai, T. et al.: "Micro Manipulation Based on Micro Physics", *Proc. of the IEEE/RSJ Int. Conf. on Intelligent Robots and Systems (IROS)*, Pittsburgh, Pennsylvania 1995, pp.236–241

[Arqu93] Arquint, Ph. and van der Schoot, B.H.: "Integrated blood-gas sensor for pO_2, pCO_2 and pH", *Sensors, Actuators & Microsystems Laboratory: Report on Research Activities*, Institute of Microtechnology, University of Neuchâtel, 1993, p.25

[Axel95] Axelrad, C. et al.: "Industrial Applications of Microtechnologies in Europe", *Proc. of the Int. Micromachine Symposium*, Tokyo, 1995, pp.37–46

[Bach91] Bacher, W. et al.: "Herstellung von Röntgenmasken für das LIGA-Verfahren", *KfK-Nachrichten, Jahrgang 23*, Forschungszentrum Karlsruhe, 1991 (2–3), S.76–83

[Bach91a] Bacher, W. et al.: "LIGA-Abformtechnik zur Fertigung von Mikrostrukturen", *KfK-Nachrichten, Jahrgang 23*, Forschungszentrum Karlsruhe, 1991 (2–3), S.84–92

[Bayer94] "Rheobay® TP AI3565 and Rheobay® TP AI3566", *Provisional Product Information*, Bayer AG, Business Group Inorganic Products, Leverkusen, 1994

[Bekey93] Bekey, G.A. and Goldberg, K.Y.: "Neural Networks in Robotics", Kluwer Academic Publishers, 1993

[Bene94] Benecke, W.: "Scaling Behaviour of Micro Actuators", *Proc. of Int. Conf. on New Actuators,* Bremen, 1994, pp. 19–24

[Bene95] Benecke, W.: "Strategy Paper on Microsystem Technology", *Proc. of the NEXUS-Workshop on Micro-Machining*, Bremen, 1995

[Benz95] Benz, M., Fatikow, S. und Großmann, B.: "Mikroaktorik für die Mikrosystemtechnik", *Reihe: Innovationen in der Mikrosystemtechnik*, VDI/VDE-IT, 1995

[Berns94] Berns, K. und Kolb, T.: "Neuronale Netze für technische Anwendungen", FZI-Berichte "Informatik", Springer-Verlag, Berlin/Heidelberg/NewYork, 1994

[Bier93] Bier, W. et al.: "Mechanische Mikrofertigung - Verfahren und Anwendungen", *1. Statuskolloquium des Projektes Mikrosystemtechnik, Tagungsband,* Forschungszentrum Karlsruhe, 1993, S.132–137

[Blatt93] Blattner, J. und Neußer, S.: "Fuzzy-Systeme und Neuronale Netze", *Mikroelektronik*, 1993 (3), Fachbeilage "Mikrosystemtechnik"

[Bleu92] Bleuler, H.: "Active Micro Levitation", *Proc. of the 3. Int. Symp. on Micro Machine and Human Science*, Nagoya, 1992, pp.129–136

[Bley91] Bley, P. und Menz, W.: "Stand und Entwicklungsziele des LIGA-Verfahrens zur Herstellung von Mikrostrukturen", *KfK-Nachrichten, Jahrgang 23*, Forschungszentrum Karlsruhe, 1991 (2–3), S.69–75

[Block92] Block, H.: "The Nature, Action and Applications of ER Fluids", *Proc. of Int. Conf. on New Actuators,* Bremen, 1992, pp.105–109

[BMFT90] "Mikrosystemtechnik", Förderungsschwerpunkt im Rahmen des Zukunftskonzeptes Informationstechnik, BMFT, Bonn, 1990

[BMFT94] "Mikrosystemtechnik 1994–1999", *BMFT-Programm im Rahmen des Zukunftskonzeptes Informationstechnik, Öffentlichkeitsarbeit,* BMFT, Bonn, 1994

[Bothe93] Bothe, H.-H.: "Fuzzy Logic: Einführung in Theorie und Anwendungen", Springer-Verlag, Berlin, 1993

[Bran91] Branebjerg, J. et al.: "A micromachined flow sensor for measuring small liquid flows", *Proc. of IEEE Int. Conf. on Transducers,* San Francisco, 1991, pp.41–46

[Brand92] Brand, S., Laux, T., und Tönshoff, H.-K.:"Piezoelektrische Aktoren", *Mikroelektronik,* 1992(6), Fachbeilage "Mikroperipherik"

[Bras95] Brasche, U.: "MST in European Industry", *mst news,* 1995 (13), p.25–29

[Breit93] Breitmeier, U.: "Qualitätskontrolle an mikromechanischen Bauteilen", *Mikroelektronik,* 1993 (4), Fachbeilage "Mikrosystemtechnik"

[Broo92] Brooks, D.: "A Practical High Speed ER Actuator", *Proc. of Int. Conf. on New Actuators,* Bremen, 1992, pp.110–115

[Brug93] Brugger, J. et al.: "Microfabricated tools for nanoscience", *Journal of Micromechanics and Microengineering,* 1993 (4), Vol. 3, pp.161–167

[Brug93a] Brugger, J. et al.: "Nanopositioniersystem", *Mikroelektronik*, 1993 (5), Fachbeilage "Mikrosystemtechnik"

[Bryze94] Bryzek, J., Petersen, K. and McCulley, W.: "Micromachines on the march", *IEEE Spectrum*, May 1994, pp.20–31

[Budig92] Budig, P.-K.: "Aktoren für den Mikrometer- und Submikrometerbereich", *Mikroelektronik*, 1992 (6), Fachbeilage "Mikroperipherik"

[Büker93] Büker, H.: "Glasfasersensorik als neue Disziplin der Mikrosensorik", *7. IAR Kolloquium, Tagungsband*, Universität Duisburg, 1993, S.11–24

[Burns94] Burns, V.: "Microelectromechanical Systems (MEMS) and SPC Market Study", *mst news*, 1994 (10), p.4

[Buser91] Buser, R. and de Rooij, N.: "ASEP: a CAD programm for silicon anisotropic etching", *Sensors and Actuators*, 1991 (28), pp.71–78

[Buser92] Buser, R., Crary, S. and Juma, O.: "Integration of the Anisotropic-Silicon-Etching Program ASEP™ within the CAEMEMS™ CAD/CAE Framework", *Proc. of the IEEE Int. Conf. on Micro Electro Mechanical Systems (MEMS)*, Travemünde, 1992, pp. 133–138

[Büst94] Büstgens, B. et al.: "Micromembrane Pump Manufactured by Molding", *Proc. of Int. Conf. on New Actuators,* Bremen, 1994, pp.86–90

[Camm92] Cammann, K.: "Biosensorik", *Mikroelektronik*, 1992 (3), Fachbeilage "Mikroperipherik"

[Camm94] Cammann, K. et al.: "Beispiele funktionstüchtiger chemischer und biochemischer Mikrosensoren für die Umweltanalytik und Medizintechnik", *1. Int. Kongress und Ausstellung für Mikrosysteme und Präzisionstechnik (Micro-Engineering 94)*, Tagungsband, Messe Stuttgart, 1994

[Camon93] Camon, H. et al.: "Electrostatic Micromotor and Millimetric Systems", *Proc. of Int. IARP Workshop on Micro Robotics and Systems*, Karlsruhe, 1993, pp.169–179

[Carl94] Carlson, J.D.: "The Promise of Controllable Fluids", *Proc. of Int. Conf. on New Actuators,* Bremen, 1994, pp.266–270

[Carp86] Carpenter, G.A. and Grossberg, S.: "Neural Dynamics of Category Learning and Recognition: Attention, Memory Consolidation, and Amnesia", in *Davis, J., Newburgh, R. and Wegmann, E. (Hrsg.): "Brain Structure, Learning, and Memory"*, AAAS Symposium Series, 1986

[Chri92] Christenson, T.R., Guckel, H. und Skrobis, K.J.: "Mikromechanische Aktoren", *Mikroelektronik*, 1992 (6), Fachbeilage "Mikroperipherik"

[CISC91] "Research in the Center for Integrated Sensors and Circuits", *Report on Research Activities*, The University of Michigan, Ann Arbor, 1991

[Clae94] Claeyssen, F., Lhermet, N. and Le Letty, R.: "State of the art in the field of magnetostrictive actuators", *Proc. of Int. Conf. on New Actuators,* Bremen, 1994, pp.203–209

[Clark92] Clark, A. E.: "High Power Magnetostrictive Transducer Materials", *Proc. of Int. Conf. on New Actuators,* Bremen, 1992, pp. 127–132

[Cocco93] Cocco, M. et al.: "An implantable neural connector incorporating microfabricated components", *Journal of Micromechanics and Microengineering*, 1993 (4), Vol. 3, pp.219–221

[Codo95] Codourey, A. et al.: "A Robot System for Automated Handling in Mirco-World", *Proc. of the IEEE/RSJ Int. Conf. on Intelligent Robots and Systems (IROS)*, Pittsburgh, Pennsylvania 1995, pp. 185–190

[Conra94] Conradt, M.: "Endoskopie", *TK aktuell*, 1994 (4), S.20–21

[Daim92] "Mikrosystems - a Key to the Future", *Öffentlichkeitsarbeit*, Daimler-Benz AG, Stuttgart, 1992

[Dario92] Dario, P. et al.: "Microactuators for Microrobots: a Critical Survey", *Journal of Micromechanics and Microengineering*, 1992 (2), pp.141–157

[Dario93] Dario, P. et al.: "Endoscopic Micromanipulator with SMA-Actuated Tip and Grippers", *Proc. of Int. Conf. on Advanced Mechatronics (ICAM)*, Tokyo, 1993, pp.490–492

[Dario94] Dario, P. et al.: "Technology and Fabrication of Hybrid Neural Interfaces for the Peripheral Nervous System", *in "Micro System Technologies 94" (Editors: H. Reichl, A. Heuberger)*, VDE-Verlag, Berlin, 1994, pp.417–426.

[Dario94a] Dario, P. et al.: "Microactuators for Microrobots: a Critical Survey", *Proc. of the IEEE Int. Conf. on Robotics and Automation*, San Diego, 1994, Tutorial S2

[DBS95] "Digitale Bildverarbeitung 1995", DBS GmbH, Bremen, 1995

[Desp93] Despont, M. et al.: "New Design of Micromachined Capacitive Force Sensor", *Journal of Micromechanics and Microengineering*, 1993 (4), Vol. 3, pp.239–242

[Dhul92] Dhuler, V.R. et al.: "A comparative study of bearing designs and operational environments for harmonic side-drive micromotors", *Proc. of the IEEE Int. Conf. on Micro Electro Mechanical Systems (MEMS)*, Travemünde, 1992, pp.171–176

[Dibb94] Dibbern, U.: "Piezoelectric Actuators in Multilayer Technique", *Proc. of Int. Conf. on New Actuators,* Bremen, 1994, pp.114–118

[Dobos90] Dobos, K.: "Integrierte Gassensoren", *Mikroelektronik*, 1990 (5), Fachbeilage "Mikroperipherik"

[Domi93] Dominguez, D. et al.: "Fabrication and characterization of a thermal flow sensor based on porous silicon technology", *Journal of Micromechanics and Microengineering*, 1993 (4), pp.247–249

[Dörs94] Dörsam, Th., Fatikow, S. and Streit, I.: "Fuzzy-Based Grasp-Force-Adaptation for Multifingered Robot Hands", *Proc. of the 3rd IEEE World Congress on Computational Intelligence*, Orlando, Florida, 1994, pp.1468–1471

[Drex91] Drexler, E. et al.: "Unbounding the Future: The Nanotechnology Revolution", William Morrow Verl., New York, 1991

[Drex92] Drexler, E. "Nanosystems: Molecular Machinery, Manufacturing, and Computation", John Wiley & Sons, New York, 1992

[Drost91] Drost, S., Endres, H.-E. und Sandmaier, H.: "Systemtechnik für Chemosensoren", *Mikroelektronik*, 1991 (3), Fachbeilage "Mikroperipherik"

[Ecke92] Eckerle, J., Andeen, G. and Kornbluh, R.: "Exploring Artificial Muscle as Robot Actuators", *Robotics Today*, 5(1), 1992

[Ecke93] Eckerle, M. et al.: "Glasfaseroptische Sensoren", *7. IAR Kolloquium, Tagungsband*, Universität Duisburg, 1993, S.119–126

[Eder91] Eder, A.: "Automatischer Entwurf in der Mikrosystemtechnik", *Mikroelektronik*, 1991 (5), Fachbeilage "Mikroperipherik"

[Egawa91] Egawa, S., Niino, T. and Higuchi, T.: "Film actuators: planar, electrostatic surface-drive actuator", *Proc. of the IEEE Int. Conf. on Micro Electro Mechanical Systems (MEMS)*, Nara, 1991, pp.9–14

[Egge96] Eggert, H. et al.: "Eine Informationsverarbeitungsumgebung zur Konstruktion und Vermessung von Mikrostrukturen", *Tagungsband zur Vortragsveranstaltung "Informationstechnik für Mikrosysteme"*, VDE/VDI-GME, Stuttgart, 1996, S.53–57

[Ehle92] Ehlers, K.: "Mikrosystemtechnik im Automobil: Abschätzung des Anwendungspotentials", *Mikroelektronik*, 1992 (2), S.62–66

[Ehrf93] Ehrfeld, W. et al.: "LIGA-Technik zur Fertigung von Mikroaktoren", *Mikroelektronik*, 1993(5), Fachbeil. "Mikrosystemtechnik"

[El-Kh93] El-Kholi, A. et al.: "Entwicklungen und Erweiterungen der Strukturierungsmöglichkeiten in der Röntgentiefenlithographie", *1. Statuskolloquium des Projektes Mikrosystemtechnik, Tagungsband,* Forschungszentrum Karlsruhe, 1993, S.114–119

[Erbel94] Erbel, R.: "Möglichkeiten der Mikrotechnik in der Kardiologie", *1. Int. Kongress und Ausstellung für Mikrosysteme und Präzisionstechnik (Micro-Engineering 94)*, Tagungsband, Stuttgart, 1994

[Esashi93] Esashi, M.: "Beispiele aus Japan", *Mikroelektronik*, 1993 (5), Fachbeilage "Mikrosystemtechnik"

[Fan93] Fan, L.-S. et al.: "Batch-Fabricated Milli-Actuators", *Proc. of the IEEE Int. Conf. on Micro Electro Mechanical Systems (MEMS)*, Fort Lauderdale, Florida, 1993, pp.179–183

[Faro95] Farooqui, M.: "Polysilicon Surface Micromachining", *Proc. of the NEXUS-Workshop on Micro-Machining*, Bremen, 1995

[Fati93] Fatikow, S. and Rembold, U.: "Principles of Micro Actuators and their Applications", *Proc. of the IARP Workshop on Micromachine Technologies and Systems*, Tokyo, 1993, pp.108–117

[Fati94] Fatikow, S., Rembold, U. and Wöhlke, G.: "A Survey of the Present State of the Art of Microsystem technology", *Journal of Design and Manufacturing*, 1994 (4), pp.293–306

[Fati94a] Fatikow, S. and Sundermann, K.: "Neural-Based Learning in Grasp Force Control of a Robot Hand", *Proc. of IFAC/IFIP/-IMACS Symposium on Artificial Intelligence in Real Time Control (AIRTC)*, Valencia, 1994, pp.139–144

[Fati95] Fatikow, S., Magnussen, B. and Rembold, U.: "A Piezoelectric Mobile Robot for Handling of Microobjects", *Proc. of the International Symposium on Microsystems, Intelligent Materials and Robots (MIMR)*, Sendai, 1995, pp.189–192

[Fati96] Fatikow, S.: "A microrobot-based automatic desk-station for assembly of micromachines", *Proc. of the 12. Int. Conference on CAD/CAM Robotics and Factories of the Future,* London, 1996, pp.174–179

[Fati96a] Fatikow, S. and Rembold, U.: "An Automated Microrobot-Based Desktop Station for Micro Assembly and Handling of Microobjects", *Proc. of the IEEE Conference on Emerging Technologies and Factory Automation (ETFA)*, Kauai, Hawaii, 1996

[Faust93] Faust, W., Michel, B. und Winkler, Th.: "Mechanisch-thermische Zuverlässigkeit", *Mikroelektronik*, 1993 (4), Fachbeilage "Mikrosystemtechnik"

[Fisch91] Fischer, K.: "Mikromechanik und Mikroelektronik vereint mit Optik", *Technische Rundschau (TR)*, 1991 (36), S.106–108

[Fleu94] Fleury, S., Herrb, M. and Chatila, R.: "Design of a Modular Architecture for Autonomous Robot", *Proc. of the IEEE Int. Conf. on Robotics and Automation*, San Diego, 1994, pp.3508–3513

[Flik94] Flik, G. et al.: "Giant Magnetostrictive Thin Film Transducers for Microsystems", *Proc. of Int. Conf. on New Actuators,* Bremen, 1994, pp.232–235

[Fricke93] Fricke, J. and Obermeier, E.: "Cantilever Beam Accelerometer Based on Surface Micromachining Technology", *Journal of Micromechanics and Microengineering*, 1993 (4), Vol. 3, pp.190–192

[Frick95] Frick, O.: "Fertigungsgerechte Montage- und Fügeverfahren für Mikrosysteme", *2. Int. Kongress und Ausstellung für Mikrosysteme und Präzisionstechnik (Micro-Engineering 95),* Tagungsband, Messe Stuttgart, 1995, S.44–51

[Früh93] Frühauf, J. et al.: "A simulation tool for orientation dependent etching" *Journal of Micromechanics and Microengineering*, 1993 (3), Vol. 3, pp.113–115

[Fuji92] Fujimasa, I.: "Future medical applications of microsystem technologies", *in "Micro System Technologies 92" (Editor: H. Reichl)*, VDE-Verlag, Berlin, 1992, pp.43–49

[Fuji93] Fujita, H.: "Group Work of Microactuators", *Proc. of the Int. IARP-Workshop on Micromachine Technologies and Systems*, Tokyo, 1993, pp.24–31

[Fuku91] Fukuda, T.: "GMA applications to micromobile robot as microactuator without power supply cables", *Proc. of the IEEE Int. Conf. on Micro Electro Mechanical Systems (MEMS)*, Nara, 1991, pp.210–215

[Fuku92] Fukuda, T. et al.: "Design and experiments of micro mobile robot using electromagnetic actuator", *Proc. of the 3. Int. Symp. on Micro Machine and Human Science*, Nagoya, 1992, pp.77–81

[Fuku93] Fukuda, T. et al.: "Active Catheter System with Multi Degrees of Freedom - Mechanism and Experimental Results of Active Catheter with Multi Units and Multi D.O.F.", *Proc. of the 4. Int. Symp. on Micro Machine and Human Science*, Nagoya, 1993, pp.155–162

[Fuku93a] Fukuda, T. et al.: "Micro Mobile Robot Using Electromagnetic Actuator", *Proc. of Int. IARP Workshop on Micro Robotics and Systems*, Karlsruhe, 1993, pp.45–50

[Fuku93b] Fukuda, T. and Ishihara, H.: "Micro Optical Robotic System with Cordless Optical Power Supply", *Proc. of Int. Conf. on Intelligent Robots and Systems (IROS)*, Yokohama, 1993, pp.1703–1708

[Fuku95] Fukuda, T. et al.: "Steering Mechanism of Underwater Micro Mobile Robot", *Proc. of Int. Conf. on Robotics and Automation*, Nagoya, 1995, pp.363–368

[Furu93] Furuhata, T. et al.: "Outer rotor surface-micromachined wobble micromotor", *Proc. of the IEEE Int. Conf. on Micro Electro Mechanical Systems (MEMS),* Fort Lauderdale, Florida, 1993, pp.161–166

[FZK93] "LIGA: Optische Komponenten", *Öffentlichkeitsarbeit,* Forschungszentrum Karlsruhe, 1993

[Gabr92] Gabriel, K.J. et al.: "Surface-Normal Electrostatic/Pneumatic Actuator", *Proc. of the IEEE Int. Conf. on Micro Electro Mechanical Systems (MEMS),* Travemünde, 1992, pp.128–132

[Gall91] Gall, M.: "The Si planar pellistor: a low power pellistor sensor in Si thin-film technology", *Sensors and Actuators B,* 1991 (4), pp. 533–538

[Gamb91] Gambert, R.: "Sensorsysteme in der Medizintechnik", *Mikroelektronik,* 1991 (3), Fachbeilage "Mikroperipherik"

[Gard94] Gardner J.W.: "Microsensors: Principles and Applications", John Wiley & Sons, Chichester, 1994

[Gass93] Gass, V., van der Schoot, B.H. and de Rooij, N.F.: "Nanofluid handling by micro-flow-sensor based on drag force measurements", *Proc. of the IEEE Int. Conf. on Micro Electro Mechanical Systems (MEMS),* Fort Lauderdale, Florida, 1993, pp.167–172

[Gemm96] Gemmeke, H.: "Kalibration, Analyse und Identifikation von Sensor-Meßdaten mit Neuronalen Netzen", *Tagungsband zur Vortragsveranstaltung "Informationstechnik für Mikrosysteme",* VDE/VDI-GME, Stuttgart, 1996, S.23–26

[Geng95] Gengenbach, U. et al.: "Ein System zur automatischen Montage von Mikrosystemen", *2. Statuskolloquium des Projektes Mikrosystemtechnik,* Tagungsband, Forschungszentrum Karlsruhe, 1995, S.62–66

[Gerh93] Gerhard, E.: "Sensorik im Wandel",*7. IAR Kolloquium, Tagungsband*, Universität Duisburg, 1993, S.3–9

[Gies92] Giesler, Th. und Meyer, J.-U.: "Plattenwellen für Biosensoren", *Mikroelektronik*, 1992 (3), Fachbeilage "Mikroperipherik"

[Gimz94] Gimzewski, J.K. et al.: "Observation of a chemical reaction using a micromechanical sensor", *Chemical Physics Letters*, 28. Januar 1994, No 5/6, Vol. 217

[Good91] Goodenough, F.: "Airbags Boom when an I.C. accelerometer sees 50G", *Electronic Design*, August 8, 1991

[Göpel94] Göpel, W.: "New Materials and Transducers for Chemical Sensors", *Sensors and Actuators*, 1994 (18)

[Grav93] Gravesen, P. et al: "Microfluidics - a review", *Journal of Micromechanics and Microengineering*, 1993 (4), Vol. 3, pp.168–182

[Gron93] Gronau, M. (Hrsg): "Technologien für Mikrosysteme", VDI-Verlag GmbH, Düsseldorf, 1993

[Guck92] Guckel, H., Christenson, T. and Skrobis, K.: "Metal micromechanisms via deep x-ray lithography, electroplating and assembly", *Proc. of Int. Conf. on New Actuators,* Bremen, 1992, pp.9–12

[Guck93] Guckel, H. et al.: "A First Functional Current Excited Planar Rotational Magnetic Micromotor", *Proc. of the IEEE Int. Conf. on Micro Electro Mechanical Systems (MEMS),* Fort Lauderdale, Florida, 1993, pp. 7–11

[Guck94] Guckel, H. et al.: "Electromagnetic, Spring Constrained Linear Actuator with Large Throw", *Proc. of Int. Conf. on New Actuators,* Bremen, 1994, pp.52–55

[Guo95] Guo, Sh. et al.: "Micro Catheter System with Active Guide Wire", *Proc. of Int. Conf. on Robotics and Automation*, Nagoya, 1995, pp.79–84

[Guth93] Guth, H. et al.: "Automatische Vermessung von 2D- und 3D-LIGA-Strukturen zur Qualitätskontrolle", *1. Statuskolloquium des Projektes Mikrosystemtechnik,* Forschungszentrum Karlsruhe, 1993, S.176–182

[Hage95] Hagena, O.F. et al.: "Erfahrungen beim Aufbau und Betrieb einer Kleinserienfertigung für LIGA-Spektrometer", *2. Statuskolloquium des Projektes Mikrosystemtechnik,* Tagungsband, Forschungszentrum Karlsruhe, 1995, S.41–44

[Haji94] Haji Babaei, J. et al.: "A New Bistable Microvalve Using an SiO_2 Beam as the Movable Part", *Proc. of Int. Conf. on New Actuators,* Bremen, 1994, pp.34–37

[Hamp94] Hampel, T. und Palotas, L.: "Simulationsgekoppelte Parameteroptimierung mittels der Evolutionsstrategie" *in M. Glesner (Hrsg.): "Aufgaben der Informatik in der Mikrosystemtechnik", Zusammenfassung der Beiträge im Verlauf der Gründung der GI-Fachgruppe 3.5.6 "Mikrosystemtechnik",* Schloß Dagstuhl, 1994

[Hash94] Hashimoto, M. et al.: "Silicon Resonant Angular Rate Sensor Using Electromagnetic Excitation and Capacitive Detection", *in "Micro System Technologies 94" (Editors: H. Reichl, A. Heuberger),* VDE-Verlag, Berlin, 1994, pp.763–771

[Hata95] Hatamura, Y., Nakao, M. and Sato, T.: "Construction of Nano Manufacturing World", *Microsystem Technologies,* 1995 (3), Vol. 1, pp.155–162

[Hata95a] Hatamura, Y.: "Realization of Integrated Manufacturing System for Functional Micromachines", Proc. of the First Int. Micromachine Symposium, Tokyo, 1995, pp.55–63

[Hatt91] Hattori, Sh. et al.: "Micro-Pump Using Polymer Gel", *Proc. of the 2. Int. Symp. on Micro Machine and Human Science,* Nagoya, 1991, pp.113–118

[Hatt92] Hattori, Sh. et al.: "Structure and Mechanism of Two Types of Micro-Pump Using Polymer Gel", *Proc. of the IEEE Int. Conf. on Micro Electro Mechanical Systems (MEMS)*, Travemünde, 1992, pp.110–115

[Haup91] Hauptmann, P.: "Sensoren: Prinzipien und Anwendungen", Carl Hanser Verlag, München/Wien, 1991

[Hawl92] Hawlitschek, G. und Laxhuber, L.: "Lichtadressierbarer potentiometrischer Sensor", *Mikroelektronik*, 1992 (3), Fachbeilage "Mikroperipherik"

[Haya91] Hayashi, I. et al.: "A Piezoelectric Cycloid Motor and Its Fundamental Characteristics", *Proc. of the 2 Int. Symp. on Micro Machine and Human Science*, Nagoya, 1991, pp.73–77

[Haze93] Hazelrigg, G.A.: "Microelectromechanical Systems Research in the United States", *Proc. of the IARP Workshop on Micromachine Technologies and Systems*, Tokyo, 1993, pp.144–153

[Hebb49] Hebb, D.: "The Organization of Behavior", Wiley Publications, New York, 1949

[Hein93] Heinzelmann, E.: "Monolithische Mikrosysteme *sur mesure*", *Technische Rundschau (TR)*, 1993 (18), S.20–25

[Hein94] Heinzelmann, E.: "Mini-Werkzeuge der Nanotechnik", *Technische Rundschau (TR)*, 1994 (6), S.22–26

[Hein94a] Heinzelmann, E.: "Erschwingliche High-Tech für KMU" *Technische Rundschau (TR)*, 1994 (16), S.18–22

[Herp94] Herpel, H.-J. und Bley, P.: "Das Forschungszentrum Karlsruhe als Partner in der Mikrosystemtechnik", *in M. Glesner (Hrsg.): "Aufgaben der Informatik in der Mikrosystemtechnik", Zusammenfassung der Beiträge im Verlauf der Gründung der GI-Fachgruppe 3.5.6 "Mikrosystemtechnik"*, Schloß Dagstuhl, 1994

[Hertz91] Hertz, J., Krogh, A. and Palmer, R.G.: "Introduction to the theory of neural networks", Addison-Wesley, 1991

[Hess92] Hesselbach, J. and Kristen, M.: "Shape Memory Actuators As Electricants", *Proc. of Int. Conf. on New Actuators,* Bremen, 1992, pp.85–91

[Heub91] Heuberger, A. (Hrsg.): "Mikromechanik: Mikrofertigung mit Methoden der Halbleitertechnologie", Springer-Verlag, Berlin, 1991

[Heur92] Heurich, M. et al.: "CO_2-sensitive Organically Modified Silicates for Application in a Gas Sensor", *in "Micro System Technologies 92" (Editor: H. Reichl)*, VDE-Verlag, Berlin, 1992, pp.359–367

[Hill94] Hilleringmann, U., Adams, St. and Goser, K.: "Micromechanic Pressure Sensors with Optical Readout and on Chip CMOS Amplifiers Based on Si-Technology", *in "Micro System Technologies 94" (Editors: H. Reichl, A. Heuberger)*, VDE-Verlag, Berlin, 1994, pp.713–722

[Hint94] Hintsche, R. et al.: "Modular microsystems made in Si-technology for chemical and biosensors", *in "Micro System Technologies 94" (Editors: H. Reichl, A. Heuberger)*, VDE-Verlag, Berlin, 1994, pp.371–379

[Hira93] Hirano, T., Furuhata, T. and Fujita, H.: "Dry Releasing of Electroplated Rotational and Overhanging Structures", *Proc. of the IEEE Int. Conf. on Micro Electro Mechanical Systems (MEMS)*, Fort Lauderdale, Florida, 1993, pp.278–283

[Hirai93] Hirai, S., Sakane, S. and Takase, K.: "Cooperative Task Execution Technology for Multiple Micro Robot Systems", *Proc. of the IARP Workshop on Micromachine Technologies and Systems*, Tokyo, 1993, pp.32–37

[Hjort95] Hjort, K.: "Micromachining of non-silicon materials", *Proc. of the NEXUS-Workshop on Micro-Machining*, Bremen, 1995

[Holl93] Holler, E. und Trapp, R.: "Ein experimenteller Telemanipulator für die Minimal-Invasive Chirurgie", *1. Statuskolloquium des Projektes Mikrosystemtechnik, Tagungsband,* Forschungszentrum Karlsruhe, 1993, S.92–99

[Horie95] Horie, M., Funabashi, H. and Ikegami, K.: "A study on micro force sensors for microhandling systems", *Microsystem Technologies,* 1995 (3), Vol. 1, pp.105–110

[Hosa93] Hosaka, H., Kuwano, H. and Yanagisawa, K.: "Electromagnetic Microrelays: Concepts and Fundamental Characteristics", *Proc. of the IEEE Int. Conf. on Micro Electro Mechanical Systems (MEMS),* Fort Lauderdale, Florida, 1993, pp.12–17

[Hoss92] Hosseini-Sianaki, A. et al.: "The High Speed Electrorheological Catch - Characteristics, Dimensional Considerations and Non Linear Operational Aspects", *Proc. of Int. Conf. on New Actuators,* Bremen, 1992, pp.118–122

[Howe90] Howe, R.T. et al.: "Silicon micromechanics: sensors and actuators on a chip", *IEEE Spectrum,* July 1990, pp.29–35

[Humb94] van Humbeeck, J., Reynaerts, D. and Stalmans, R.: "Shape Memory Alloys: Functional and Smart", *Proc. of Int. Conf. on New Actuators,* Bremen, 1994, pp.312–316

[Hund94] Hund, H.: "Mikrosystemtechnik in der mittelständischen Industrie", *SENSOR report,* 1994 (2), S.31–33

[Idog93] Idogaki, T.: "Micromachine Technology: Advanced Maintenance Technologies for Electric-Power Plants", *Proc. of the IARP Workshop on Micromachine Technologies and Systems,* Tokyo, 1993, pp.50–58

[Ikuta92] Ikuta, K., Aritomi, S. and Kabashima, T.: "Tiny Silent Linear Cybernetic Actuator Driven by Piezoelectric Device with Electromagnetic Clamp", *Proc. of the IEEE Int. Conf. on Micro Electro Mechanical Systems (MEMS),* Travemünde, 1992, pp.232–237

[Ikuta92a] Ikuta, K., Aritomi, S. and Kabashima, T.: "Silent cybernetic actuator", *Proc. of the 3. Int. Symp. on Micro Machine and Human Science*, Nagoya, 1992, pp.143–148

[IMC94] "Micro Mechanics", *Report on R&D Activities*, Industrial Microelectronics Center, Kista, Sweden, 1994

[Inde95] Indermühle P.-F. et al.: "AFM imaging with an xy-micropositioner with integrated tip", *Sensors and Actuators A*, 1995 (46–47), pp.562–565

[INFO95] "Modellbibliothek für komplexe analoge Bauelemente", *MST-Infobörse*, 1995 (1), VDI/VDE-IT, Teltow

[Inoue95] Inoue, T. et al.: "Micromanipulation Using Magnetic Field", *Proc. of Int. Conf. on Robotics and Automation*, Nagoya, 1995, pp. 679–684

[Ishi95] Ishihara, H. et al.: "Approach to Distributed Micro Robotic Systems", *Proc. of Int. Conf. on Robotics and Automation*, Nagoya, 1995, pp.375–380

[Itoh92] Itoh, T. and Okamura, M.: "Development of Ultra Small DC Motor", *Proc. of the 3. Int. Symp. on Micro Machine and Human Science*, Nagoya, 1992, pp.27–33

[Itoh93] Itoh, T., Itoh, K. and Okamura, M.: "Electromagnetic Ultra Small Motor", *Proc. of Int. Conf. on Advanced Mechatronics (ICAM)*, Tokyo, 1993, pp.82–85

[Jaeck92] Jaecklin, V.P. et al.: "Micromechanical comb actuators with low driving voltage", *Proc. of Int. Conf. on New Actuators,* Bremen, 1992, pp.40–45

[Jaeck93] Jaecklin, V.P. et al.: "Optical Microshutters and Torsional Micromirrors for Light Modulator Arrays", *Proc. of the IEEE Int. Conf. on Micro Electro Mechanical Systems (MEMS)*, Fort Lauderdale, Florida, 1993, pp.124–127

[Jano92] Janocha, H. (Hrsg.): "Aktoren - Grundlagen und Anwendungen", Springer-Verlag, Berlin Heidelberg New York, 1992

[Jano93] Janocha, H.: "Aktoren in Systemen der Automatisierungstechnik", *VDE-Fachseminar "Moderne Aktoren und Sensoren in der Automatisierungstechnik"*, Tagungsband, Universität des Saarlandes, Saarbrücken, 1993

[Jano94] Janocha, H., Schäfer, J. and Jendritza, D.J.: "Design criteria for the application of solid-state actuators", *Proc. of Int. Conf. on New Actuators,* Bremen, 1994, pp.246–250

[Jean92] Jeanerret, S., de Rooij, N. und van der Schoot, B.: "Chemische Analysesysteme", *Mikroelektronik*, 1992 (2), Fachbeilage "Mikroperipherik"

[Jend93] Jendritza, D. J.: "Piezoelektrische Aktoren in der Mechatronik", *VDE-Fachseminar "Moderne Aktoren und Sensoren in der Automatisierungstechnik"*, Tagungsband, Universität des Saarlandes, Saarbrücken, 1993

[Jend93a] Jendritza, D. J. und Janocha, H.: "Große Zukunft für neue Aktoren", *Elektronik*, 06.04.93 (7), Sonderdruck, Franzis-Verlag, München

[Jend94] Jendritza, D. J. und Schröder, J.: "Possible Applications and Limitations of Piezoelectric Actuators in Hydraulic Systems", *Proc. of Int. Conf. on New Actuators,* Bremen, 1994, pp.138–143

[Joha93] Johansson, St.: "Hybrid Techniques in Microrobotics", *Proc. of Int. IARP Workshop on Micro Robotics and Systems*, Karlsruhe, 1993, pp.72–83

[Joha95] Johansson, St.: "Micromanipulation for Micro- and Nano-Manufacturing", *Proc. of the INRIA/IEEE Conference on Emerging Technologies and Factory Automation (ETFA)*, Paris, 1995, Tome 3, pp.3–8

[Josw92] Joswig, J.: "Active micromechanic valve", *Proc. of Int. Conf. on New Actuators,* Bremen, 1992, pp.183–185

[Kahl93] Kahlert, J. und Frank, H.: "Fuzzy-Logik und Fuzzy-Control", Vieweg Verlag, Braunschweig/Wiesbaden, 1993

[Kalb94] Kalb, H. et al.: "Electrostatically Driven Linear Stepping Motor in LIGA Technique", *Proc. of Int. Conf. on New Actuators,* Bremen, 1994, pp.83–85

[Kall94] Kallenbach, E., Albrecht, A. and Birli, O.: "Design of microactuators", *in "Micro System Technologies 94" (Editors: H. Reichl, A. Heuberger),* VDE-Verlag, Berlin, 1994, pp.979–988

[Kasp94] Kasper, M. und Reichl, H.: "Optimierungsmethoden für die Systemintegration", in *VDI/VDE-IT (Hrsg.): "Untersuchungen zum Entwurf von Mikrosystemen", Reihe: Innovationen in der Mikrosystemtechnik,* Teltow, 1994, S.73–86

[Kato91] Kato, H. et al.: "Photoelectric Inclination Sensor for Controlling the Attitude of Milli-machine", *Proc. of the 2. Int. Symp. on Micro Machine and Human Science,* Nagoya, 1991, pp.93–101

[Kell91] Keller, W. et al.: "Aufbau- und Verbindungstechnik", *KfK-Nachrichten, Jahrgang 23,* Forschungszentrum Karlsruhe, 1991 (2–3), S.143–147

[Kies88] Kiesewetter, L.: "Terfenol in linear motor", *Proc. of the Conf. on Giant Magnetostrictive Alloys,* Marbella, 1988

[Kiku93] Kikuya, Y. et al.: "Micro Alignment Machine for Optical Coupling", *Proc. of the IEEE Int. Conf. on Micro Electro Mechanical Systems (MEMS),* Fort Lauderdale, Florida, 1993, pp.36–41.

[Kimu91] Kimura, M. und Fujiyoshi, M.: "Force Measurement and Control of Electrostatic Microactuator", *Proc. of the 2. Int. Symp. on Micro Machine and Human Science,* Nagoya, 1991, pp.119–124

[Klaa94] Klaaßen, B. und Paap, K.L.: "Methoden zur Multilevel-Analogsimulation", in *VDI/VDE-IT (Hrsg.): "Untersuchungen zum Entwurf von Mikrosystemen", Reihe: Innovationen in der Mikrosystemtechnik,* Teltow, 1994, S.15–24

[Klei92] Kleinschmidt, P und Schmidt, F.: "Des Sensors liebstes Kind", *Technische Rundschau (TR)*, 1992 (44), S.46–48

[Klein95] Klein, St. et al.: "A New Large-Chamber SEM and Its Application in Micromechanical Assembly Processes", Proc. of the Int. Conf. on Flexible Automation and Intelligent Manufacturing, Stuttgart, 1995

[Knob96] Knobloch, J.: "Fuzzy-Mikrosysteme in der Sensorik", *Tagungsband zur Vortragsveranstaltung "Informationstechnik für Mikrosysteme"*, VDE/VDI-GME, Stuttgart, 1996, S.11–15

[Kohl94] Kohl, M. et al.: "Characterization of NiTi Shape Memory Microdevices Produced by Microstructuring of Etched Sheets or Sputter Deposited Films", *Proc. of Int. Conf. on New Actuators,* Bremen, 1994, pp.317–320

[Kohl94a] Kohl, M. und Menz, W.: "Konzepte der Mikro-/Makroankopplung beim Entwurf von Mikrosystemen", in *VDI/VDE-IT (Hrsg.): "Untersuchungen zum Entwurf von Mikrosystemen", Reihe: Innovationen in der Mikrosystemtechnik,* Teltow, 1994, S.123–184

[Kohl94b] Kohl, F. et al.: "A micromachined flow sensor for liquid and gaseous fluids", *Sensors and Actuators A*, 1994 (41–42), pp.293–299

[Kohl96] Kohl, D. et al.: "Examples of signal conditioning in gas sensor systems", *Tagungsband zur Vortragsveranstaltung "Informationstechnik für Mikrosysteme"*, VDE/VDI-GME, Stuttgart, 1996, S.5–10

[Koho87] Kohonen, T.: "Self-Organization and Associative Memory", Springer-Verlag, Berlin, 1987

[Kram92] Kramer, W.: "Piezoelectric Actuators for Automotive Application: Current Issues and Future Prospects", *Proc. of Int. Conf. on New Actuators,* Bremen, 1992, pp.47–50

[Krat91] Kratzer, K.P.: "Unüberwachte Adaption mit dem Kosinus-Klassifikator", *Informationstechnik,* 1991 (4)

[Krat93] Kratzer, K.P.: "Neuronale Netze: Grundlagen und Anwendungen", Carl Hanser Verlag, München/Wien, 1993

[Krev91] Krevet, B.: "Neuronale Netze und Mikrosystemtechnik", *KfK-Nachrichten, Jahrgang 23,* Forschungszentrum Karlsruhe, 1991 (2–3), S.158–164

[Kröm95] Krömer, O. et al.: "Intelligentes triaxiales Beschleunigungssensorsystem", *2. Statuskolloquium des Projektes Mikrosystemtechnik, Tagungsband,* Forschungszentrum Karlsruhe, 1995, S.75–80

[Krull93] Krull, F. und Endres, H.-E.: "Spürnasen: Grundbauelemente der chemischen Mikrosensorik", *Technische Rundschau (TR),* 1993 (18), S.28–34, (20), S.46–52

[Krull95] Krull, F.: "Nutzen der dritten Dimension", *Technische Rundschau (TR),* 1995 (11), S.10–16

[Kruse94] Kruse, R. et al. (Ed.): "Fuzzy Systems in Computer Science", Vieweg Verlag, 1994

[Kumar92] Kumar, S. and Cho, D.: "Electrostatically Levitated Microactuators", *Journal of Micromechanics and Microengineering,* 1992 (2), pp.96–103

[Kuri93] Kuribayashi, K. et al.: "Trial Fabrication of Micron Sized Arm Using Reversible TiNi Alloy Thin Film Actuators", *Proc. of the IEEE/RSJ Int. Conf. on Intelligent Robots and Systems (IROS),* Yokohama, 1993, pp.1697–1702

[Lamm93] Lammerink, T., Elwenspoek, M. and Fluitman, J.: "Integrated

Micro-liquid Dosing System", *Proc. of the IEEE Int. Conf. on Micro Electro Mechanical Systems (MEMS),* Fort Lauderdale, Florida, 1993, pp.254–259

[Lawr92] Lawrence, J.: "Neuronale Netze: Computersimulation biologischer Intelligenz", Systhema Verlag, München, 1992

[Lee92] Lee, A., Ljung, P. and Pisano, A.: "Polysilicon Micro Vibromotors", *Proc. of the IEEE Int. Conf. on Micro Electro Mechanical Systems (MEMS)*, Travemünde, 1992, pp.177–182

[Lehr92] Lehr, H. et al.: "Application of the LIGA-Technique for the Development of Microactuators Based on Electromagnetic Principles", *Proc. of Int. Conf. on New Actuators,* Bremen, 1992, pp.13–18

[Lehr94] Lehr, H. et al.: "Fertigung von Mikroaktoren mittels LIGA-Technik", *1. Int. Kongress und Ausstellung für Mikrosysteme und Präzisionstechnik,* Tagungsband, Messe Stuttgart, 1994

[Lher92] Lhermet, N. et al.: "Actuators based on biased magnetostrictive rare earth-iron alloys", *Proc. of Int. Conf. on New Actuators,* Bremen, 1992, pp.133–137

[Lin93] Lin, J., Obermeier, E. und Schlichting, V.: "Elektrostatisch aktivierte Mikroblende", *Mikroelektronik,* 1993 (5), Fachbeilage "Mikrosystemtechnik"

[Lind93] Linders, J.: "Mikromaschinen", *Mikroelektronik,* 1993 (5), Fachbeilage "Mikrosystemtechnik"

[Lisec94] Lisec, T. et al.: "A Fast Switching Silicon Valve for Pneumatic Control Systems", *Proc. of Int. Conf. on New Actuators,* Bremen, 1994, pp.30–33

[Löch94] Löchel, B. et al.: "Electroplated Electromagnetic Components for Actuators", *Proc. of Int. Conf. on New Actuators,* Bremen, 1994, pp.109–113

[Lore93] Lorenz, Th.: "Faseroptische Temperatursensoren mit flüssigen Kristallen", *7. IAR Kolloquium, Tagungsband,* Universität Duisburg, 1993, S.127–139

[M2S293] "M2S2 Zwischenbericht", Laborverbund für Mikromechanik auf Silizium in der Schweiz, Neuchâtel, 1993

[Magn94] Magnussen, B., Fatikow, S. und Rembold, U.: "Micro Actuators: Principles and Applications", *in M. Glesner (Hrsg.): "Aufgaben der Informatik in der Mikrosystemtechnik", Zusammenfassung der Beiträge im Verlauf der Gründung der GI-Fachgruppe 3.5.6 "Mikrosystemtechnik"*, Schloß Dagstuhl, 1994

[Magn95] Magnussen, B., Fatikow, S. and Rembold, U.: "Actuation in Microsystems: Problem Field Overview and Practical Example of the Piezoelectric Robot for Handling of Microobjects", *Proc. of the INRIA/IEEE Conference on Emerging Technologies and Factory Automation (ETFA)*, Paris, 1995, Tome 3, pp.21–27

[Magn96] Magnussen, B.: "Infrastruktur für Steuerungs- und Regelungssysteme von robotischen Miniatur- und Mikrogreifern", *Dissertation,* Universität Karlsruhe, Fakultät für Informatik, 1996

[Mast93] Mastrangelo, C. and Saloka, G.: "A dry-release method based on polymer columns for microstructure fabrication", *Proc. of the IEEE Int. Conf. on Micro Electro Mechanical Systems (MEMS)*, Fort Lauderdale, Florida, 1993, pp.77–81

[Mato94] Matoba, H., Kim, C. J. and Muller, R. S.: "Fabrication of a Bistable Snapping Microactuator", *in "Micro System Technologies 94" (Editors: H. Reichl, A. Heuberger)*, VDE-Verlag, Berlin, 1994, pp.1005–1013

[Mats93] Matsuoka, T. et al.: "Mechanical Analysis For Micro Mobile Machine With Piezoelectric Element", *Proc. of the IEEE/RSJ Int. Conf. on Intelligent Robots and Systems (IROS)*, Yokohama, 1993, pp.1685–1690

[Mehl92] Mehlhorn, T. et al.: "CMOS-compatible Capacitive Silicon Pressure Sensor", *in "Micro System Technologies 92" (Editor: H. Reichl)*, VDE-Verlag, Berlin, 1992, pp.277–285

[Meiss94] von Meiss, P.: "Automated Assembly Units for Microsystems", *mst news*, 1994 (9), p.2–3

[Menz93] Menz, W. und Bley, P.: "Mikrosystemtechnik für Ingenieure", VCH Verlag, Weinheim, 1993

[Menz93a] Menz, W.: "Die LIGA-Technik und ihr Potential für die industrielle Anwendung", *1. Statuskolloquium des Projektes Mikrosystemtechnik, Tagungsband,* Forschungszentrum Karlsruhe, 1993, S.19–28

[Menz95] Menz, W.: "Physikalische Mikrokomponenten für die Mikrosystemtechnik", *2. Statuskolloquium des Projektes Mikrosystemtechnik, Tagungsband,* Forschungszentrum Karlsruhe, 1995, S.11–14

[Meyer95] Meyer, J.-U.: "Mikrosysteme zur Ankopplung an das Nervensystem", *Tagungsband des 2. Workshops "Methoden- und Werkzeugentwicklung für den Mikrosystementwurf" im Rahmen des 3. Statusseminars zum BMBF-Verbundprojekt METEOR*, Karlsruhe, 1995, S.19–26

[Micr91] "Three-dimensional microstructures made from metals, plastics and ceramics", *Öffentlichkeitsarbeit,* MicroParts Gesellschaft für Mikrostrukturtechnik mbH Karlsruhe, 1991

[Mins69] Minsky, M. and Papert, S.: "Perceptrons: An Introduction to Computational Geometry", MIT Press, Massachusetts, 1969

[Mits93] Mitsuishi, M. et al.: "Development of Tele-Operated Micro-Handling/Machining System Based on Information Transformation", *Proc. of the IEEE/RSJ Int. Conf. on Intelligent Robots and Systems (IROS)*, Yokohama, 1993, pp.1473–1478

[MIT92] "Research in Microsystems Technology", *Annual Report*, MIT, Cambridge, 5/1992

[MITI91] "Micromachine Technology", *Introduction to a new project for the national research and development programm*, MITI, Tokyo, 1991

[Mizu93] Mizuno, T. et al.: "Light Driven Microactuator for a Micropump", *Proc. of the IARP Workshop on Micromachine Technologies and Systems*, Tokyo, 1993, pp.84–89

[Mohr91] Mohr, J. et al.: "Herstellung von beweglichen Mikrostrukturen mit dem LIGA-Verfahren", *KfK-Nachrichten, Jahrgang 23*, Forschungszentrum Karlsruhe, 1991 (2–3), S.110–117

[Mohr92] Mohr, J. et al.: "Microactuators Fabricated by the LIGA Process", *Proc. of Int. Conf. on New Actuators,* Bremen, 1992, pp.19–23

[Moil92] Moilanen, H. et al.: "Laser Ablation Deposition for Fabrication of Low Voltage Actuator Using Bimorph Structures of ND-Doped PZT Thin Films", *Proc. of Int. Conf. on New Actuators,* Bremen, 1992, pp.191–195

[Mokw93] Mokwa, W.: "Intelligente Sensorsysteme auf Siliziumbasis", *7. IAR Kolloquium, Tagungsband*, Universität Duisburg, 1993, S.25–75

[Mori93] Morishita, H. and Hatamura, Y.: "Development of Ultra Precise Manipulator System for Future Nanotechnology", *Proc. of Int. IARP Workshop on Micro Robotics and Systems*, Karlsruhe, 1993, pp.34–42

[Mull90] Muller, R.S.: "Microdynamics", *Sensors and Actuators A*, 1990 (21–23), pp. 1–8

[Müll93] Müller, C., Hein, H. und Mohr, J.: "Mikrospektrometer für spektrale Analyseaufgaben im sichtbaren und nahen Infrarotbereich",

1. Statuskolloquium des Projektes Mikrosystemtechnik, Tagungsband, Forschungszentrum Karlsruhe, 1993, S.103–108

[Müll95] Müller, J.: "Schichten für die Mikrosystemtechnik", *Technische Rundschau (TR)*, 1995 (33), S.26–29

[Muro92] Muro, H. et al.: "Integrated Piezoresistive Accelerometers with Oil-Damping", *in "Micro System Technologies 92" (Editor: H. Reichl)*, VDE-Verlag, Berlin, 1992, pp.233–241

[Nach92] "Elektronik puscht Automobilentwicklung", *VDI-Nachrichten*, 10.04.92

[Nach94] "Bild aus über 400000 Mikrospiegeln", *VDI-Nachrichten*, 25.02.94

[Nase93] Nase, R.: "Reluktanz-Antriebe", *VDE-Fachseminar "Moderne Aktoren und Sensoren in der Automatisierungstechnik"*, Tagungsband, Universität des Saarlandes, Saarbrücken, 1993

[Nauck94] Nauck, D., Klawonn, F. und Kruse, R.: "Neuronale Netze und Fuzzy-Systeme", Vieweg Verlag, Braunschweig/Wiesbaden, 1994

[Nent92] Nentwig, J., Scheller, F.W. et al.: "Elektrochemische Biosensoren", *Mikroelektronik*, 1992 (3), Fachbeilage "Mikroperipherik"

[NEXU95] "NEXUS Market 2002: An Opinion Survey", *in Proc. of the NEXUS-Workshop on Micro-Machining*, Bremen, 1995

[NU-Te93] NU-Tech GmbH (Deutschland): "Komplettsimulation piezoelektrischer Systeme", *Sensor Magazin*, 1993 (1–2), S.21–22

[Oliv93] Olivier, M. et al.: "A Tele-Nano-Positioning-Robot For Micro-Fabrication And Micro-Assembly", *Proc. on the IARP Workshop on Micromachine Technologies and Systems*, Tokyo, 1993, pp.16–23

[Pavl94] Pavlicek, H.: "Entwurf und Simulation in der Mikrotechnik", *1. Int. Kongress und Ausstellung für Mikrosysteme und Präzisionstechnik (Micro-Engineering 94)*, Tagungsband, Messe Stuttgart, 1994

[Peet93] Peeters, E. et al.: "Developments in etch stop techniques", *MicroMechanics Europe, Workshop Digest*, Neuchâtel, 1993, pp.35–49

[Pelr92] Pelrine, R., Eckerle, J. and Chiba, S.: "Review of Artificial Muscle Approaches", *Proc. of the 3. Int. Symp. on Micro Machine and Human Science*, Nagoya, 1992, pp.1–17

[Popp91] Poppinger, M.: "Entwurf piezoresistiver Drucksensoren", *Mikroelektronik*, 1991 (5), Fachbeilage "Mikroperipherik"

[Prod95] "Wende durch steigende Investitionen", *Produktion,* Nr. 12, 23.3.95, S.4

[Quan93] Quandt, E.: "Magnetostriktive Schichten als Aktoren in der Mikrosystemtechnik", *1. Statuskolloquium des Projektes Mikrosystemtechnik, Tagungsband,* Forschungszentrum Karlsruhe, 1993, S.151–156

[Quen94] Quenzer, H.J. et al.: "Fabrication of Relief-Topographic Surfaces with a One-Step UV-Lithographic Process", *in "Micro System Technologies 94" (Editors: H. Reichl, A. Heuberger)*, VDE-Verlag, Berlin, 1994, pp.163–172

[Quick92] "Torpedo-Pille gegen Darm-Entzündung", *Quick*, 20.8.92, S.24

[Ramp94] Rampersad, H.K.: "Integrated and Simultaneous Design for Robotic Assembly", John Wiley & Sons, Chichester, 1994

[Rech93] Rech, B.: "Aktoren mit elektrorheologischen Flüssigkeiten", *VDE-Fachseminar "Moderne Aktoren und Sensoren in der Automatisierungstechnik"*, Tagungsband, Universität des Saarlandes, Saarbrücken, 1993

[Recke95] Recke, C., Vogel, Th. und Kasper, M.: "Modellgenerierung für mechanische Mikrosystem-Komponenten", *Tagungsband des 2. Workshops "Methoden- und Werkzeugentwicklung für den Mikrosystementwurf" im Rahmen des 3. Statusseminars zum BMBF-Verbundprojekt METEOR*, Karlsruhe, 1995, S.100–109

[Reig94] Reignier, P.: "Fuzzy logic techniques for mobile robot obstacle avoidance", *Robotics and Autonomous Systems*, 1994 (12), pp.143–153

[Remb95] Rembold, U. et al.: "The Use of Actuation Principles for Micro Robots", *in: "The Ultimate Limits of Fabrication and Measurement" (Editors: M. Welland, J. Gimzewski)*, Kluwer, Dordrecht, 1995, pp.33–40

[Rich92] Richter, A. and Zengerle, R.: "Properties and applications of a micro membrane pump with electrostatic drive", *Proc. of Int. Conf. on New Actuators,* Bremen, 1992, pp.28–33

[Rich92a] Richter, A.: "Mikropumpen für Dosiersysteme", *Mikroelektronik,* 1992 (2), Fachbeilage "Mikroperipherik"

[Rieg91] Riegel, J.: "Gasanalyse mit Wärmetönungssensoren", *Mikroelektronik,* 1991 (4), Fachbeilage "Mikroperipherik"

[Ritt91] Ritter, H., Martinetz, Th. und Schulten, K.: "Neuronale Netze: Eine Einführung in die Neuroinformatik selbstorganisierender Netzwerke", Addison-Wesley, Bonn/München, 1991

[Rogn93] Rogner, A. et al.: "The LIGA technique: What are the new opportunities?", *SUSS report,* 1993 (3.Quarter)

[Rojas93] Rojas, R.: "Theorie der neuronalen Netze: Eine systematische Einführung", Springer-Verlag, Berlin, 1993

[Roth92] Roth, R.C.: "The elastic wave motor - a versatile terfenol driven, linear actuator with high force and great precision", *Proc. of Int. Conf. on New Actuators,* Bremen, 1992, pp.138–141

[Rück93] Rückert, U., Spaanenburg, L. und Anlauf, J.: "Hardwareimplementierung Neuronaler Netze", *Automatisierungstechnische Praxis*, 1993 (7), S.414–420

[Russ94] Russel, R.A.: "A robotic system for performing sub-millimetre grasping and manipulation tasks", *Robotics and Autonomous Systems*, 1994 (3), pp.209–218

[Ruth95] Ruther, P., Feit, K. und Bacher, W.: "Hydraulischer MIkroaktor", *2. Statuskolloquium des Projektes Mikrosystemtechnik*, Tagungsband, Forschungszentrum Karlsruhe, 1995, S.189

[Salo93] Salomon, P.: "Anwendungspotential der Integrierten Optik", *F&M*, 1993 (6)

[SAML93] "Sensors, Actuators & Microsystems Laboratory", *Report on Research Activities*, Institute of Microtechnology, University of Neuchâtel, 1993

[Sato92] Sato, K. "Electrostatic Film Actuator with a Large Vertical Displacement", *Proc. of the IEEE Int. Conf. on Micro Electro Mechanical Systems (MEMS)*, Travemünde, 1992, pp.1–5

[Sato95] Sato, T. et al.: "Hand-Eye System in Nano Manipulation World", *Proc. of Int. Conf. on Robotics and Automation*, Nagoya, 1995, pp.59–66

[Scha94] Schaller, Th. et al.: "Mechanische Mikrostrukturierung metallischer Oberflächen", *F&M*, 1994 (5–6), S.274–278

[Schä93] Schäfer, J.: "Entwurf und Einsatz von magnetostriktiven Aktoren", *VDE-Fachseminar "Moderne Aktoren und Sensoren in der Automatisierungstechnik"*, Tagungsband, Saarbrücken, 1993

[Schä94] Schäfer, W. et al.: "Perspektiven der Fertigungsgeräteindustrie in der industriellen Produktion von Mikrosystemen", *1. Int. Kongress und Ausstellung für Mikrosysteme und Präzisionstechnik (Micro-Engineering 94)*, Tagungsband, Messe Stuttgart, 1994

[Sche92] Scheinbeim, J. et al.: "Electrostrictive response of elastomeric polymers", *ACS Polymer Preprints*, 33(2), 1992, pp.385–386

[Schod93] Schoder, D.: "Adaptive Regler", *Mikroelektronik*, 1993 (3), Fachbeilage "Mikrosystemtechnik"

[Schom93] Schomburg, W. et al.: "Mikromembranpumpen als Elemente eines optochemischen Mikroanalysesystems", *1. Statuskolloquium des Projektes Mikrosystemtechnik, Tagungsband,* Forschungszentrum Karlsruhe, 1993, S.76–82

[Schrö93] Schröer, B.: "Aktoren in der Mikrosystemtechnik", *Mikroelektronik*, 1993 (6), Fachbeilage "Mikrosystemtechnik"

[Schra94] Schrage, J. et al.: "Störsignaleinkopplungen in mikrosystemspezifische Sensoranordnungen", in *VDI/VDE-IT (Hrsg.): "Untersuchungen zum Entwurf von Mikrosystemen", Reihe: Innovationen in der Mikrosystemtechnik,* Teltow, 1994, S.59–72

[Schu95] Schulze, S.: "Silicon bonding in microsystem technology", *Proc. of the NEXUS-Workshop on Micro-Machining*, Bremen, 1995

[Schur95] Schurr, M. O. et al.: "Interdisciplinary Technology Development for Minimally Invasive Therapy", *mst news*, 1995 (13), p.2–4

[Schwa94] Schwarzenbach, H.U., Roos, M. and Anderheggen, E.: "Numerical Modelling and Simulation of Actuator Operation – State of the Art and Requirements", *Proc. of Int. Conf. on New Actuators,* Bremen, 1994, pp.96–99

[Schwe93] Schweizer, M. et al.: "Umgang mit atomar kleinen Strukturen", *Technische Rundschau (TR)*, 1993 (34), S.20–23

[SSEL91] "Solid-State Electronics Laboratory", *Report on Research Activities*, The University of Michigan, Ann Arbor, 1991

[Stei94] Steiner, K. et al.: "Thin-film SnO_2 Gas Sensors on Si Substrates for CO and CO_2 Detection", *in "Micro System Technologies 94"*

(Editors: H. Reichl, A. Heuberger), VDE-Verlag, Berlin, 1994, pp.429–437

[Stix92] Stix, G.: "Trends in Micromechanics: Micron Machinations", *Scientific American*, Nov. 1992, pp.72–80

[Stöck92] Stöckel, D.: "Status and Trends in Shape Memory Technology", *Proc. of Int. Conf. on New Actuators,* Bremen, 1992, pp.79–84

[Stro93] Strohrmann, M. et. al.: "LIGA-Sensoren und intelligente Sensorsysteme zur Messung von Beschleunigungen", *1. Statuskolloquium des Projektes Mikrosystemtechnik, Tagungsband,* Forschungszentrum Karlsruhe, 1993, S.65–70

[Suzu91] Suzumori, K., Iikura, Sh. and Tanaka, H.: "Applying a Flexible Microactuator to Robotic Mechanisms", *Proc. of Int. Conf. on Robotics and Automation*, Sacramento, CA, 1991, pp.1622–1627

[Suzu91a] Suzumori, K., Iikura, Sh. and Tanaka, H.: "Flexible Microactuator for Miniature Robots", *Proc. of the IEEE Int. Conf. on Micro Electro Mechanical Systems (MEMS)*, Nara, 1991, pp.204–209

[Suzu91b] Suzumori, K., Kondo, F. and Tanaka, H.: "Miniature Walking Robot Using Flexible Microactuators", *Proc. of the 2. Int. Symp. on Micro Machine and Human Science*, Nagoya, 1991, pp.29–36

[Suzu94] Suzumori, K., Koga, A. and Haneda, R.: "Microfabrication of Integrated FMAs using Stereo Lithography", *Proc. of the IEEE Int. Conf. on Micro Electro Mechanical Systems (MEMS)*, Oiso, 1994, pp.136–141

[Tana95] Tanaka, Sh.: "Pacemakers and Implantable Cardioverter Defibrillators", *Proc. of the Int. Micromachine Symposium*, Tokyo, 1995, pp.11–21

[Tilm93] Tilmans, H. and Bouwstra, S.: "A novel design of a highly sensitive low differential-pressure sensor using built-in resonant strain gauges", *Journal of Micromechanics and Microengi-*

neering, 1993 (4), Vol. 3, pp.198–202

[Tin92] Tinschert, F.: "Oil Pressure Control with Shape Memory Springs in Hydraulic Systems", *Proc. of Int. Conf. on New Actuators,* Bremen, 1992, pp.92–96

[Tiro93] Tirole, N. et al.: "3D Silicon Electrostatic Microactuator", *Journal of Micromechanics and Microengineering*, 1993 (3), Vol. 3, pp.155–157

[Tiro93a] Tirole, N. et al.: "Microfabrication tools for the design of 3D microdevices", *MicroMechanics Europe, Workshop Digest,* Neuchâtel, 1993, pp.97–100

[Tisch93] Tischhauser, H.: "Microaccelerometers", *in Staufert, G., Reber, A. and Hieber, H.: "Packaging"*, UETP MEMS, Swiss Found. for Research in Microtechnology, Neuchâtel, 1993, pp.153–166

[Torii93] Torii, A. et al.: "Adhesive Force of the Microstructures Measured by the Atomic Force Microscope", *Proc. of the IEEE Int. Conf. on Micro Electro Mechanical Systems (MEMS)*, Fort Lauderdale, Florida, 1993, pp.111–116.

[Trau93] Trauboth, H.: "Aufgaben der Informationsverarbeitung in der Mikrosystemtechnik", *1. Statuskolloquium des Projektes Mikrosystemtechnik, Tagungsband,* Forschungszentrum Karlsruhe, 1993, S.36–47

[Tschu92] Tschulena, G.: "Micromechanics Business Opportunities", *in "Micro System Technologies 92" (Editor: H. Reichl)*, VDE-Verlag, Berlin, 1992, pp.51–58

[Tschu95] Tschulena G.R.: "Innovative Products and Future Markets of Micromachines/MST in Europe", *Proc. of the Int. Micromachine Symposium*, Tokyo, 1995, pp.23–30

[Ueno91] Ueno, Y. et al.: "One-Chip Hall Sensor Array for High Resolution Angle Measurement", *Proc. of the 2. Int. Symp. on Micro Ma-*

chine and Human Science, Nagoya, 1991, pp.85–92

[Urban95] Urban, G. et al.: "Development of a Micro Flow-System with Integrated Biosensor Array", *in "Micro Total Analysis Systems" (Editors: A. van den Berg and P. Bergveld)*, Kluwer, Dordrecht, 1995, pp.259–262

[VDI92] VDI/VDE-IT (Hrsg.): "Statusbericht zur indirekt-spezifischen Maßnahme", Berlin, 1992

[VDI93] VDI/VDE-IT (Hrsg.): "Aufbau- und Verbindungstechnik für faser- und integriert-optische Sensoren", Reihe: Innovationen in der Mikrosystemtechnik, Teltow, 1993

[VDI94] VDI/VDE-IT (Hrsg.): "Technologien und Werkzeuge zum Entwurf und zur Realisierung anwendungsspezifischer Mikrosensoren für mechanische Größen", Reihe: Innovationen in der Mikrosystemtechnik, Teltow, 1994

[Wagn92] Wagner, B., Kreutzer, M. and Benecke, W.: "Linear and Rotational Magnetic Micromotors Fabricated Using Silicon Technology", *Proc. of the IEEE Int. Conf. on Micro Electro Mechanical Systems*, Travemünde, 1992, pp.183–189

[Wall92] Wallrabe, U. et al.: "Theoretical and experimental results of an electrostatic micro motor with large gear ratio fabricated by the LIGA process", *Proc. of the IEEE Int. Conf. on Micro Electro Mechanical Systems (MEMS)*, Travemünde, 1992, pp.139–140

[Wall93] Wallrabe, U. et al.: "Design and Test of Electrostatic Micromotors made by the LIGA-Process", *Proc. of Int. IARP Workshop on Micro Robotics and Systems*, Karlsruhe, 1993, pp.89–97

[Wall95] Wallrabe, U. et al.: "Möglichkeiten der Mikrosystemtechnik zur Herstellung von Mikrokomponenten für einen Herzkatheter" *2. Statuskolloquium des Projektes Mikrosystemtechnik*, Tagungsband, Forschungszentrum Karlsruhe, 1995, S.123–127

[Weick94] Weickmann, M et al.: "Simulation eines Mikrosystems am Beispiel eines Mikrolasers", in *VDI/VDE-IT (Hrsg.): "Untersuchungen zum Entwurf von Mikrosystemen", Reihe: Innovationen in der Mikrosystemtechnik*, Teltow, 1994, S.1–13

[Wein94] Weiner, M.: "Neurotechnologie: Belebende Impulse für tote Nerven", *Der Fraunhofer*, 1994(2), S.22–24

[Weng94] Wengelink, J. et al.: "Generation of Relief-Type Surface Topographies for Integrated Microoptical Elements", *in "Micro System Technologies 94" (Editors: H. Reichl, A. Heuberger)*, VDE-Verlag, Berlin, 1994, pp.209–217

[Wiel93] van der Wiel, A.J.: "A flow sensor for gases and liquids", *Sensors, Actuators & Microsystems Laboratory: Report on Research Activities*, Institute of Microtechnology, University of Neuchâtel, 1993, p.6

[Witte92] Witte, M. and Gu, H.: "Force and Position Sensing Resistors: an Emerging Technology", *Proc. of Int. Conf. on New Actuators*, Bremen, 1992, pp.168–170

[Wolff94] Wolff, C. and Wendt, E.: "Application of Electrorheological Fluids in Hydraulic Systems", *Proc. of Int. Conf. on New Actuators*, Bremen, 1994, pp.284–287

[Xie94] Xie, B. et al.: "Simultaneous determination of multiple analytes using a thermal micro-biosensor fabricated by micromachining", *in "Micro System Technologies 94" (Editors: H. Reichl, A. Heuberger)*, VDE-Verlag, Berlin, 1994, pp.391–398

[Yamad91] Yamada, Y. et al.: "Tactile Sensor Fabricated on a Flexible Film Sheet", *Proc. of the 2. Int. Symp. on Micro Machine and Human Science*, Nagoya, 1991, pp.79–84

[Yamag93] Yamaguchi, M. et al.: "Distributed Electrostatic Micro Actuator", *Proc. of the IEEE Int. Conf. on Micro Electro Mechanical Systems (MEMS)*, Fort Lauderdale, Florida, 1993, pp.18–23

[Yama95] Yamagata, Y. and Higuchi, T.: "A Micropositioning Device for Precision Automatic Assembly Using Impact Force of Piezoelectric Elements", *Proc. of Int. Conf. on Robotics and Automation,* Nagoya, 1995, pp.666–671

[Zache95] Zacheja, J. et al.: "Implantable Telemetric Endosystem for Minimal Invasive Pressure Measurements", *Proc. of MEDTECH'95,* Berlin, 1995

[Zdeb94] Zdeblick, M. J. et al.: "Thermopneumatically Actuated Microvalves And Integrated Electro-Fluidic Circuits", *Proc. of Int. Conf. on New Actuators,* Bremen, 1994, pp.56–60

[Zeng92] Zengerle, R., Richter, A. and Sandmaier, H.: "A micro membran pump with electrostatic actuation", *Proc. of the IEEE Int. Conf. on Micro Electro Mechanical Systems (MEMS),* Travemünde, 1992, pp.19–24

[Zesch95] Zesch, W. et al.: "Inertial Drives for Micro- and Nanorobots: Two Novel Mechanisms", Proc. of Int. Symp. on Microrobotics and Micromechanical Systems, Philadelphia, Pennsylvania, 1995, pp.80–88

[Ziad94] Ziad, H., Spirkovitch, S. and Rigo, S.: "MMIC Applications for Electrostatic Micromotors", *Proc. of Int. Conf. on New Actuators,* Bremen, 1994, pp.46–51

[Zimm91] Zimmerman, H.-J.: "Fuzzy Sets, Decision Making and Expert Systems", Kluwer, Boston, 1991

[Zimm94] Zimmermann, H.-J. und von Altrock, C. (Hrsg.): "Fuzzy Logic: Anwendungen", Oldenbourg, München/Wien, 1994

[Zum93] Zum Gahr K.-H.: "Materialforschung für Mikrosysteme", *1. Statuskolloquium des Projektes Mikrosystemtechnik, Tagungsband,* Forschungszentrum Karlsruhe, 1993, S.48–54

Subject Index

MIX
Papier aus verantwortungsvollen Quellen
Paper from responsible sources
FSC® C105338

If you have any concerns about our products,
you can contact us on
ProductSafety@springernature.com

In case Publisher is established outside the EU,
the EU authorized representative is:
Springer Nature Customer Service Center GmbH
Europaplatz 3, 69115 Heidelberg, Germany

Printed by Libri Plureos GmbH
in Hamburg, Germany